Undergraduate Texts in Mathematics

Editors
S. Axler
K.A. Ribet

Undergraduate Texts in Mathematics

Gerald Edgar

Measure, Topology, and Fractal Geometry

Second Edition

 Springer

Gerald Edgar
Department of Mathematics
The Ohio State University
Columbus, OH 43210-1174, USA
edgar@math.ohio-state.edu

ISBN 978-0-387-74748-4   ISBN 978-0-387-74749-1 (eBook)

Library of Congress Control Number: 2007934922

Mathematics Subject Classification (2000): 28A80, 28A78, 28-01, 54F45

Printed on acid-free paper

9 8 7 6 5 4 3 2 1

springer.com

Dedication

Arnold Ross
(1906–2002)
[37]

Preface

What is a fractal? Benoit Mandelbrot coined the term in 1975. There is (or should be) a mathematical definition, specifying a basic idea in geometry. There is also a figurative use of the term to describe phenomena that approximate this mathematical ideal. Roughly speaking, a fractal set is a set that is more "irregular" than the sets considered in classical geometry. No matter how much the set is magnified, smaller and smaller irregularities become visible. Mandelbrot argues that such geometric abstractions often fit the physical world better than regular arrangements or smooth curves and surfaces. On page 1 of his book, *The Fractal Geometry of Nature*, he writes, "Clouds are not spheres, mountains are not cones, coastlines are not circles, and bark is not smooth, nor does lightning travel in a straight line." [44, p. 1]*

To define *fractal*, Mandelbrot writes: "A fractal is by definition a set for which the Hausdorff–Besicovitch dimension strictly exceeds the topological dimension." [44, p. 15] It might be said that this book is a meditation on that verse. Study of the Hausdorff dimension requires measure theory (Chap. 5); study of topological dimension requires metric topology (Chap. 2). Note, however, that Mandelbrot later expressed some reservations about this definition: "In science its [the definition's] generality was to prove excessive; not only awkward, but genuinely inappropriate ... This definition left out many 'borderline fractals', yet it took care, more or less, of the frontier 'against' Euclid. But the frontier 'against' true geometric chaos was left wide open." [44, p. 159] We will discuss in Chap. 6 a way (proposed by James Taylor) to repair the definition. Mandelbrot himself, in the second printing of [44, p. 459], proposes "... to use 'fractal' without a pedantic definition, to use 'fractal dimension' as a generic term applicable to *all* the variants in Chapter 39, and to use in each specific case whichever definition is the most appropriate." I have not adopted the first of these suggestions in this book, since a term without a "pedantic definition" cannot be discussed mathematically. I have, however, used the term "fractal dimension" as suggested.

---

* Bracketed numbers like this refer to the references collected on p. 257.

This is a *mathematics* book. It is not about how fractals come up in nature; that is the topic of Mandelbrot's book [44]. It is not about how to draw fractals on your computer. (I did have a lot of fun using a Macintosh to draw the pictures for the book, however. There will be occasional use of the Logo programming language for illustrative purposes.) Complete proofs of the main results will be presented, whenever that can reasonably be done. For some of the more difficult results, only the easiest non-trivial case of the proof (such as the case of two dimensions) is included here, with a reference to the complete proof in a more advanced text.

The main examples that will be considered are subsets of Euclidean space (in fact, usually two-dimensional Euclidean space); but as we will see, it is helpful to deal with the more abstract setting of "metric spaces".

This book deals with *fractal geometry*. It does not cover, for example, chaotic dynamical systems. That is a separate topic, although it is related. Another book of the same size could be written* on it. This book does not deal with the Mandelbrot set. Some writing on the subject has left the impression that "fractal" is synonymous with "Mandelbrot set"; that is far from the truth. This book does not deal with stochastic (or random) fractals; their rigorous study would require more background in probability theory than I have assumed for this book.

Prerequisites

(1) Experience in reading (and, preferably, writing) mathematical proofs is essential, since proofs are included here. I will say "necessary and sufficient" or "contrapositive" or "proof by induction" without explanation. Readers without such experience will only be able to read the book at a more superficial level by skipping many of the proofs. (Of course, no mathematics student will want to do that!)
(2) Basic abstract set theory will be needed. For example, the abstract notion of a function; finite vs. infinite sets; countable vs. uncountable sets.
(3) The main prerequisite is calculus. For example: What is a continuous function, and why do we care? The sum of an infinite series. The limit of a sequence. The least upper bound axiom (or property) for the real number system. Proofs of the important facts are included in many of the modern calculus texts.

Advice for the reader

Here is some advice for those trying to read the book without guidance from an experienced instructor. The most difficult (and tedious) parts of the book are probably Chaps. 2 and 5. But these chapters lead directly to the most important parts of the book, Chaps. 3 and 6. Most of Chap. 2 is independent

---

* For example, [16].

of Chap. 1, so to ease the reading of Chap. 2, it might be reasonable to intersperse parts of Chap. 2 with parts of Chap. 1. In a similar way, Chaps. 3, 4, and 5 are mostly independent of each other, so parts of these three chapters could be interspersed with each other.

There are many exercises scattered throughout the text. Some of them deal with examples and supplementary material, but many of them deal with the main subject matter at hand. Even though the reader already knows it, I must repeat: Understanding will be greatly enhanced by work on the exercises (even when a solution is not found). Answers to some of the exercises are given elsewhere in the book, but in order to encourage the reader to devote more work to the exercises, I have not attempted to make them easy to find. When an exercise is simply a declarative statement, it is to be understood that it is to be proved. (Professor Ross points out that if it turns out to be false, then you should try to salvage it.) Some exercises are easy and some are hard. I have even included some that I do not know how to solve. (No, I won't tell you which ones they are.)

Take a look at the Appendix. It is intended to help the reader of the book. There is an index of the main terms defined in the book; an index of notation; and a list of the fractal examples discussed in the text.

Some illustrations are not referred to in the main text. An instructor who knows what they are may use them for class assignments at the appropriate times.

Some of the sections that are more difficult, or deal with less central ideas, are marked with an asterisk (*). They should be considered optional. A section of "Remarks" is at the end of each chapter. It contains many miscellaneous items, such as: references for the material in the chapter; more sophisticated proofs that were omitted from the main text; suggestions for course instructors.

Notation

Most notation used here is either explained in the text, or else taken from calculus and elementary set theory. A few reminders and additional explanations are collected here.

Integers: $\mathbb{Z} = \{\cdots, -2, -1, 0, 1, 2, \cdots\}$.

Natural numbers or positive integers: $\mathbb{N} = \{1, 2, 3, \cdots\}$.

Real numbers: $\mathbb{R} = (-\infty, \infty)$.

Intervals of real numbers:

$$
\begin{aligned}
(a, b) &= \{\, x : a < x < b \,\}, \\
(a, b] &= \{\, x : a < x \le b \,\}, \quad \text{etc.}
\end{aligned}
$$

The notation (a, b) also represents an ordered pair, so the context must be used to distinguish.

Set difference or relative complement: $X \setminus A = \{\, x \in X : x \notin A \,\}$.

If $f\colon X \to Y$ is a function, and $x \in X$, I will use parentheses $f(x)$ for the value of the function at the point x; if $C \subseteq X$ is a set, I will use square brackets for the image set $f[C] = \{\, f(x) : x \in C \,\}$.

The union of a family $(A_i)_{i \in I}$ of sets, written

$$\bigcup_{i \in I} A_i,$$

consists of all points that belong to at least one of the sets A_i. The intersection

$$\bigcap_{i \in I} A_i$$

consists of the points that belong to all of the sets A_i. The family $(A_i)_{i \in I}$ is said to be **disjoint** iff $A_i \cap A_j = \varnothing$ for any $i \neq j$ in the index set I.

The **supremum** (or least upper bound) of a set $A \subseteq \mathbb{R}$ is written $\sup A$. By definition $u = \sup A$ satisfies (1) $u \geq a$ for all $a \in A$, and (2) if $y \geq a$ for all $a \in A$, then $y \geq u$. Thus, if A is not bounded above, we write $\sup A = \infty$, and if $A = \varnothing$, we write $\sup A = -\infty$. The **infimum** (or greatest lower bound) is $\inf A$. The **upper limit** of a sequence $(x_n)_{n=1}^{\infty}$ is

$$\limsup_{n \to \infty} x_n = \lim_{n \to \infty} \sup_{k \geq n} x_k.$$

And, if $\alpha(r)$ is defined for real $r > 0$,

$$\limsup_{r \to 0} \alpha(r) = \lim_{s \to 0} \sup_{0 < r < s} \alpha(r).$$

Similar notation is used for the **lower limit** or $\liminf$.

The sign $\square$ signals the end of a proof.

The origin of the book

I offered a course in 1987 at The Ohio State University on fractal geometry. It was intended for graduate students in mathematics, and it was based on the books of Hurewicz and Wallman [35] and Falconer [23]. I tried to keep the prerequisites at a low enough level that, for example, a graduate student in physics could take the course. The prerequisites listed were: metric topology and Lebesgue measure.* When the course was announced, I began getting inquiries from many other types of students, who were interested in studying fractal geometry more rigorously, but did not have even this minimal background. For example, a student in computer science with a strong background in calculus would still have required two more years of mathematics

---

* Then I found, to my surprise, that Lebesgue integration is not considered necessary for physics students. I suppose the fact that I find this incredible is an illustration of my ignorance of how mathematics is applied in practice.

study ("Advanced Calculus" and "Introductory Real Analysis") before being prepared for the course. This book is intended to fit this sort of student. Only a small part of those two courses is actually required for the study of fractal geometry, at least at the most elementary level. The required topics from metric topology and measure theory are covered in Chaps. 2 and 5. (Mathematics students may be able to skip much of these two chapters.)

This book is directly derived from notes prepared for use in a course offered in 1988 in connection with the program for talented high school students that is run here at The Ohio State University by Professor Arnold Ross for eight weeks every summer. The influence of these young students can be seen in many small ways in the book. (In particular, 1.5.7 and 1.6.1.) Past practice in the Ross program suggested the fruitful use of ultrametric spaces.

Parts of the manuscript were read by Manav Das, Don Leggett, William McWorter, Lorraine Rellick, and Karl Schmidt. Their comments led to many improvements in the manuscript. I would like to thank the many people at Springer-Verlag New York, especially mathematics editor Rüdiger Gebauer, mathematics assistant editor Susan Gordon, mathematics editor Ulrike Schmickler-Hirzebruch (editor of the "Undergraduate Texts in Mathematics" series), production editor Susan Giniger, and Kenneth Dreyhaupt.

. . .

Columbus, Ohio   *Gerald A. Edgar*
                                                        March 1990

Second Edition

Arnold Ross retired from his summer program in 2000 at age 94. He passed away on September 25, 2002. But the summer program has continued, with Professor Daniel Shapiro as director.

Although this Second Edition is substantially the same as the First, there were many changes. Most important was an increased emphasis on the packing measure (Sect. 6.2), so that now it is often treated on a par with the Hausdorff measure. The topological dimensions were rearranged (Chap. 3), so that the covering dimension is the major one, and the inductive dimensions are the variants. A "reduced cover class" notion was introduced to help in proofs for Method I or Method II measures (p. 149). Research results since 1990 that affect these elementary topics have been taken into account. Some examples have been removed (golden rectangle, Barnsley wreath). Other examples have been added (Barnsley leaf, Li lace, Julia set). Terminology has been changed (addressing function, self-referential equation, Lipschitz α, constituent). Notation has been changed (spectral radius). Most of the figures have been re-drawn.

Many readers of the first edition of this book provided comments and suggestions which have improved the second edition. Among these are: Richard B. Darst, Manav Das, Paolo Facchi & Saverio Pascazio, Julia Genyuk, Na Peng, Lorraine Rellick, John F. Rossi, M. Szyszliowicz, and Yang Ke. Springer has been helpful in the preparation, especially pure mathematics editor Mark Spencer and editorial assistant Charlene Cerdas. In addition to the software provided with Mac OS X, preparation of the book involved: ASC Logo, InkScape, Photoshop, TeXShop.

Columbus, Ohio   *Gerald A. Edgar*
                                                          August 2007

His days and times are past,
And my reliances on his fracted dates
Have smit my credit: I love and honor him
—W. Shakespeare, *Timon of Athens*

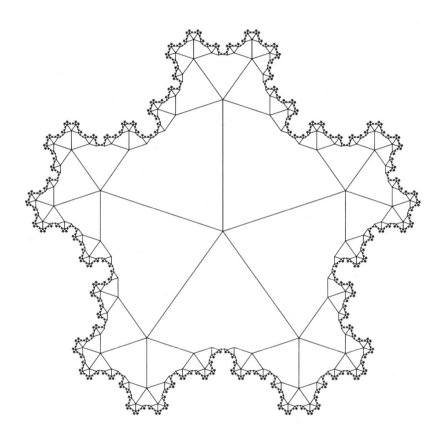

Contents

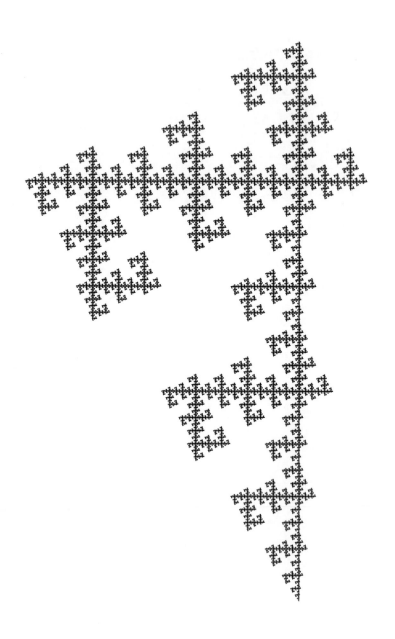

1

Fractal Examples

A few basic mathematical examples of fractals will be introduced in this chapter. Their analysis, and especially the question of what makes them "fractals", must be postponed until much later in the book.

One of the surprising ideas in the subject is the contention that the "dimension" of a set might be a real number that is not an integer. If we say that a set C in the plane has dimension 1.7, then we mean that its properties are "between" those of a curve (dimension 1) and an open region (dimension 2). The technical aspects of such "fractal" dimensions are discussed in Chap. 6.

Another characteristic feature of fractals emphasizes their difference from the sets of classical geometry. Typically, a fractal looks irregular; but more importantly, after it is magnified it still looks irregular. A typical set from classical geometry becomes very simple looking if it is magnified enough. One of the ways in which this behavior under magnification can be specified is through the use of an "iterated function system". Several of the examples in this chapter have descriptions in terms of iterated function systems, but the more detailed discussion is in Chap. 4.

1.1 The Triadic Cantor Dust

We will begin with an example. It will be treated in several different ways. Examples that we study later will often have something in common with one or more of the aspects treated for this example. This example is known as "the Cantor set". Mandelbrot has called it the "Cantor dust"; this descriptive name shows what kind of set it is. (The term "dust" refers to the fact that the set is zero-dimensional; this will be discussed in Chap. 3.)

Construction by Tremas

The *triadic Cantor dust* is a subset of the line $\mathbb{R}$. A sequence of approximations is first defined. Start with the closed interval $C_0 = [0, 1]$. Then the set C_1

is obtained by removing the middle third from $[0, 1]$, leaving $[0, 1/3] \cup [2/3, 1]$. The next set C_2 is defined by removing the middle third of each of the two intervals of C_1. This leaves

$$C_2 = [0, 1/9] \cup [2/9, 1/3] \cup [2/3, 7/9] \cup [8/9, 1].$$

And so on. (See Fig. 1.1.1.) The ***triadic Cantor dust*** is the "limit" C of the sequence C_n of sets. The sets decrease: $C_0 \supseteq C_1 \supseteq C_2 \supseteq \cdots$. So we will define the "limit" to be the intersection of the sets,

$$C = \bigcap_{k \in \mathbb{N}} C_k.$$

We will later do other constructions in a similar way. The parts that are removed are called ***tremas***. (Mandelbrot coined this word from a Latin word meaning "hole".)

The sequence of sets is defined recursively. This means that it will often be easy to prove facts about the sets by induction. For example, the set C_k consists of 2^k disjoint closed intervals, each of length $(1/3)^k$. So the ***total length*** of C_k, the sum of the lengths, is $(2/3)^k$. The limit is

$$\lim_{k \to \infty} \left(\frac{2}{3}\right)^k = 0.$$

So the total length of the Cantor dust itself is zero. (The mathematical version of "total length" is called "Lebesgue measure". It will be dealt with in Chap. 5). So total length is not a very useful way to compute the size of C. We will see that this is related to the fact that the fractal dimension of C is strictly less than 1.

Let us consider more carefully what points constitute the Cantor set. If $[a, b]$ is one of the closed intervals that makes up one of the approximations C_k, then the endpoints a and b belong to *all* of the future sets C_m, $m \geq k$, and therefore belong to the intersection C. (Again, prove this by induction.)

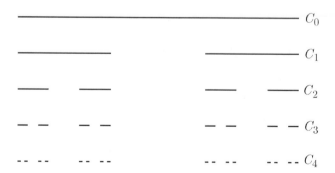

Fig. 1.1.1. The Triadic Cantor Dust

Taking all the endpoints of all the intervals of all the approximations C_k, we get an infinite set of points, all belonging to C. (This set of endpoints is only a countable set, however.)

It is important to note that there are points in C other than these end-points.

Exercise 1.1.2. The point $1/4$ is not an endpoint of any interval of any set C_k. But the point $1/4$ belongs to C.

It is also important to note that C is not like the usual sets of elementary geometry. At first, it is likely to tax your powers of geometrical visualization. If you can't imagine it today, try again tomorrow. Here are a few tidbits to help:

Exercise 1.1.3. (a) The set C contains no interval of positive length.
(b) The set C has no isolated points: that is, if $a \in C$, then for every $\varepsilon > 0$, no matter how small, the interval $(a - \varepsilon, a + \varepsilon)$ contains points of C in addition to a.
(c) The set C is closed: that is, if $a \in \mathbb{R}$ has the property that every interval of the form $(a - \varepsilon, a + \varepsilon)$ intersects C, then $a \in C$.

Exercise 1.1.4. Let r be a positive number. How many tremas are there with length $\geq r$?

Coordinates

There is a convenient way to characterize the elements of the triadic Cantor dust in terms of their expansions in base 3.

First we will review the standard facts concerning base 3. You know how expansions in the usual base 10 work. Base 3 is of course similar. Every positive integer x has a unique representation

$$x = \sum_{j=0}^{M} a_j 3^j,$$

where the "digits" a_j are chosen from the list $0, 1, 2$. For example,

$$15 = 0 \times 3^0 + 2 \times 3^1 + 1 \times 3^2.$$

We will sometimes write simply $15 = (120)_3$. It is understood that the sub-script specifying the base is always written in base ten. Similarly, we may represent fractions: Every number x between 0 and 1 has a representation in the form

$$x = \sum_{j=-\infty}^{-1} a_j 3^j,$$

with digits $0, 1, 2$. These are written with a "radix point" (or "ternary point" in this case):

$$7/9 = (0.21)_3,$$
$$1/4 = (0.02020202\cdots)_3 \quad \text{(a repeating expansion)},$$
$$\sqrt{2} = (1.1020112\cdots)_3 \quad \text{(non-repeating)}.$$

Some numbers (rational numbers of the form $a/3^k$) admit two different expansions. For example,

$$1/3 = (0.1000000\cdots)_3 = (0.0222222\cdots)_3.$$

Proposition 1.1.5. *Let $x \in [0, 1]$. Then x belongs to the triadic Cantor dust C if and only if x has a base 3 expansion using only the digits 0 and 2.*[*]

Proof. The first place to the right of the ternary point is a 1 if and only if x is between

$$(0.1000000\cdots)_3 = 1/3 \quad \text{and} \quad (0.1222222\cdots)_3 = 2/3.$$

The first trema is the interval $(1/3, 2/3)$. After this trema is removed, we have C_1. (The numbers $1/3$ and $2/3$ each have two expansions, one with a 1 in the first place, and one without. So they should not be removed.) So C_1 contains exactly the numbers in $[0, 1]$ that have a base 3 expansion not using 1 in the first place. The second place of a number x in C_1 is a 1 if and only if x belongs to one of the second-level tremas $(1/9, 2/9)$ or $(7/9, 8/9)$. When these tremas are removed, we have C_2. So C_2 contains exactly the numbers in $[0, 1]$ that have a base 3 expansion using 1 neither in the first place nor the second place. Continuing in this way, we see that the points remaining in $C = \bigcap_{k \in \mathbb{N}} C_k$ are exactly the numbers in $[0, 1]$ that have a base 3 expansion not using 1 at all. □

The Cantor dust is uncountable. This follows from the representation just proved, together with the observation that each real number has at most two representations base 3. (Actually, for numbers in the Cantor dust, two different sequences of 0's and 2's always represent different real numbers. See Exercise 1.6.2.)

Construction by Translations

Suppose L is a subset of $\mathbb{R}$ and s is a real number. The **translate** of L by s is the set

$$\{\, x + s : x \in L \,\}.$$

---

[*] This proposition will show that $1/4 \in C$, which is part of Exercise 1.1.2. But I hope you already solved it yourself. Was your solution for the special number $1/4$ simpler than this general case?

Fig. 1.1.6. Translation Construction

That is, we add s to each element of L. This is sometimes written $L + s$.

We construct recursively a sequence (L_k) of subsets of the line $\mathbb{R}$, together with a sequence (s_k) of real numbers. (Fig. 1.1.6.) Begin with the number $s_0 = 2/3$, and a starting set L_0 consisting of the one point 0. The next set L_1 is obtained by combining L_0 with its translate by s_0. So $L_1 = \{0, 2/3\}$. The next term in the sequence (s_k) will be $1/3$ times the previous one; $s_1 = (1/3)(2/3) = 2/9$. It will be used to obtain L_2 from L_1:

$$L_2 = L_1 \cup (L_1 + s_1) = \{0, 2/9, 2/3, 8/9\}.$$

Then $s_2 = (1/3)s_1$ and $L_3 = L_2 \cup (L_2 + s_2)$. And so on. The set of interest is the "limit" L of the sequence L_n. This sequence is increasing: $L_0 \subseteq L_1 \subseteq L_2 \subseteq \cdots$, so one reasonable definition for a limit would be the union:

$$L = \bigcup_{k \in \mathbb{N}} L_k.$$

But in fact, we will see later another (and, for our purposes, better) way to define the "limit" of a sequence of sets such as (L_k).

What is the connection between this construction and the previous construction of the Cantor dust C? The set L_k consists of 2^k points. They are exactly the left endpoints of the intervals that make up the set C_k. Or, they are the right endpoints of the tremas removed from $[0, 1]$ to construct C_k (plus the one point 0).

The points of L_k are the numbers in $[0, 1]$ that have a base 3 representation with k digits involving only 0's and 2's. (Prove this by induction on k.) For example, L_2 consists of

$$(0.00)_3 = 0,$$
$$(0.02)_3 = 2/9,$$
$$(0.20)_3 = 2/3,$$
$$(0.22)_3 = 8/9.$$

Certainly the set $L = \bigcup_{k \in \mathbb{N}} L_k$ is not equal to the Cantor dust C. The number $1/4$ does not belong to L. But L is "close" to C in the following sense.

Proposition 1.1.7. *If $x \in C$, then x is the limit of a sequence of points of L.*

Proof. Since $x \in C$, we know that x has a base 3 representation:

$$x = \sum_{j=1}^{\infty} a_j 3^{-j}, \qquad \text{each } a_j = 0 \text{ or } 2.$$

If this representation is truncated after only k terms, we get a number

$$x_k = \sum_{j=1}^{k} a_j 3^{-j},$$

which is an element of L_k. Now

$$|x - x_k| = \sum_{j=k+1}^{\infty} a_j 3^{-j} \le \sum_{j=k+1}^{\infty} 2 \cdot 3^{-j} = 3^{-k}.$$

Since $\lim_{k \to \infty} 3^{-k} = 0$, we may conclude that $\lim_{k \to \infty} x_k = x$. Thus $x \in C$ is the limit of the sequence $x_k \in L$.

The set L is **dense** in the Cantor set C. This means that $L \subseteq C$, and every point of C is the limit of a sequence of points of L.

Iterated Function System

Let $r > 0$ and $a \in \mathbb{R}$. The **dilation** on $\mathbb{R}$ with **ratio** r and **center** a is the function $f \colon \mathbb{R} \to \mathbb{R}$ given by $f(x) = rx + (1 - r)a$.

Consider the two dilations on $\mathbb{R}$ defined by:

$$f_1(x) = \frac{x}{3}, \qquad f_2(x) = \frac{x+2}{3}.$$

They both have ratio $1/3$. The first has center 0 and the second has center 1.

Proposition 1.1.9. *The triadic Cantor dust C satisfies the self-referential equation*

$$C = f_1[C] \cup f_2[C].$$

Fig. 1.1.8. Two dilations

Proof. It follows by induction that

$$C_{k+1} = f_1[C_k] \cup f_2[C_k]$$

for $k = 0, 1, \cdots$.

First I will prove that $C \subseteq f_1[C] \cup f_2[C]$. Suppose $x \in C$. Then $x \in C_1$. So either $x \in [0, 1/3]$ or $x \in [2/3, 1]$. We take the case in which $x \in [2/3, 1]$; the other case is similar. Now for any k, we know $x \in C_{k+1} = f_1[C_k] \cup f_2[C_k]$. But

$$f_1[C_k] \subseteq f_1[[0,1]] = \left[0, \frac{1}{3}\right],$$

so in fact $x \in f_2[C_k]$, or $3x - 2 \in C_k$. This is true for all k, so $3x - 2 \in \bigcap_{k \in \mathbb{N}} C_k = C$. Thus, $x \in f_2[C]$. In the other case, $x \in f_1[C]$. So in any case, we have $x \in f_1[C] \cup f_2[C]$.

Next I will prove that

$$C \supseteq f_1[C] \cup f_2[C].$$

Suppose $x \in f_1[C] \cup f_2[C]$. Either $x \in f_1[C]$ or $x \in f_2[C]$. We take the case $x \in f_2[C]$; the other case is similar. Thus, $3x - 2 \in C$. Now for any k, we know $3x - 2 \in C_k$, or $x \in f_2[C_k] \subseteq C_{k+1}$. Thus

$$x \in \bigcap_{k \in \mathbb{N}} C_{k+1} = \bigcap_{k \in \mathbb{N}} C_k = C.$$

This completes the proof that $C \supseteq f_1[C] \cup f_2[C]$.   □

We will call the pair (f_1, f_2) an ***iterated function system***, and we will say that C is the[*] ***invariant set*** (or ***attractor***) of the iterated function system (f_1, f_2).

Exercise 1.1.10. There are sets $A \neq C$ also satisfying $A = f_1[A] \cup f_2[A]$. How many can you find?

1.2 The Sierpiński Gasket

The next example is a set in the plane known as the ***Sierpiński gasket***.

Construction by tremas

Start with a filled-in equilateral triangle with side length 1 (the triangle together with the region inside). Call it S_0. It may be subdivided into four smaller triangles, using lines joining the midpoints of the sides. The smaller

---

[*] In Chap. 4, extra conditions will be added that will make the invariant set of an iterated function system unique, justifying the word "the".

triangles have side length $1/2$. The middle triangle is rotated 180 degrees compared to the others. The trema to be removed is the middle triangle (the "open triangle"—remove the interior but leave boundary of the triangle, the edges and vertices). After it is removed, the remaining set is S_1, a subset of S_0. Now each of the three remaining triangles should be subdivided into smaller triangles with edge length $1/4$, and the three middle triangles removed. The result is S_2, a subset of S_1. We should continue in the same way, to obtain a sequence S_k of sets. The **Sierpiński gasket** is the limit S of this sequence of sets. (Fig. 1.2.1.) The sequence is decreasing $(S_0 \supseteq S_1 \supseteq S_2 \supseteq \cdots)$, so by the "limit" we mean the intersection $S = \bigcap_{k \in \mathbb{N}} S_k$.

The set S_k consists of 3^k triangles, with side 2^{-k}. So the **total area** of S_k is $3^k \cdot (2^{-k})^2 \cdot \sqrt{3}/4$. This converges to 0 as $k \to \infty$. The total area of the Sierpiński gasket itself is therefore 0. Thus "area" is not very useful in measuring the size of S. Area is used to measure the size of a set of dimension 2. A line segment, which has dimension 1, has area 0. In a similar way, we will see that the Sierpiński gasket S can be said to have dimension less than 2.

The line segments that make up the boundary of one of the triangles of S_n remain in all the later approximations S_k, $k \geq n$. So the set S contains at least all of these line segments. In S_k there are 3^k triangles, each having 3 sides of length 2^{-k}. So the "total length" of S is at least $3^k \cdot 3 \cdot 2^{-k}$. This goes to ∞ as $k \to \infty$. So it makes sense to say that the total length of S is infinite. So "length" is not very useful to measure the size of S. Length is used to measure the size of a set of dimension 1. A square (with its inside), which has dimension 2, has infinite length, since it contains as many disjoint line segments as you like. In a similar way, we will see that the Sierpiński gasket S can be said to have dimension greater than 1.

So S supposedly has dimension greater than 1 but also less than 2. There is no integer between 1 and 2. The way around this dilemma, proposed by Hausdorff in 1919, is to allow the dimension of a set to be a fraction. According

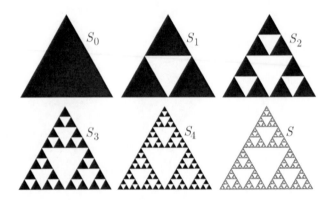

Fig. 1.2.1. The Sierpiński gasket

to Hausdorff's definition (Chap. 6), the Sierpiński gasket has dimension approximately 1.58.

Iterated Function System

Let $r > 0$ be a real number and let a be a point in the plane. The ***dilation*** with ratio r and center a is a map f of the plane to itself such that each point x is mapped to a point $f(x)$, which is on the ray from a through x, and the distance from a to $f(x)$ is r times the distance from a to x. By convention, $f(a) = a$ also. (Fig. 1.2.2. If $r < 1$, as shown, distances decrease; if $r > 1$, distances increase.)

Exercise 1.2.3. A dilation f maps lines to lines: that is, if L is a line, then the set $f[L] = \{ f(x) : x \in L \}$ is also a line. A dilation f preserves angles:* that is, if lines L_1 and L_2 meet at angle θ, then lines $f[L_1]$ and $f[L_2]$ also meet at angle θ.

The Sierpiński gasket was constructed above using approximations S_k. Let f_1, f_2, f_3 be the three dilations with ratio $1/2$ and centers at the three vertices of the triangle S_0. Now it follows by induction that

$$S_{k+1} = f_1[S_k] \cup f_2[S_k] \cup f_3[S_k].$$

Then, in much the same way as in Proposition 1.1.9, we can see that

$$S = f_1[S] \cup f_2[S] \cup f_3[S]$$

by showing each side is a subset of the other. This self-referential equation means that S is the ***invariant set*** of the iterated function system (f_1, f_2, f_3).

Coordinates

There is a description of the Sierpiński gasket in terms of coordinates.

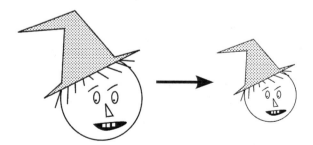

Fig. 1.2.2. A dilation of the plane

---

* We say that f is ***conformal***.

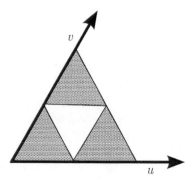

Fig. 1.2.5. Coordinate system

Exercise 1.2.4. Let coordinates (u, v) be defined in the plane with origin at one corner of the triangle S_0, and axes along two of the sides of S_0. Then coordinates (u, v) with $0 \leq u \leq 1$, $0 \leq v \leq 1$ represent a point of the Sierpiński gasket if and only if the base 2 expansions of u and v never have 1 in the same place.

Another description of the condition on the base 2 expansions of u and v is to say that the sum $u + v$ can be computed (base 2) without carrying. You will also need to take into account the numbers with two different expansions in base 2.

The preceding exercise may suggest a "translation" type construction for the Sierpiński gasket. Start with a single point. Choose two directions, at a 60 degree angle with each other. Start with a set L_0 containing that one point, and a number $s_0 = 1/2$. The next set is the union of three sets: L_0 and the translates of L_0 through distance s_0 in the two chosen directions. Then let $s_1 = (1/2)s_0$, and let L_2 consist of L_1 together with the translates of L_1 through distance s_1 in the two directions. And so on. (See Fig. 1.2.7.)

We say that a plane set L is **_dense_** in a set S iff $L \subseteq S$ and every point of S is the limit of a sequence of points of L.

Exercise 1.2.6. Prove that the union $L = \bigcup_{k \in \mathbb{N}} L_k$ is dense in the Sierpiński gasket S.

Consider Pascal's triangle:

$$
\begin{array}{ccccccccc}
& & & & 1 & & & & \\
& & & 1 & & 1 & & & \\
& & 1 & & 2 & & 1 & & \\
& 1 & & 3 & & 3 & & 1 & \\
1 & & 4 & & 6 & & 4 & & 1 \\
\end{array}
$$

etc.

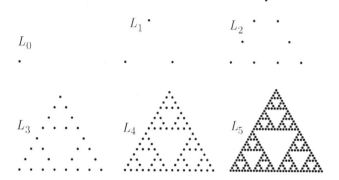

Fig. 1.2.7. Translation construction

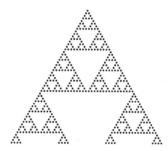

Fig. 1.2.8. Pascal's triangle modulo 2

Now, if we make a black dot wherever there is an odd number, and leave blank wherever there is an even number, we will get a geometric arrangement in the plane (Fig. 1.2.8).

Exercise 1.2.9. Why does the figure look like the Sierpiński gasket?

1.3 A Space of Strings

An *infinite binary tree* is pictured in Fig. 1.3.1. It is supposed to continue indefinitely at the top. Each *node* has two nodes immediately above it, called its *children*. (Sometimes it might be convenient to distinguish them as the *left child* and the *right child*.) The node at the bottom, with no parent, is called the *root* of the tree. (Sometimes the tree is drawn the other side up; then it looks less like a "tree", but terminology such as "child" is more reasonable.)

What is a more concrete model of this structure; a model that could be used to investigate the properties of an infinite binary tree?

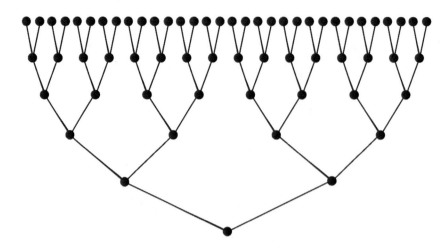

Fig. 1.3.1. Infinite Binary Tree

We consider two symbols, say 0 and 1. Then we consider finite **strings** (or **words**) made up of these symbols. For example

$$001010011$$

The number of symbols in a string α is called the **length** of the string, and written $|\alpha|$. The string above has length 9. How many strings of length n are there? By convention, we say that there is a unique string of length 0, called the **empty string**, which will be denoted Λ.

If α and β are two strings, then we may form a string $\alpha\beta$, called the **concatenation**, by listing the symbols of the string α followed by the symbols of the string β.

The set of all such finite strings (from the alphabet $E = \{0, 1\}$) can be identified with the infinite binary tree: The root of the tree corresponds to Λ; if α is a string, then the left child of α is $\alpha 0$ and the right child of α is $\alpha 1$.

We will write $E^{(n)}$ for the set of all strings of length n from the alphabet E. (Recall that $E^{(0)}$ has one element Λ.) We will write

$$E^{(*)} = E^{(0)} \cup E^{(1)} \cup E^{(2)} \cup \cdots$$

for the set of all finite strings. Because of the correspondence with the nodes of the infinite binary tree, we may sometimes refer to $E^{(*)}$ itself as the infinite binary tree. String α represents an ancestor of string β in the tree if and only if α is an **initial segment** of β, that is, $\beta = \alpha\gamma$ for some string γ. Or, α is a **prefix** of β. We will write $\alpha \leq \beta$ in this case. If $|\alpha| \geq n$, we write $\alpha{\restriction}n$ for the **initial segment** of α of length n; it consists of the first n symbols of the string α. Thus, $(001010011){\restriction}4 = 0010$.

Let A be a subset (finite or infinite) of the infinite binary tree $E^{(*)}$. A node $\gamma \in E^{(*)}$ is a **lower bound** for the set A iff $\gamma \leq \alpha$ for all $\alpha \in A$. (When we think of strings, we might say that γ is a "common prefix" for the set A.) A node β is a **greatest lower bound** for the set A iff β is a lower bound for A and $\gamma \leq \beta$ for any other lower bound γ for A. (In string language, β is the "longest common prefix" for the strings in the set A.)

Proposition 1.3.2. *Every nonempty subset A of the infinite binary tree $E^{(*)}$ has a unique greatest lower bound.*

Proof. First, choose some $\gamma \in A$. This is possible since A is not empty. Let $n = |\gamma|$ be the length of γ. Consider the integers k with $0 \leq k \leq n$. Some of them, for example 0, have the property that $\gamma{\restriction}k$ is a lower bound for A. Let k_0 be the largest such k. (A finite nonempty set of integers has a largest element.) I claim that $\gamma{\restriction}k_0$ is the desired greatest lower bound.

First, $\gamma{\restriction}k_0$ is a lower bound for A. Let β be any other lower bound for A. Now $\gamma \in A$, so $\beta \leq \gamma$. That means $\beta = \gamma{\restriction}k$ for some k with $0 \leq k \leq n$. By the definition of k_0, we know that $k \leq k_0$. So $\beta \leq \gamma{\restriction}k_0$. Thus $\gamma{\restriction}k_0$ is the greatest lower bound of A.

If α and β are both greatest lower bounds of the set A, then each is $\leq$ the other, so they are equal. □

Next, let $E^{(\omega)}$ be the set of *infinite* strings from the alphabet E. For $\sigma \in E^{(\omega)}$ there are initial segments $\sigma{\restriction}n$ of all sizes; we can think of an infinite string σ in terms of its initial segments, beginning with the empty string Λ:

$$\Lambda = \sigma{\restriction}0$$
$$< \sigma{\restriction}1$$
$$< \sigma{\restriction}2$$
$$< \cdots .$$

Another way to describe an element $\sigma \in E^{(\omega)}$ is as an infinite sequence of letters from $E = \{0, 1\}$. (The Greek letter omega, ω, is used in set theory to represent the least infinite ordinal. Here is shows the order type that is used for our infinite strings.)

If $\sigma \in E^{(\omega)}$ and $\alpha \in E^{(*)}$, then concatenation $\alpha\sigma \in E^{(\omega)}$ still makes sense. If $\sigma \in E^{(\omega)}$ and $n \in \mathbb{N}$, then the prefix $\sigma{\restriction}n \in E^{(n)}$ makse sense.

We will be interested in certain subsets of $E^{(\omega)}$. If $\alpha \in E^{(*)}$ is a finite string, let

$$[\alpha] = \left\{ \sigma \in E^{(\omega)} : \alpha \leq \sigma \right\},$$

the set of all infinite strings that begin with α. We may call this the **cylinder** defined by α. Now the two children of α are $\alpha 0$ and $\alpha 1$. The corresponding sets satisfy

$$[\alpha] = [\alpha 0] \cup [\alpha 1], \qquad [\alpha 0] \cap [\alpha 1] = \varnothing.$$

We can think of the usual base 2 system as defining a function h from strings $E^{(\omega)}$ to real numbers, by adding a binary point on the left. For example, the periodic string $\sigma = 001001001 \cdots$ corresponds to the real number

$$h(\sigma) = (0.001001001 \cdots)_2 = \frac{1}{7}.$$

The set $h\left[E^{(\omega)}\right]$ of values of h is exactly $[0, 1]$.

I have made assertions before about when two strings correspond to the same real number. Can you prove them?

Exercise 1.3.3. What are necessary and sufficient conditions on infinite strings $\sigma, \tau \in E^{(\omega)}$ so that $h(\sigma) = h(\tau)$?

A situation like the one just described will be called a "string model": the set of interest, such as $[0, 1]$, is related to a set $E^{(\omega)}$ of strings, called the "model", by a function, such as $h\colon E^{(\omega)} \to \mathbb{R}$, called the "model map". The model can be used to study the set of interest. If the model map h is understood, sometimes we may say that the infinite string σ is the **address** of the point $h(\sigma)$. Alternate terminology calls $E^{(\omega)}$ a **code space** and h the **addressing function**.

Here is a second example of a string model. The set of interest is the triadic Cantor dust C. The space of strings is again the set $E^{(\omega)}$ of infinite strings from the two-letter alphabet $\{0, 1\}$. But now the map $h\colon E^{(\omega)} \to \mathbb{R}$ is slightly different. Basically what we want to do is to have the letters 0 and 1 correspond to digits 0 and 2 respectively, and write the numbers in base 3.

So, for example, the periodic string $\sigma = 001001001 \cdots$ corresponds to the real number

$$h(\sigma) = (0.002002002 \cdots)_3 = \frac{1}{13}.$$

According to Proposition 1.1.5, the range of h is exactly the Cantor dust: $h\left[E^{(\omega)}\right] = C$. The model map h is related to the two dilations associated with C above:

Exercise 1.3.4. Let $h\colon E^{(\omega)} \to \mathbb{R}$ be the model map just defined. Then for all strings $\sigma \in E^{(\omega)}$,

$$h(0\sigma) = \frac{h(\sigma)}{3}$$

$$h(1\sigma) = \frac{h(\sigma) + 2}{3}.$$

1.4 Turtle Graphics

Many of the examples we will discuss are defined recursively. At least the finite approximations to the sets can be drawn by a computer graphics program.

When it seems useful, the Logo programs for the sets will be included in the book. In this section, the few elements of the Logo language are discussed. The main part of Logo that we will be concerned with is drawing with "turtle graphics".

Other computer languages could be used in place of Logo. The main requirement is the existence of graphics commands. Abelson and diSessa [1, Appendix B] discuss how to implement turtle graphics in other languages (BASIC, Pascal, APL, Lisp, Smalltalk). For example, in Pascal we would define routines FORWARD, BACK, LEFT, RIGHT, and then use the Logo programs with appropriate changes in syntax (parentheses and commas).

Logo

We should think of drawing in the plane, as represented by the computer screen. Our drawing instrument, known as a turtle, is pictured in Fig. 1.4.1(a). (On some versions of Logo, it may be simplified to a triangle.) Its properties include a "position" (a point in the plane) and a "heading" (the direction the turtle faces).

The commands forward and back make the turtle move. The argument is the distance to move, measured in some convenient units. As it moves, the turtle draws a line.

The commands left and right turn the turtle (change the heading). The angle is measured in degrees.

Between penup and pendown, no drawing occurs.

The command repeat can be used for repetition. Its first argument is the number of times to repeat, and the second argument is the list of commands to be repeated.

Variables can be used to store values. A colon preceding the name of the variable means that we want to refer to the value of the variable. A double-quote preceding the name of the variable means that we want to refer to the name of the variable itself. One of the Logo assignment statements is make. To assign a value to the variable, make needs to know its name, not its old

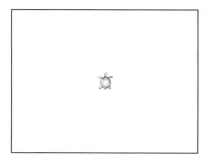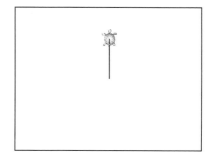

Fig. 1.4.1. (a) Turtle   (b) forward 50

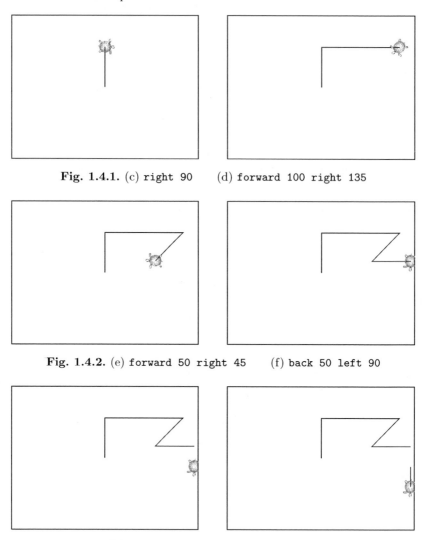

Fig. 1.4.1. (c) `right 90`   (d) `forward 100 right 135`

Fig. 1.4.2. (e) `forward 50 right 45`   (f) `back 50 left 90`

Fig. 1.4.3. (g) `penup forward 25`   (h) `pendown forward 25`

value. But on the other hand, `repeat` needs to know the value 5, not the name of the variable.

```
make "n 5
repeat :n [forward 50 left 360/:n]
```

The usual arithmetic can be performed: addition `:x + :y`, subtraction `:x - :y`, multiplication `:x * :y`, division `:x / :y`, square root `sqrt :x`. Some versions of Logo also have powers `:x ∧ :y`.

Commands can be combined to make new commands. Here is a definition of a command `polygon`. It will have as arguments the length of each side, and

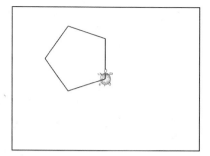

Fig. 1.4.4. `repeat 5 [forward 50 left 72]`

the number of sides. The key words `to` and `end` show such a definition. (Some versions of Logo do not use `to` and `end`, but have other methods of defining commands.)

```
to polygon :size :n
    repeat :n [forward :size left 360 / :n]
end
```

After `polygon` has been defined, it may be used like any other command (Fig. 1.4.5). Polygons with more and more sides (of shorter and shorter lengths) converge to a circle. (Convergence for sets is discussed in Sections 2.3 and 2.4.) Since the graphics screen of the computer (and the ink on this page, as well) has only finite resolution, drawing a regular polygon with enough sides is the same as drawing a circle.

Exercise 1.4.6. Write an ellipse program, using only the Logo commands discussed above.

The first argument of the `if` command is a condition to be tested. If it is true, then the second argument, a list of commands, is executed. If not, the third argument is executed. The `stop` command ends the execution of the routine. Control returns to wherever the routine was called from. Recursive

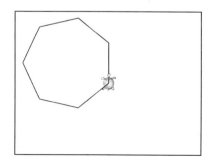
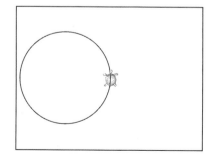

Fig. 1.4.5. (a) `polygon 50 7`   (b) `polygon 1 360`

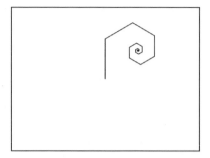

Fig. 1.4.7. `hideturtle spiral 50`

routines will be used frequently. In the following example, the routine `spiral` calls itself to draw a spiral at a smaller size. The `if` condition insures that it does not run indefinitely.

```
to spiral :size
    if :size < 1 [stop] [
        forward :size
        right 60
        spiral :size * 0.8]
end
```

1.5 Sets Defined Recursively

We will consider several sets that are defined recursively. Some of the ways used for defining the triadic Cantor dust and the Sierpiński gasket are recursive. Most of the examples in this section are "dragon" curves. They can be described well using a recursive Logo program.

The Koch Curve

The first construction is of "trema" type. (Fig. 1.5.1.) It begins with a triangle L_0 (including the interior) having angles of 120, 30, 30 degrees. This triangle can be subdivided into three smaller triangles: two isosceles triangles, angles 120, 30, 30, with long sides along the short sides of the original triangle; and one equilateral triangle. For the next approximation L_1, the trema to be removed is the equilateral triangle. Then we repeat, again and again. The Koch curve is the "limit". (Why is it called a "curve"? See Proposition 2.4.10.)

A second construction for the Koch curve is of "dragon" type. It involves approximations that are polygons. Here is a program in the Logo language. This is a good example of a "recursive program". The procedure `Koch` calls itself four times to draw four copies of the curve at 1/3 the size. But each of

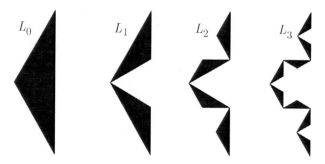

Fig. 1.5.1. Koch curve

these four calls of Koch calls four more. And so on. The variable depth is used to end the recursion at a certain number of levels.

```
to Koch :depth :size
   if :depth = 0 [forward :size stop] [
     Koch :depth - 1 :size / 3
     left 60
     Koch :depth - 1 :size / 3
     right 120
     Koch :depth - 1 :size / 3
     left 60
     Koch :depth - 1 :size / 3]
end
```

The first set P_0 is a line segment: its picture (Fig. 1.5.2) is obtained by executing Koch 0 200. The value 200 is simply a convenient size. The next approximation P_1 is obtained by executing Koch 1 200. And so on. The set

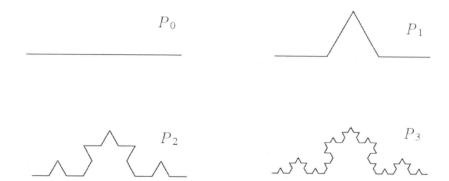

Fig. 1.5.2. Results of program Koch

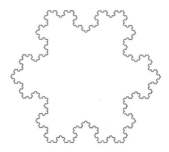

Fig. 1.5.3. Snowflake

P_k consists of 4^k line segments of length 3^{-k}. The **Koch curve** is the "limit" P of the sequence P_k.

The Koch curve can be obtained as an iterated function system construction. For example, it is made up of 4 parts, each similar to the whole. See Plate 8. (Similarities are discussed in Sect. 2.1.)

Three copies of the Koch curve, originating from three sides of an equilateral triangle, form a simple closed curve, often known as the **snowflake curve** (Fig. 1.5.3).

Heighway's Dragon

(See Fig. 1.5.4.) Heighway's dragon is a set in the plane. The approximation P_0 is a line segment of length 1. The next approximation is P_1; it is obtained

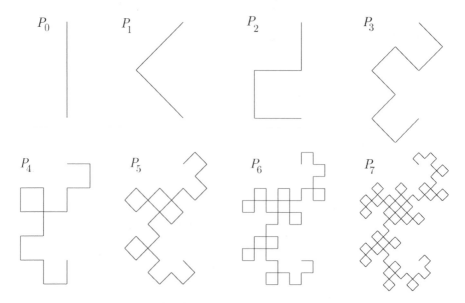

Fig. 1.5.4. Heighway's dragon

from P_0 by replacing the line segment by a polygon with two segments, each of length $1/\sqrt{2}$, joined at a right angle. The two ends are the same as before. (There are two choices of how this can be done. We choose the one on the "left" side.) For P_2, each line segment in P_1 is replaced by a polygon with two segments, each having length $1/\sqrt{2}$ times the length of the segment that is replaced. The choices alternate between left and right, starting with left, counting from the endpoint on the bottom. **_Heighway's dragon_** is the "limit" P of this sequence P_n of polygons.

The program below is shorter than the description given above, and (with a little study) is less prone to be misinterpreted.* This is a good reason for using Logo to describe the construction.

```
; Heighway's Dragon
make "factor 1 / sqrt 2
to heighway :depth :size :parity
   if :depth = 0 [forward :size stop] [
     left :parity * 45
     heighway :depth - 1 :size * :factor 1
     right :parity * 90
     heighway :depth - 1 :size * :factor (-1)
     left :parity * 45]
end
```

The third argument is supposed to be either 1 or -1, depending on whether the next corner is supposed to go to the left or to the right. To generate P_0, execute `heighway 0 200 1`. To generate P_1, execute `heighway 1 200 1`.

Proposition 1.5.5. *All of the approximations P_n remain in some bounded region of the plane.*

Proof. Every point of P_0 has distance at most 1 from the endpoint. Every point of P_1 has distance at most $1/2$ from some point of P_0. We can see by induction, that every point of P_k has distance at most $(1/\sqrt{2})^{k+1}$ from some point of P_{k-1}. Therefore, any point of P_k has distance from the endpoint at most

$$1 + \sum_{j=1}^{k} \left(\frac{1}{\sqrt{2}}\right)^{j+1} < 1 + \sum_{j=1}^{\infty} \left(\frac{1}{\sqrt{2}}\right)^{j+1}.$$

This is a geometric series with ratio < 1, so it converges to a finite value. $\square$

The following exercise may be easier to approach after the discussion of similarities (Sect. 2.1) and convergence of a sequence sets (Sects. 2.4 and 2.5).

Exercise 1.5.6. There are two similarities f_1 and f_2 of the plane onto itself, with ratio $1/\sqrt{2}$, so that Heighway's dragon P satisfies the equation $P = f_1[P] \cup f_2[P]$.

---

* Did you figure out what I was saying about "left" and "right"?

Proposition 1.5.7. *In an approximation P_k of Heighway's dragon, the polygon never crosses itself.*

Proof. It is possible that P_k visits some point twice. But I will show that it does not traverse any line segment twice, and when it visits a point twice, it does not cross itself. The vertices of the polygon P_k lie on a square lattice L_k with edge length $(1/\sqrt{2})^k$. Suppose P_k visits a point twice without visiting an entire line segment twice; then that point must be a vertex of the square lattice. At each vertex, the polygon has a right-angle corner. So it does not cross itself there.

I must only show that P_k does not include a line segment more than once. If it does, it must include a complete edge of the square lattice. Let k be an integer such that P_k does not include any line segment more than once. Let S be some square of the square lattice L_k. I will show that P_{k+1} does not include more than once any of the four line segments of L_{k+1} inside S.

Let t be an edge of L_{k+1} inside S. If t were traversed by P_{k+1} more than once, then the two sides of S adjacent to t (e_1 and e_2 in the figure) must both be traversed by P_k. Now let us color the squares of L_k in a checker-board pattern. The square to the left of the first line segment of P_k will be colored black, and the squares will alternate white and black. Since there is a right-angle turn between two consecutive segments of P_k, the square to the left is always the black square. By convention, let us say that the first line segment is "vertical". The edges alternate between vertical and horizontal. Now when P_{k+1} is constructed, the new pairs of edges are placed alternately to the left and to the right of the old edges. So the new edges corresponding to a vertical edge of P_k will be in the black square bordered by the edge, and the new edges corresponding to a horizontal edge of P_k will be in the white square bordered by the edge. So not both of the edges e_1 and e_2 will produce new edges inside S. $\square$

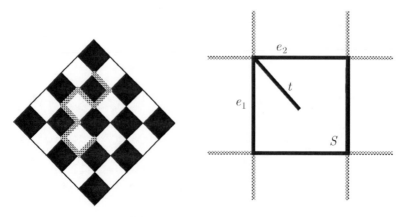

Fig. 1.5.7. (a) Square Lattice.   (b) The Square S

Fudgeflake

Using an angle other than 90 degrees, we can obtain variant forms of Heighway's dragon. In Fig. 1.5.8, the angle is 120 degrees. (For this version, left and right have been reversed). Three copies of this curve, joined in an equilateral triangle, surround a set known as the **fudgeflake**. A fudgeflake is made up of 3 small fudgeflakes; so the fudgeflake tiles the plane (p. 39).*

Sierpiński dragon

Here is another dragon. (See Fig. 1.5.9.)

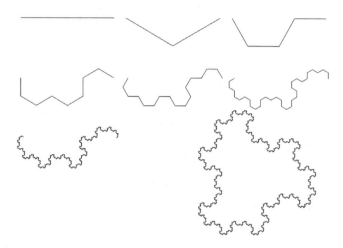

Fig. 1.5.8. 120-degree dragon and fudgeflake

Fig. 1.5.9. Sierpiński dragon

---

* A set with fractal boundary that tiles the plane is known as a **fractile**.

```
to SD :depth :size :parity
  if :depth = 0 [forward :size stop] [
    left 60 * :parity
    SD :depth - 1 :size / 2 (-:parity)
    right 60 * :parity
    SD :depth - 1 :size / 2 :parity
    right 60*:parity
    SD :depth - 1 :size / 2 (-:parity)
    left 60 * :parity]
end
```

Exercise 1.5.10. What does this dragon construction have to do with the Sierpiński gasket?

McWorter's Pentigree

Fig. 1.5.11 illustrates another dragon; we will call it *McWorter's pentigree*. It is a subset of the plane. The first approximation P_0 is a line segment. In any future stage P_n, each line segment of P_n is replaced by six line segments in a particular pattern to form P_{n+1}. Here is a Logo program. Why was the value $(3 + \sqrt{5})/2$ chosen for the value of the variable shrink?

```
; McWorter's pentigree
make "shrink (3 + sqrt 5) / 2
to pent :depth :size
  if :depth = 0 [forward :size stop] [
    left 36
    pent :depth - 1 :size / :shrink
    left 72
    pent :depth - 1 :size / :shrink
```

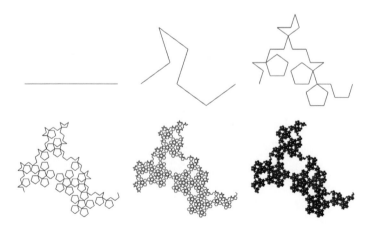

Fig. 1.5.11. McWorter's Pentigree

```
      right 144
      pent :depth - 1 :size / :shrink
      right 72
      pent :depth - 1 :size / :shrink
      left 72
      pent :depth - 1 :size / :shrink
      left 72
      pent :depth - 1 :size / :shrink
      right 36]
  end
```

Exercise 1.5.12. Does an approximation P_n for McWorter's pentigree ever cross itself?

Five copies of the pentigree fit together to form a set with five-fold rotational symmetry (Plate 1). This set will also be called "the second form of McWorter's pentigree". It can be thought of as made up of 6 sets similar to the whole, with ratio $2/(3 + \sqrt{5}) = (3 - \sqrt{5})/2$ (Plate 2).

Consider the "translation" construction illustrated in Fig. 1.5.13. Set L_0 is a single point. Set L_1 is obtained from L_0 by translating L_0 in 5 equally spaced directions by some distance s_0 together with L_0 itself. Set L_2 is obtained from L_1 by translating it in 5 directions (the opposites of the previous directions)

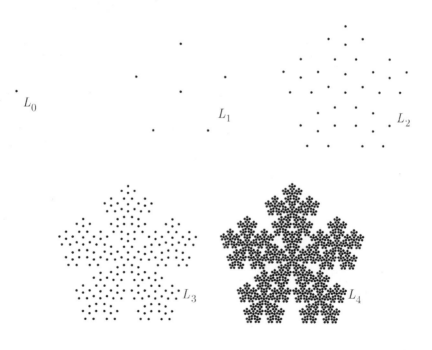

Fig. 1.5.13. Translation Construction

by distance $s_1 = rs_0$, where $r = (3 - \sqrt{5})/2$. Set L_3 is obtained by translating L_2 in the original 5 directions by distance $s_2 = rs_1$. And so on.

Exercise 1.5.14. What does this construction have to do with McWorter's pentigree?

Barnsley's Leaf

Here is a set defined recursively in Logo form. If `:depth` is large enough so that you can watch it being drawn, you may see that most line-segments are re-traced many times. Some results are shown in Fig. 1.5.15. The limit B of these approximations will be called **Barnsley's leaf**.

```
make "stwo sqrt 2
to leaf :depth :size
  if :depth < 1 [forward 2 * :size back 2 * :size] [
    forward :size
    leaf :depth - 2 :size / 2
    back :size
    left 45
    leaf :depth - 1 :size / :stwo
    right 90
    leaf :depth - 1 :size / :stwo
    left 45]
end
```

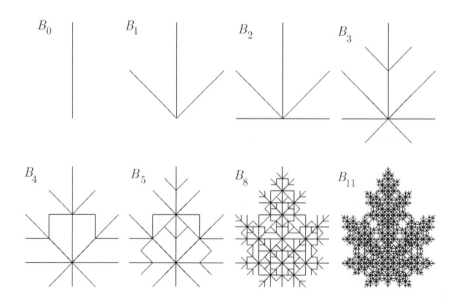

Fig. 1.5.15. Barnsley's leaf construction

Notice that one recursive copy is shrunk by factor $1/2$, and two copies are shrunk by factor $1/\sqrt{2}$.

Exercise 1.5.16. Determine three similarities f_1, f_2, f_3 of the plane into itself so that Barnsley's leaf B satisfies the self-referential equation $B = f_1[B] \cup f_2[B] \cup f_3[B]$.

A Julia Set

The maps in an iterated function system used to construct a set need not be similarities. Write $\mathbb{C}$ for the set of complex numbers, geometrically thought of as a Euclidean plane. We briefly discuss the Julia set for a function $\varphi \colon \mathbb{C} \to \mathbb{C}$ given by $\varphi(z) = z^2 + c$. Here, c is a fixed complex number. (In the pictures, I used $c = -0.15 + 0.72i$ because I liked the look of the result.) The map φ has inverse f defined by $f(z) = \sqrt{z - c}$. Actually, a complex number (except zero) has two square-roots, so there are two inverses, say $f_0(z) = \sqrt{z - c}$ and $f_1(z) = -\sqrt{z - c}$. It won't matter how we choose one of the two square-roots for f_0 because we will use them both. The Julia set in this case is a nonempty compact set $J \subseteq \mathbb{C}$ that satisfies the self-referential equation[*]

$$J = f_0[J] \cup f_1[J].$$

A construction for J will work much as in the previous examples. See Fig. 1.5.17. Start with a set J_0. (I used a parallelogram connecting four points which turn out to be the extremities of J.) Then recursively define

$$J_{n+1} = f_0[J_n] \cup f_1[J_n].$$

The limit of the sequence J_n is J. The limit of a sequence of sets will be discussed in Sect. 2.5.

Another possibility starts with J_0 as the circle $|z| = 2$. This was chosen so that the image J_1 is a curve that lies inside that circle. And subsequent images J_n will curves, each be inside its predecessor. Plate 14 shows all of these drawn in the same picture, with the bands between successive curves colored in different colors. Points inside all the curves are black, this is the filled-in Julia set. This picture is the "escape time" coloring often seen (for example [3, plate 13]) but this time constructed using the inverses of φ.

Reflections in a Circle (on a Sphere)

Some fractal examples lie on the surface of a sphere. To represent them in drawings on a flat page, we sometimes use "stereographic projection" to set

---

[*] To get uniqueness, we exclude the point c where the two square-roots coincide, thus requiring that $J \subseteq \mathbb{C} \setminus \{c\}$. Otherwise, the filled-in Julia set would be a second possibility. Compact sets are discussed in Sect. 2.3.

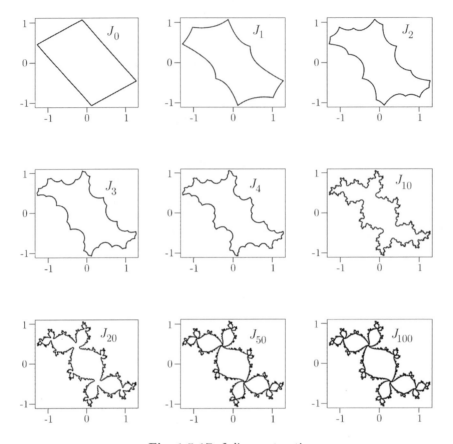

Fig. 1.5.17. Julia construction

up a correspondence between the sphere Σ and a plane Π. In Fig. 1.5.18, the horizontal plane Π is tangent to the "south pole" of the sphere Σ. Each point S on Σ (except the north pole itself) together with the north pole N determines a line NS that meets Π in a unique point P. The correspondence $S \leftrightarrow P$ is called **stereographic projection**.

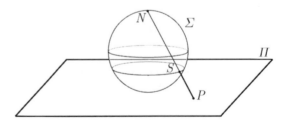

Fig. 1.5.18. Stereographic projection

Fig. 1.5.20. Reflection in a circle

Exercise 1.5.19. Under stereographic projection, a circle C on the sphere Σ maps to either (1) a circle on Π or (2) a line on Π.

Maps of the sphere that are easy to visualize, such as rotations about a diameter, may not be as easy to visualize as maps of the plane. And vice versa. If C is a circle on Σ, we want to define a map of Σ to itself described as "reflection in the circle" C. When C is a great circle, that is easy to imagine as a map of Σ to itself. When the circle passes through the north pole N, then reflection in C becomes reflection in a line in Π, which is easy to imagine. For a general circle we may proceed as follows: rotate the sphere to map the circle C to a circle C' through N. Then reflect in C' which is reflection in a line in the stereographic projection. Finally, rotate back to the starting orientation.

Exercise 1.5.21. Let the set $\mathbb{C}$ of complex numbers be identified with the plane Π in such a way that the equator of Σ corresponds to the "unit circle" $K = \{ z \in \mathbb{C} : |z| = 1 \}$ of the complex plane. The reflection in K corresponds to reflection in the equator. Show that this reflection, in terms of complex numbers, is described by

$$z \mapsto \frac{1}{\bar{z}}.$$

Of course the point $z = 0$ corresponds to the south pole, and that reflects to the north pole, so the image of 0 under the reflection in K is not a point of $\mathbb{C}$.

Figure 1.5.22(a) depicts **Pharaoh's breastplate**, which is a fractal constructed on a sphere. First we choose circles (called "generators") as in Fig. 1.5.22(b): six circles, each tangent to four of the others. The picture is the stereographic projection—the dashed circle is the equator (not one of the six); the vertical line down the middle is of course a circle through the

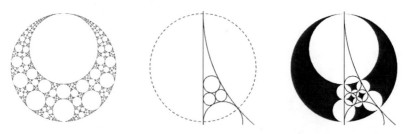

Fig. 1.5.22. (a) Pharaoh's breastplate   (b) Generators   (c) Osculating basis

north pole. These six circles define six reflection maps $f_1, \cdots, f_6$ of the sphere to itself. The fractal P is the smallest nonempty closed* set invariant under all six of these reflections: $P = f_1[P] = f_2[P] = \cdots = f_6[P]$.

There is an efficient way to draw the fractal. We define an "osculating basis" of eight disks, each with bounding circle orthogonal to three of the generators and not meeting the other three generators. In Fig. 1.5.22(c), these disks are in white. One of them is the northern hemisphere (in the stereographic projection, the outside of the unit circle).

The first approximation P_0 is the entire sphere. The first tremas U_0 to be removed are the open disks of the osculating basis. This leaves the the approximation P_1. The next tremas to be removed are the images of U_0 under the six reflections. (Some of the images are repeats.) This leaves the set P_2. We continue in this way recursively, and the limit P is the intersection of the approximations P_n.

A colorful image is contructed as follows. Color the disks of the osculating basis in up to eight colors, then when you take an image of a trema, use the same color. The complement of P will then be colored in the colors chosen. See Plate 9.

Fig. 1.5.23. Construction by tremas

---

* *Closed set* is defined on p. 47.

1.6 Number Systems

Let us consider the usual decimal representation of real numbers, and how it can be generalized. Suppose we have a number b for the base (or "radix") of our number system, and a finite set $D = \{d_1, d_2, \cdots, d_k\}$ of numbers, called "digits". We will always assume that 0 is one of the digits. A "whole number" for this system will have the form

$$\sum_{j=0}^{M} a_j b^j \tag{1}$$

where each $a_j \in D$. Write W for the set of whole numbers. A "fraction" for this system will have the form

$$\sum_{j=-\infty}^{-1} a_j b^j \tag{2}$$

where each $a_j \in D$. Write F for the set of fractions. The general number represented by this system is the sum of one of each:

$$\sum_{j=-\infty}^{M} a_j b^j. \tag{3}$$

In order for the representations (2) to converge, we must have $|b| > 1$. Our usual decimal number system has

$$b = 10 \quad \text{and} \quad D = \{0, 1, 2, 3, 4, 5, 6, 7, 8, 9\}.$$

The binary number system has

$$b = 2 \quad \text{and} \quad D = \{0, 1\}.$$

In these two cases, the numbers of the form (1) are exactly the nonnegative integers, and the numbers of the form (2) are the elements of the interval $[0, 1]$. So the numbers of the form (3) are exactly the nonnegative real numbers, $[0, \infty)$. In both of these cases, there are numbers that have no representation in the form (3), namely the negative numbers.

Exercise 1.6.1. Let $b = -2$ and $D = \{0, 1\}$. The numbers of the form (1) are exactly the integers. The numbers of the form (2) constitute the closed interval $[-2/3, 1/3]$. Every real number has the form (3).

Both the decimal and binary systems have the fault that some numbers have two different representations. For example

$$0 \times 10^{-1} + \sum_{j=2}^{\infty} 9 \times 10^{-j} = 1 \times 10^{-1} + \sum_{j=2}^{\infty} 0 \times 10^{-j}.$$

A system related to the Cantor set avoids this fault.

Exercise 1.6.2. Let $b = 3$ and $D = \{0, 2\}$. Then no number has two different representations in the form (3).

But in this case, some numbers have no representation at all (such as $1/2$). Can we avoid both problems at once?

Exercise 1.6.3. Let b be a real number with $|b| > 1$, and let D be a finite set of real numbers including 0. Then either some real number has no expansion in the form (3) or some real number has more than one expansion in the form (3).

Next, let us consider a number system to represent complex numbers. Now the base b may be a complex number, and the digit-set D is a finite set of complex numbers (including 0). We are interested in representing complex numbers in the form

$$\sum_{j=-\infty}^{M} a_j b^j, \tag{3}$$

where all $a_j \in D$.

Exercise 1.6.4. Let b be a complex number with $|b| > 1$, and let D be a finite set of complex numbers including 0. Then either some complex number has no expansion in the form (3) or some complex number has more than one expansion in the form (3).

Exercise 1.6.5. If every complex number has an expansion of the form (3), then in fact there is a complex number with at least three different expansions in the form (3).

One useful property that a number system for complex numbers might have is that the set W of whole numbers is the set of "algebraic integers" of a number field.* Two examples: The complex numbers of the form $u + iv$, where u and v are integers; they are called the **Gaussian integers**. For a second example, let

$$\omega = \frac{-1 + i\sqrt{3}}{2} = \cos\frac{2\pi}{3} + i\sin\frac{2\pi}{3}.$$

Note that $\omega^3 = 1$, and $\omega^2 = \overline{\omega} = 1/\omega$. The complex numbers of the form $u + v\omega$ (u and v integers) are sometimes known as the **Eisenstein integers**.[†]

Exercise 1.6.6. Let $b = -1 + i$ and $D = \{0, 1\}$. Describe the set W of whole numbers (1) for this number system.

---

[*] An algebraic integer is a complex number that is a zero of a polynomial such that the coefficients are integers and the coefficient of the highest-degree term is 1. The number field $\mathbb{Q}(i)$ is made up of the complex numbers of the form $u + vi$, where u and v are rational numbers. It turns out that such a number is an algebraic integer if and only if u and v are both integers.

[†] They are the algebraic integers of the number field $\mathbb{Q}(\sqrt{-3})$.

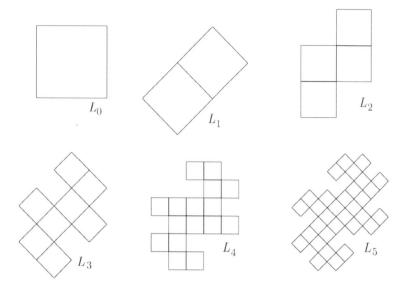

Fig. 1.6.7. Twindragon Construction

A construction of the set F of "fractions" for the number system with base $b = -1 + i$ and digit set $D = \{0, 1\}$ is illustrated in Fig. 1.6.7. For the first approximation, we consider the set W of "whole numbers", in this case the Gaussian integers; they form a regular square lattice S_0 in the complex plane. The points that are closer to 0 than to any other Gaussian integer form a square. Call this square L_0. Next, consider the numbers representable in our system using at most one place to the right of the radix point. They, too, form a square lattice S_1 in the complex plane, but with shorter sides, and at an angle. The set of points closer to 0 than to any other element of S_1 is a square, and the set of points closer to $(.1)_{-1+i}$ than to any other element of S_1 is also a square. These two squares constitute a set L_1, the next approximation of F. When we take two digits to the right of the radix point, we get a set L_2 made up of four squares. Continuing in this way, we obtain F as the limit of a sequence (L_k) of approximations.

Fig. 1.6.8. Twindragon

The set obtained in this way is the **twindragon**. Plate 4 suggests that it is made up of two copies of Heighway's dragon. Since every complex number can be represented in this number system, the plane is covered by countably many twindragons, namely the sets $w + F$, one for each Gaussian integer w. The sets $w + F$ overlap only in their boundaries, so this constitutes a tiling of the plane. The twindragon (we will see later) has a fractal boundary. Therefore the twindragon is a "fractile".

This set F can be seen from the point of view of an iterated function system. The set F is the union of two parts, namely the set F_0 of all numbers of the form (2) with $a_{-1} = 0$, and the set F_1 of all numbers of the form (2) with $a_{-1} = 1$. Now the elements of F_0 are exactly b^{-1} times the elements of F; and the elements of F_1 are of the form $b^{-1} + b^{-1}x$, where $x \in F$. So if we write

$$f_0(x) = b^{-1}x,$$
$$f_1(x) = b^{-1} + b^{-1}x,$$

then we have a self-referential equation

$$F = f_0[F] \cup f_1[F].$$

So F is the invariant set for this iterated function system. See Plate 5.

Exercise 1.6.9. Find a complex number with three representations in this system.

Let us turn next to the **Eisenstein** number system.

Exercise 1.6.10. Let $b = -2$ and $D = \{0, 1, \omega, \omega^2\}$. Describe the set W of whole numbers.

For calculations in this system, it may be easier to write $A = \omega$ and $B = \omega^2$. The set F of fractions is pictured in Fig. 1.6.11. See also Plate 6.

Fig. 1.6.11. Eisenstein fractions

Exercise 1.6.12. Describe an iterated function system for the Eisenstein fractions.

The Eisenstein fractions make up another "fractile".

1.7 *Remarks

Fractal sets, such as those we have seen in this chapter, have been used by mathematicians over the years. Only since Mandelbrot's book has there been interest in them beyond mathematics. Even among mathematicians, these sets had often been considered to have little interest.

I recall a discussion I had with one of my colleagues some years ago. We were talking about a problem dealing with abstract integration. In the course of the discussion, it became apparent that I considered the Cantor set a more natural setting for the problem than the interval. The colleague was surprised by that. He had thought that the Cantor set was only a pathological counterexample. I wonder what he would have said about the Sierpiński gasket.

In 1883, Georg Cantor published a description of the set that today bears his name. It is known as the "Cantor discontinuum", the "Cantor middle-thirds set", or simply the "Cantor set". Over the years it has come to occupy a special place at the heart of descriptive set theory. It was the most important example used by Hausdorff in his paper [32] on fractional dimension.

The "iterated function system" was named and popularized by Michael Barnsley [2], [3].

Wacław Sierpiński published his description of the "Sierpiński gasket" in 1915. This name was assigned to it by Mandelbrot. The pre-Mandelbrot literature calls it something like "Sierpiński's other curve". (It was called a curve, despite its appearance, because it has topological dimension 1; see Chap. 3. "Sierpiński's curve" was already used to refer to another example, called Sierpiński's carpet—*dywan Sierpińskiego* [44, p. 144].)

The use of "strings" in this book may be a bit unusual. The more conventional terminology involves "sequences" from the set $\{0, 1\}$, so there is no essential difference. Many students will be familiar with strings from their computer-related courses (but probably not infinite strings). Sequences are used in other ways in this book, so I hope this terminology will reduce the confusion a little.

The description of dragon curves is done naturally using recursive computer programs. Instead of using a pseudo-code to formulate such programs, or (worse yet) inventing another ad-hoc notation, I have chosen to use a real computer language. One of my favorite languages, especially for drawing pictures, is Logo. Some programmers have negative opinions concerning Logo, because it has been used to teach small children. But that should not be held against it. There are programming tasks that I would not use Logo for, but the simple recursive pictures that are of concern in this book seem suited to Logo.

A reference for turtle graphics is [1]. It contains some interesting ideas on plane geometry. It contains a solution for Exercise 1.4.6:

```
to ellipse :s :e
    make "n 1
    repeat 360 [right :n forward :s left :n
        left :n forward (:s * :e) right :n
        make "n :n + 1]
end
```

If you show the turtle, and ask him to do `ellipse 1 0.5`, you will see that he is doing a lot of work.*

H. von Koch's curve dates from 1904. It is a continuous curve that has a tangent line nowhere. The closed "snowflake" version of Fig. 1.5.3 is sometimes used as an example of a curve of infinite length surrounding a finite area. We will see later that the snowflake curve has fractal dimension strictly larger than 1. This is a much more precise assertion than merely saying the curve has infinite length.

Heighway's dragon dates from about 1967; according to Martin Gardner [28], it was discovered by physicist John E. Heighway, and studied by Heighway together with physicists Bruce A. Banks and William G. Harter. This dragon was publicized in [13], which contains a wealth of information about the polygonal approximations. Proposition 1.5.7, which states that a polygonal approximation does not cross itself, was observed by Harter and Heighway; a proof was published by Davis and Knuth [13]. The proof of Proposition 1.5.7 that appears here contains elements of a proof submitted by Brian Conrad, Jon Grantham, and Roger Lee during the Ross summer program. The name "Heighway" is spelled "Heightway" in some of the references on the subject.

The fudgeflake is found in [44, p. 72]. Mandelbrot considered this shape to be derived by "fudging" the snowflake (Fig. 1.5.3) using alternating left and right. William McWorter described the pentigree in [50]. "Pentigree" is from "pentagon-filligree". The first edition of this book contains the first published analysis of this interesting dragon. Slightly changing the definition yields another interesting result, the pentadendrite (see Plate 3).

```
make "offangle 11.82
make "shrink 1 / 2.87
to dend :depth :size
    if :depth = 0 [forward :size stop] [
        left :offangle
        dend :depth - 1 :size * :shrink
        left 72
        dend :depth - 1 :size * :shrink
        right 72
```

---

* This was true in 1990. But nowadays computers may be so fast that you won't see this.

```
dend :depth - 1 :size * :shrink
right 144
dend :depth - 1 :size * :shrink
left 72
dend :depth - 1 :size * :shrink
left 72
dend :depth - 1 :size * :shrink
right :offangle]
end
```

Michael Barnsley's leaf is found in [4, p. 330 and Fig. 4.8]. This is part of his discussion of methods to concoct an iterated function system that approximates a previously given picture.

Julia sets are named for Gaston Julia. He and Pierre Fatou are both credited with creating the theory of iteration in the complex plane at about the same time (1918). There was a sometimes bitter priority dispute between the two of them about this material.

The fractal called Pharaoh's breastplate was described by Mandelbrot in [46, p. 126].

The fractal sets associated with complex number systems are discussed, for example, in [13], [29], [30]. The statement of Exercise 1.6.1 was improved by a proof submitted by Dan Bernstein, Keith Conrad, and Paul Lefelhocz during the Ross summer program. Exercises 1.6.3 and 1.6.4 are stated as "either/or". It is possible that *both* alternatives occur: there is some number with more than one expansion, while there is another number with no expansion. For example, modify the Eisenstein number system (Exercise 1.6.10) by using base $b = -2$, but using only 3 of the digits, say $D = \{0, 1, \omega\}$. (Thanks to Peter Hinow for this example.)

Here is a hint for Exercise 1.1.2. To see that $1/4$ is not an endpoint, prove by induction that all endpoints are of the form $m/3^k$, for nonnegative integers m, k. Then use unique factorization to show that $1/4$ is not of this form. To show that $1/4 \in C$, prove by induction on k that $1/4 \in C_k$ for all k. In fact, it is easier if you prove more: both $1/4$ and $3/4$ are in C_k for all k.

> Si elle était douée de vie,
> il ne serait pas possible de l'anéantir,
> sans la supprimer d'emblée
> car elle renaîtrait sans cesse
> des profondeurs de ses triangles,
> comme la vie dans l'univers.
> —E. Cesàro, 1905

> *If it [the Koch curve] were given life,*
> *it would not be possible to destroy it*
> *except by doing away with it all at once,*
> *since it would be endlessly reborn*
> *of the depths of its triangles,*
> *like life in the universe.*

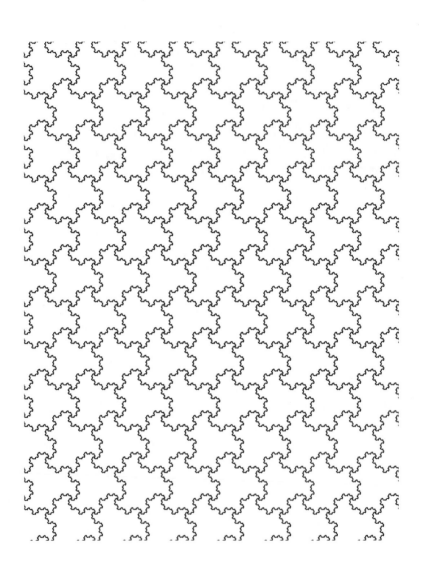

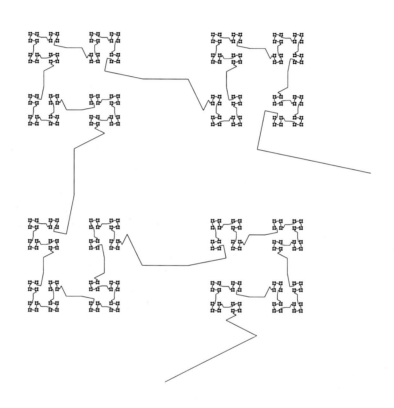

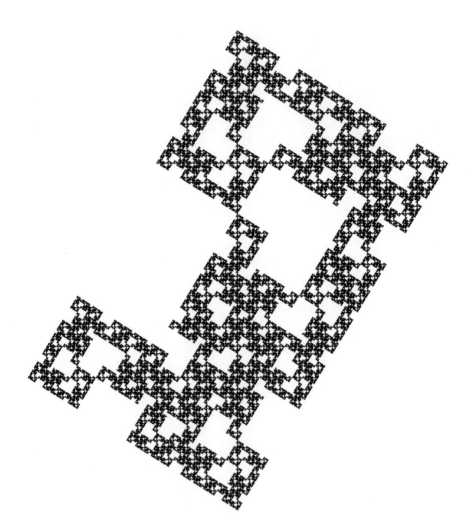

2

Metric Topology

This chapter contains the mathematical background for much of the rest of the book. If the book were organized in strictly logical fashion, then this would be the first chapter of the book; but I included instead some more fractal-like material as Chap. 1. Chapter 2 is a more technical chapter. Have patience! It really is useful for the understanding of the rest of the book.

Mathematics students will eventually learn almost everything in this chapter in the normal course of their studies. So many readers may be able to skip this chapter completely; but it is here for those who need it. Many of the proofs (and exercises) are merely the usual real-number proofs adapted to the setting of metric spaces. So a student who has experience dealing with the proofs of ordinary calculus will see many familiar ideas. A reader who does not care about being mathematically rigorous* could skip the proofs in this chapter. Metric topology is, in fact, important for a lot of modern mathematics. The selection of topics for this chapter was determined by what is required later in the book; so this chapter is a bit peculiar as an introduction to metric topology.

2.1 Metric Space

A **metric space** is a set S together with a function $\varrho \colon S \times S \to [0, \infty)$ satisfying

$$\varrho(x, y) = 0 \quad \Longleftrightarrow \quad x = y;$$
$$\varrho(x, y) = \varrho(y, x);$$
$$\varrho(x, z) \leq \varrho(x, y) + \varrho(y, z).$$

The last inequality is known as the **triangle inequality**: in Euclidean geometry, it says that the sum of the lengths of two sides of a triangle is at least

---

* "A simple man believes every word he hears; a clever man understands the need for proof." (Proverbs 14:15, *New English Bible*)

equal to the length of the third side. The nonnegative real number $\varrho(x, y)$ is called the **distance** between x and y. The function ϱ itself is called a **metric** on the set S. A metric space may be written as a pair (S, ϱ), but if the metric is understood, it will be referred to simply as S.

Examples

Let us consider a few examples of metric spaces.

Theorem 2.1.1. *The set $\mathbb{R}$ of real numbers, with the function*

$$\varrho(x, y) = |x - y|$$

is a metric space.

Proof. First, note that $|x - y| \geq 0$. Also, $|x - y| = 0$ if and only if $x = y$. Next, $\varrho(x, y) = |x - y| = |-(x - y)| = |y - x| = \varrho(y, x)$. For the triangle inequality, let us consider several cases:

(1) $x \leq y \leq z$: Then $\varrho(x, y) + \varrho(y, z) = (y - x) + (z - y) = z - x = \varrho(x, z)$.
(2) $x \leq z \leq y$: Then $\varrho(x, y) + \varrho(y, z) = (y - x) + (y - z) \geq y - x \geq z - x = \varrho(x, z)$.
(3) $y \leq x \leq z$: Then $\varrho(x, y) + \varrho(y, z) = (x - y) + (z - y) \geq z - y \geq z - x = \varrho(x, z)$.
(4) $y \leq z \leq x$: Then $\varrho(x, y) + \varrho(y, z) = (x - y) + (z - y) \geq x - y \geq x - a = \varrho(x, z)$.
(5) $z \leq y \leq x$: Then $\varrho(x, y) + \varrho(y, z) = (x - y) + (y - z) = x - z = \varrho(x, z)$.
(6) $z \leq x \leq y$: Then $\varrho(x, y) + \varrho(y, z) = (y - x) + (y - z) \geq y - z \geq x - z = \varrho(x, z)$. $\square$

Exercise 2.1.2. When is the triangle inequality actually an equality in the metric space $\mathbb{R}$?

If d is a positive integer, then $\mathbb{R}^d$ is the set of all ordered d-tuples of real numbers. We can define several operations in this setting. For $x = (x_1, x_2, \ldots, x_d) \in \mathbb{R}^d$, $y = (y_1, y_2, \ldots, y_d) \in \mathbb{R}^d$, and $s \in \mathbb{R}$, define

$$sx = (sx_1, sx_2, \ldots, sx_d),$$
$$x + y = (x_1 + y_1, x_2 + y_2, \ldots, x_d + y_d),$$
$$x - y = x + (-1)y,$$
$$|x| = \sqrt{x_1^2 + x_2^2 + \cdots + x_d^2}.$$

We define d-**dimensional Euclidean space** to be the set $\mathbb{R}^d$ with the metric $\varrho(x, y) = |x - y|$.

In order to show that this is a metric space , I will prove two basic inequalities.

Theorem 2.1.3 (Cauchy's inequality). *Let x_1, $x_2, \cdots, x_d$, y_1, $y_2, \cdots, y_d$ be 2d real numbers. Then*

$$\left(\sum_{j=1}^{d} x_j y_j \right)^2 \le \left(\sum_{j=1}^{d} x_j^2 \right) \left(\sum_{j=1}^{d} y_j^2 \right).$$

Proof. If λ is any real number, then

$$\sum_{j=1}^{d} (x_j - \lambda y_j)^2 \ge 0.$$

Multiplying this out and collecting terms, we see

$$\left(\sum_{j=1}^{d} y_j^2 \right) \lambda^2 - 2 \left(\sum_{j=1}^{d} x_j y_j \right) \lambda + \left(\sum_{j=1}^{d} x_j^2 \right) \ge 0.$$

This is true for all real numbers λ. But in order for a quadratic polynomial $A\lambda^2 + B\lambda + C$ to be nonnegative for all λ, it is necessary that $B^2 - 4AC \le 0$. In this case, it means

$$4 \left(\sum_{j=1}^{d} x_j y_j \right)^2 - 4 \left(\sum_{j=1}^{d} y_j^2 \right) \left(\sum_{j=1}^{d} x_j^2 \right) \le 0,$$

which is equivalent to the inequality to be proved. $\qquad\square$

Theorem 2.1.4 (Minkowski's inequality). *Let $x, y \in \mathbb{R}^d$. Then $|x + y| \le |x| + |y|$.*

Proof. Write $x = (x_1, x_2, \cdots, x_d)$ and $y = (y_1, y_2, \cdots, y_d)$. Then

$$|x + y|^2 = \sum_{j=1}^{d} (x_j + y_j)^2$$

$$= \sum_{j=1}^{d} x_j^2 + 2 \sum_{j=1}^{d} x_j y_j + \sum_{j=1}^{d} y_j^2$$

$$\le \sum_{j=1}^{d} x_j^2 + 2 \left(\sum_{j=1}^{d} x_j^2 \right)^{1/2} \left(\sum_{j=1}^{d} y_j^2 \right)^{1/2} + \sum_{j=1}^{d} y_j^2$$

$$= \left(\left(\sum_{j=1}^{d} x_j^2 \right)^{1/2} + \left(\sum_{j=1}^{d} y_j^2 \right)^{1/2} \right)^2$$

$$= (|x| + |y|)^2.$$

By taking the square root of the extremes, we may conclude (since both of these terms are nonnegative)

$$|x + y| \leq |x| + |y|.$$ $\square$

Corollary 2.1.5. *The space $\mathbb{R}^d$ is a metric space with the metric $\varrho(x,y) = |x - y|$.*

Proof. Write $x = (x_1, x_2, \cdots, x_d)$ and $y = (y_1, y_2, \cdots, y_d)$. First,

$$\varrho(x,y) = \sqrt{(y_1 - x_1)^2 + (y_2 - x_2)^2 + \cdots + (y_d - x_d)^2} \geq 0.$$

If $\varrho(x,y) = 0$, then $(y_1 - x_1)^2 + (y_2 - x_2)^2 + \cdots + (y_d - x_d)^2 = 0$. But a square is nonnegative, so this means that all terms must be 0. That is, $x_j = y_j$ for all j, so that $x = y$. The equation $\varrho(x,y) = \varrho(y,x)$ is clear. For the triangle inequality, we apply Minkowski's inequality:

$$\varrho(x,y) + \varrho(y,z) = |x - y| + |y - z|$$
$$\geq |(x - y) + (y - z)| = |x - z| = \varrho(x,z).$$ $\square$

Exercise 2.1.6. Find necessary and sufficient conditions for Cauchy's inequality to be an equality in $\mathbb{R}^d$.

Exercise 2.1.7. Find necessary and sufficient conditions for Minkowski's inequality to be an equality in $\mathbb{R}^d$.

Next we will consider the set $E^{(\omega)}$ of infinite strings from the two-letter alphabet $E = \{0, 1\}$. We will define a metric $\varrho_{1/2}$ for this space. The basic idea is that two strings should be considered "close" if they begin in the same way.

So let σ and τ be two infinite strings. If $\sigma = \tau$, then of course the distance must be zero:

$$\varrho_{1/2}(\sigma, \sigma) = 0.$$

If $\sigma \neq \tau$, then there is a first time they disagree: we can write $\sigma = \alpha\sigma'$, $\tau = \alpha\tau'$, where α is a (possibly empty) finite string, and the first character of σ' is different than the first character of τ'. (In the language used before, α is the longest common prefix of σ and τ.) If k is the length of α, then define

$$\varrho_{1/2}(\sigma, \tau) = \left(\frac{1}{2}\right)^k.$$

Proposition 2.1.8. *The set $E^{(\omega)}$ is a metric space under the metric $\varrho_{1/2}$.*

Proof. Clearly $\varrho_{1/2}(\sigma, \tau) \geq 0$. If $\sigma \neq \tau$, then $\varrho_{1/2}(\sigma, \tau) = (1/2)^k > 0$. The equation $\varrho_{1/2}(\sigma, \tau) = \varrho_{1/2}(\tau, \sigma)$ is also clear.

So all that remains is the triangle inequality. Let σ, τ, θ be three strings. I will prove $\varrho_{1/2}(\sigma, \tau) \leq \max\{\varrho_{1/2}(\sigma, \theta), \varrho_{1/2}(\theta, \tau)\}$. If two of the strings are equal, then this is clear. So suppose they are all different. Let k be the length of the longest common prefix of σ and θ, and let m be the length of the longest common prefix of θ and τ. If $n = \min\{k, m\}$, we know that the first n letters of σ agree with the first n letters of θ; and the first n letters of θ agree with the first n letters of τ. Therefore, the first n letters of σ agree with the first n letters of τ. So the longest common prefix of σ and τ has length at least n. Therefore:

$$\begin{aligned} \varrho_{1/2}(\sigma, \tau) \leq (1/2)^n &= (1/2)^{\min\{k, m\}} \\ &= \max\{(1/2)^k, (1/2)^m\} \\ &= \max\{\varrho_{1/2}(\sigma, \theta), \varrho_{1/2}(\theta, \tau)\}. \end{aligned}$$

Finally, this "ultra-triangle" inequality implies the ordinary triangle inequality, since

$$\max\{\varrho_{1/2}(\sigma, \theta), \varrho_{1/2}(\theta, \tau)\} \leq \varrho_{1/2}(\sigma, \theta) + \varrho_{1/2}(\theta, \tau). \qquad \square$$

Related Definitions

If S is a metric space with metric ϱ, and $T \subseteq S$, then T is also a metric space with metric ϱ_T defined by

$$\varrho_T(x, y) = \varrho(x, y) \qquad \text{for } x, y \in T.$$

In the future, we will usually write simply ϱ rather than ϱ_T.

The **diameter** of a subset A of a metric space S is

$$\operatorname{diam} A = \sup\{\varrho(x, y) : x, y \in A\}.$$

The diameter of A is the distance between the two most distant points of A, if such points exist. But, for example, if $A = [0, 1)$, the diameter is 1. Even though no two points of A have distance exactly 1, there are pairs x, y of points of A with distance as close as we like to 1; and there are no pairs x, y of points of A with distance greater than 1.

If A and B are nonempty sets in a metric space S, define the **distance** between them by

$$\operatorname{dist}(A, B) = \inf\{\varrho(x, y) : x \in A, y \in B\}.$$

Note that this is not a metric, for example because the triangle inequality fails. If $A = 0$ and $B = (0, 1]$ in $\mathbb{R}$, then $\operatorname{dist}(A, B) = 0$ even though $A \neq B$.

Let S be a metric space, $x \in S$, and $r > 0$. The **open ball** with center x and radius r is the set $B_r(x) = \{y \in S : \varrho(y, x) < r\}$. The **closed ball** with center x and radius r is the set $\overline{B}_r(x) = \{y \in S : \varrho(y, x) \leq r\}$.

Let S be a metric space, and let A be a subset. An **interior point** of A is a point x so that $B_\varepsilon(x) \subseteq A$ for some $\varepsilon > 0$. A set A is called an **open** set iff every point of A is an interior point.

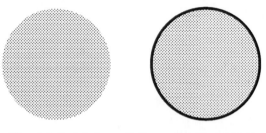

Fig. 2.1.9. (a) Open Ball   (b) Closed Ball

Proposition 2.1.10. *An open ball $B_r(x)$ is an open set.*

Proof. Let $y \in B_r(x)$. Then $\varrho(x, y) < r$, so that $\varepsilon = r - \varrho(x, y)$ is positive. The triangle inequality shows that $B_\varepsilon(y) \subseteq B_r(x)$. So y is an interior point of $B_r(x)$. □

Theorem 2.1.11. *Let S be a metric space. Then $\varnothing$ and S are open sets. If U and V are open sets, so is $U \cap V$. If $\mathcal{U}$ is any family of open sets, then the union*

$$\bigcup_{U \in \mathcal{U}} U$$

is also open.

Proof. Certainly every point of $\varnothing$ has whatever property I choose to name, such as being an interior point. So $\varnothing$ is an open set.

Let $x \in S$. Then certainly $B_1(x) \subseteq S$. So S is an open set.

Suppose U and V are both open. Let $x \in U \cap V$. Then x is an interior point of U, so there is $\varepsilon_1 > 0$ with $B_{\varepsilon_1}(x) \subseteq U$. Also, x is an interior point of V, so there is $\varepsilon_2 > 0$ with $B_{\varepsilon_2}(x) \subseteq V$. Therefore, if ε is the minimum of ε_1 and ε_2, then we have $B_\varepsilon(x) \subseteq U \cap V$. So $U \cap V$ is an open set.

Let $\mathcal{U}$ be a family of open sets, and write

$$V = \bigcup_{U \in \mathcal{U}} U.$$

Let $x \in V$. Then $x \in U$ for some $U \in \mathcal{U}$. So there is $\varepsilon > 0$ with $B_\varepsilon(x) \subseteq U \subseteq V$. Therefore V is an open set. □

Let S be a metric space, and let $A \subseteq S$. A point $x \in S$ is an **accumulation point** of A iff, for every $\varepsilon > 0$, the ball $B_\varepsilon(x)$ contains points of A other than x. A set A is **closed** iff it contains all of its accumulation points. Comparing this to the definition of "open set", we can easily see that a set A is closed if and only if its complement $S \setminus A$ is open.

Exercise 2.1.12. Let S be a metric space. Then $\varnothing$ and S are closed sets. If A and B are closed sets, so is $A \cup B$. If $\mathcal{C}$ is any family of closed sets, then the intersection

$$\bigcap_{A \in \mathcal{C}} A$$

is also closed.

Proposition 2.1.13. *A closed ball $\overline{B_r}(x)$ is a closed set.*

Proof. Suppose $y \notin \overline{B_r}(x)$. Then $\varrho(x, y) > r$, so that $\varepsilon = \varrho(x, y) - r$ is positive. The triangle inequality shows that $B_\varepsilon(y) \cap \overline{B_r}(x) = \varnothing$. Therefore y is not an accumulation point of $\overline{B_r}(x)$. This shows that $\overline{B_r}(x)$ is a closed set.   □

A family $\mathcal{B}$ of open subsets of a metric space S is called a **base for the open sets** of S iff, for every open set $A \subseteq S$, and every $x \in A$, there is $U \in \mathcal{B}$ such that $x \in U \subseteq A$. What would be a good definition for a "base for the closed sets"?

Exercise 2.1.14. A family $\mathcal{B}$ of open subsets of a metric space S is a base for the open sets if and only if every open set T is of the form

$$T = \bigcup_{A \in \mathcal{A}} A$$

for some $\mathcal{A} \subseteq \mathcal{B}$.

Of course the definition of "open set" shows that the collection of all open balls is a base for the open sets of S.

An **ultrametric** space S is a metric space for which the metric ϱ satisfies the **ultra-triangle inequality**:

$$\varrho(x, z) \leq \max\{\varrho(x, y), \varrho(y, z)\}.$$

Note that in Proposition 2.1.8, we proved that $(E^{(\omega)}, \varrho_{1/2})$ is an ultrametric space.

The properties of an ultrametric space may seem strange if you are familiar only with Euclidean space and its subsets. Here are a few examples to help you understand the situation.

Exercise 2.1.15. Let S be an ultrametric space. Prove:

(1) Every triangle is isosceles: that is, if $x, y, z \in S$, then at least two of $\varrho(x, y), \varrho(y, z), \varrho(x, z)$ are equal.
(2) A ball $B_r(x)$ of radius r has diameter at most r.
(3) Every point of a ball is a center: that is, if $y \in B_r(x)$, then $B_r(x) = B_r(y)$.
(4) A closed ball is an open set.
(5) An open ball is a closed set.

2.2 Metric Structures

Metric spaces support many of the concepts that are well-known from Euclidean space. We will discuss functions on metric spaces and sequences in metric spaces.

Functions on Metric Spaces

Suppose S and T are metric spaces. A function $h\colon S \to T$ is an **isometry** iff

$$\varrho_T(h(x), h(y)) = \varrho_S(x, y)$$

for all $x, y \in S$. Two metric spaces are **isometric** iff there is an isometry of one onto the other. A "property" is called a **metric** property iff it is preserved by isometry, that is: if S and T are isometric, and one has the property, then so does the other.

What are the isometries of the Euclidean plane $\mathbb{R}^2$ into itself? Some examples are pictured in Fig. 2.2.3. In fact, the maps of these types are the only isometries of $\mathbb{R}^2$. (But, of course, there are infinitely many isometries of each of the four types.) Here is an outline of the proof:

(1) Let ABC and $A'B'C'$ be two congruent triangles in the plane. Any point P determines a corresponding point P' such that

$$|A - P| = |A' - P'|, |B - P| = |B' - P'|, \text{ and } |C - P| = |C' - P'|.$$

(2) Likewise another point Q yields Q', and $|P - Q| = |P' - Q'|$.
(3) Any two congruent triangles are related by a unique isometry.
(4) Two given congruent line segments AB, $A'B'$ are related by just two isometries: one direct and one opposite.
(5) Any isometry with an invariant point is a rotation or a reflection.
(6) An isometry with no invariant point is a translation or a glide-reflection.

A detailed argument along these lines may be found in [12, Chap. 3].

A function $h\colon S \to T$ is a **similarity** iff there is a positive number r such that

$$\varrho(h(x), h(y)) = r\varrho(x, y)$$

for all $x, y \in S$. (An alternate term is **similitude**.) The number r is the **ratio** of h. Two metric spaces are **similar** iff there is a similarity of one onto the other.

Exercise 2.2.1. Let f be the dilation of $\mathbb{R}^2$ with center a and ratio $r > 0$. Then

$$|f(x) - f(y)| = r\,|x - y|$$

for all $x, y \in \mathbb{R}^2$.

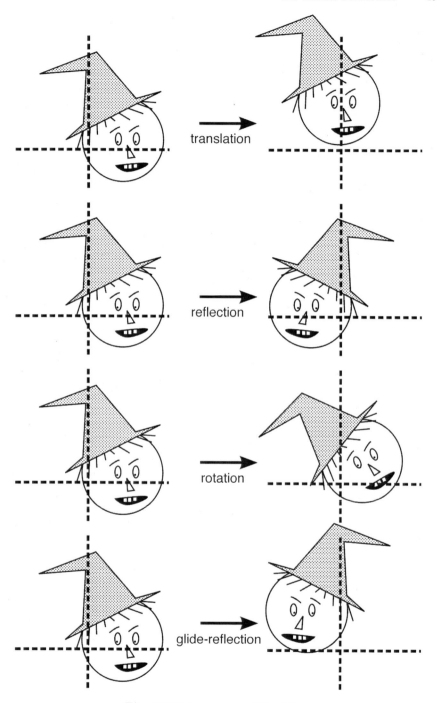

Fig. 2.2.3. Isometries of the Plane

Exercise 2.2.2. Describe all of the similarities of two-dimensional Euclidean space $\mathbb{R}^2$ onto itself.

Let S and T be metric spaces. Let $x \in S$. A function $h \colon S \to T$ is **continuous** at x iff, for every $\varepsilon > 0$, there is $\delta > 0$ such that for all $y \in S$

$$\varrho(x, y) < \delta \quad \Longrightarrow \quad \varrho(h(x), h(y)) < \varepsilon.$$

The function h is simply called **continuous** iff it is continuous at every point $x \in S$. This is one of the ten most important definitions in all of mathematics. A thorough understanding of it will be useful to you not only in the study of fractal geometry, but also in much of the other mathematics you will study.

Exercise 2.2.4. Let $h \colon S \to T$, let $x \in S$, and let $\varepsilon, \delta > 0$. Then we have

$$\varrho(x, y) < \delta \quad \Longrightarrow \quad \varrho(h(x), h(y)) < \varepsilon$$

for all $y \in S$ if and only if

$$h\big[B_\delta(x)\big] \subseteq B_\varepsilon\big(h(x)\big).$$

Exercise 2.2.5. Isometries and similarities are continuous functions.

Continuity can be phrased in terms of open sets.

Theorem 2.2.6. *A function $h \colon S \to T$ is continuous if and only if $h^{-1}[V]$ is open in S for all V open in T.*

Proof. First, suppose that h is continuous. Let V be an open set in T. I must show that $h^{-1}[V]$ is an open set in S. So let $x \in h^{-1}[V]$. Then $h(x) \in V$, which is open, so there is $\varepsilon > 0$ with $B_\varepsilon(h(x)) \subseteq V$. By the continuity of h, there is $\delta > 0$ such that $h[B_\delta(x)] \subseteq B_\varepsilon(h(x)) \subseteq V$. Therefore $B_\delta(x) \subseteq h^{-1}[V]$. So $h^{-1}[V]$ is an open set.

Conversely, suppose that $h^{-1}[V]$ is open in S whenever V is open in T. Let $x \in S$. I must show that h is continuous at x. Let $\varepsilon > 0$. then $B_\varepsilon(h(x))$ is an open set in T. So $W = h^{-1}[B_\varepsilon(h(x))]$ is an open set in S. Now $x \in W$, so there is $\delta > 0$ with $B_\delta(x) \subseteq W$. Therefore $h[B_\delta(x)] \subseteq h[W] \subseteq B_\varepsilon(h(x))$. So h is continuous. $\qquad\square$

Exercise 2.2.7. Let $\mathcal{B}$ be a base for the open sets of T. A function $h \colon S \to T$ is continuous if and only if $h^{-1}[V]$ is open in S for all $V \in \mathcal{B}$.

A function $h \colon S \to T$ is a **homeomorphism** of S onto T iff it is bijective, and both h and h^{-1} are continuous. Two metric spaces are **homeomorphic** iff there is a homeomorphism of one onto the other. A "property" is known as a **topological** property iff it is preserved by homeomorphism.

Sequences in a Metric Space

Let S be a set. A **_sequence_** in S is, strictly speaking, a function $f\colon \mathbb{N} \to S$. It is defined by the infinite list of values $f(1), f(2), f(3), \cdots$. Often we will write something like

$$(x_n)_{n \in \mathbb{N}}$$

and understand that the function is specified by $f(1) = x_1$, $f(2) = x_2$, and so on. We may even write simply (x_n).

A sequence (x_n) in a metric space S **_converges_** to the point $x \in S$ iff for every $\varepsilon > 0$, there is $N \in \mathbb{N}$ so that $\varrho(x_n, x) < \varepsilon$ for all $n \geq N$. If this happens, we write $\lim_{n \to \infty} x_n = x$, or $x_n \to x$. Also, x is called the **_limit_** of the sequence (x_n). We say that the sequence is **_convergent_** iff it converges to some point.

In order for the definitions to make sense the way they have been worded here, we need to know that limits are unique:

Exercise 2.2.8. Let (x_n) be a sequence in a metric space S. If $x_n \to a$ and $x_n \to b$, then $a = b$.

Many of the definitions of this section have equivalent formulations in terms of sequences.

Theorem 2.2.9. *Let S and T be metric spaces, and let $h\colon S \to T$ be a function. Then h is continuous if and only if, for every sequence (x_n) in S,*

$$x_n \to x \quad \Longrightarrow \quad h(x_n) \to h(x).$$

Proof. First, suppose h is continuous. Let (x_n) be a sequence in S, and suppose $x_n \to x$. I must prove that $h(x_n) \to h(x)$. So let $\varepsilon > 0$. Since h is continuous at x, there is $\delta > 0$ with $h\left[B_\delta(x)\right] \subseteq B_\varepsilon(h(x))$. Since $x_n \to x$, there is $N \in \mathbb{N}$ so that $x_n \in B_\delta(x)$ for all $n \geq N$. But then $h(x_n) \in B_\varepsilon(h(x))$ for all $n \geq N$. This shows that $h(x_n) \to h(x)$.

For the other direction, I will prove the contrapositive. Suppose h is not continuous. I must prove that the convergence property

$$x_n \to x \quad \Longrightarrow \quad h(x_n) \to h(x)$$

fails. Since h is not continuous, there exist $x \in S$ and $\varepsilon > 0$ such that for all $\delta > 0$, there exists $y \in S$ with $\varrho(x, y) < \delta$ but $\varrho(h(x), h(y)) \geq \varepsilon$. In particular, for $\delta = 1/n$, there is $x_n \in S$ with $\varrho(x_n, x) < 1/n$ but $\varrho(h(x_n), h(x)) \geq \varepsilon$. This means that the sequence (x_n) converges to x, but the image sequence $(h(x_n))$ does not converge to $h(x)$. So the convergence property fails. $\square$

Exercise 2.2.10. If $f\colon S_1 \to S_2$ is continuous and $g\colon S_2 \to S_3$ is continuous, then the composition $g \circ f\colon S_1 \to S_3$ is also continuous.

Let $(x_n)_{n \in \mathbb{N}}$ be a sequence. Suppose we choose an infinite subset of the positive integers, and list them in order:

$$k_1 < k_2 < k_3 < \cdots .$$

Then we may form a new sequence

$$(x_{k_i})_{i \in \mathbb{N}} .$$

This is called a **subsequence** of (x_n).

Let (x_n) be a sequence in a metric space S and let $x \in S$. We say that x is a **cluster point** of the sequence (x_n) iff for every $\varepsilon > 0$, and every $N \in \mathbb{N}$, there exists $n \geq N$ with $\varrho(x_n, x) < \varepsilon$.

Proposition 2.2.11. *The point x is a cluster point of the sequence (x_n) if and only if x is the limit of some subsequence of (x_n).*

Proof. Suppose x is a cluster point of (x_n). We will define integers $k_1 < k_2 < \cdots$ recursively. Now $1 > 0$, so there is n with $\varrho(x_n, x) < 1$. Let k_1 be such an n. Then $1/2 > 0$, so there is $n \geq k_1 + 1$ with $\varrho(x_n, x) < 1/2$. Let k_2 be such an n. Suppose k_j has been defined. Then $1/(j+1) > 0$, so there is $n \geq k_j + 1$ with $\varrho(x_n, x) < 1/(j+1)$. Let k_{j+1} be such an n. So, we get a sequence $k_1 < k_2 < \cdots$ such that $\varrho(x_{k_j}, x) < 1/j$ for all j. Thus x is the limit of the subsequence (x_{k_j}) of (x_n).

Conversely, suppose x is the limit of the subsequence (x_{k_j}) of (x_n). Let $\varepsilon > 0$. Then there is $J \in \mathbb{N}$ so that $\varrho(x_{k_j}, x) < \varepsilon$ for $j \geq J$. If $N \in \mathbb{N}$, then there is k_j with both $j \geq J$ and $k_j \geq N$. So we have $\varrho(x_{k_j}, x) < \varepsilon$. Therefore x is a cluster point of the sequence (x_n). $\qquad\square$

Exercise 2.2.12. Let A be a subset of a metric space S. Then the following are equivalent:

(1) A is closed;
(2) If $x \in S$ and there is a sequence (x_n) in A such that $x_n \to x$, then $x \in A$.

Exercise 2.2.13. Let A be a subset of a metric space S. Then the following are equivalent:

(1) A is open;
(2) For every $x \in A$ and every sequence (x_n) in S such that $x_n \to x$, there exists $N \in \mathbb{N}$ so that $x_n \in A$ for all $n \geq N$.

Completeness

A **Cauchy sequence** in a metric space S is a sequence (x_n) satisfying: for every $\varepsilon > 0$ there is $N \in \mathbb{N}$ so that $\varrho(x_n, x_m) < \varepsilon$ for all n, m with $n \geq N$ and $m \geq N$. This is an important definition for the real line $\mathbb{R}$. A sequence in $\mathbb{R}$ is convergent if and only if it is a Cauchy sequence. For a general metric space, only one direction remains true in general:

Proposition 2.2.14. *Every convergent sequence is a Cauchy sequence.*

Proof. Suppose $x_n \to x$. I will show that (x_n) is a Cauchy sequence. Let $\varepsilon > 0$. Then also $\varepsilon/2 > 0$. Since $x_n \to x$, there is $N \in \mathbb{N}$ such that $\varrho(x_n, x) < \varepsilon/2$ for all $n \geq N$. Then, if $n, m \geq N$, we have

$$\varrho(x_n, x_m) \leq \varrho(x_n, x) + \varrho(x, x_m) < \frac{\varepsilon}{2} + \frac{\varepsilon}{2} = \varepsilon.$$

Therefore (x_n) is a Cauchy sequence.   □

Consider the metric space S consisting of all the rational numbers. The number $\sqrt{2}$ is irrational. But if its decimal expansion is truncated after n places, the result is a rational number:

$$x_1 = 1.4$$
$$x_2 = 1.41$$
$$x_3 = 1.414$$
$$x_4 = 1.4142$$

and so on. This is a Cauchy sequence in the metric space S that does not converge in S. The theorem of Cauchy is so useful, however, that we will single out those metric spaces where it is true: A metric space S is called **complete** iff every Cauchy sequence in S converges (in S).

Exercise 2.2.15. Three-dimensional Euclidean space $\mathbb{R}^3$ is complete.

Exercise 2.2.16. Suppose S is an ultrametric space. Then a sequence (x_n) in S is a Cauchy sequence if and only if $\varrho(x_n, x_{n+1}) \to 0$.

Proposition 2.2.17. *The space $E^{(\omega)}$ of infinite strings from the alphabet $\{0, 1\}$ is complete under the metric $\varrho_{1/2}$.*

Proof. Let (σ_n) be a Cauchy sequence in $E^{(\omega)}$. We first define a "candidate" τ for the limit, then we prove that it is, in fact, the limit. For each k, there is $n_k \in \mathbb{N}$ so that for all $n, m \geq n_k$, we have

$$\varrho_{1/2}(\sigma_n, \sigma_m) < \left(\frac{1}{2}\right)^k.$$

That means that $\sigma_{n_k} \lceil k = \sigma_m \lceil k$ for all $m \geq n_k$. Define τ as follows: The kth letter of τ is the kth letter of σ_{n_k}. So τ satisfies

$$\tau \lceil k = \sigma_{n_k} \lceil k$$

for all k.

To see that $\sigma_n \to \tau$, let $\varepsilon > 0$ be given. Choose k so that $(1/2)^k < \varepsilon$. Then for $m > n_k$, we have $\sigma_m \lceil k = \tau \lceil k$, so $\varrho_{1/2}(\sigma_m, \tau) \leq (1/2)^k < \varepsilon$. This shows that $\sigma_n \to \tau$.   □

Exercise 2.2.18. Completeness is a metric property, but not a topological property.

The **closure** of a set A is the set $\overline{A}$, consisting of A together with all of its accumulation points. It is a closed set. A set A is **dense** in a set B iff $\overline{A} = B$. It may be useful to verify that this agrees with the definitions given above for $\mathbb{R}$ (p. 6) and $\mathbb{R}^2$ (p. 10).

A point x that belongs both to the closure of the set A and to the closure of the complementary set $S \backslash A$ is called a **boundary point** of A. The **boundary** of A is the set of all boundary points of A. We will write it as ∂A.

A set has empty boundary if and only if it is clopen.* The boundary of the interval (a, b) in the metric space $\mathbb{R}$ is the two-point set $\{a, b\}$. The boundary of the ball $B_r(x)$ in $\mathbb{R}^d$ is the sphere $\{\, y \in \mathbb{R}^d : |x - y| = r \,\}$. In a general metric space S, however, we can only say that the boundary of the ball $B_r(x)$ is a subset of the sphere $\{\, y \in S : \varrho(x, y) = r \,\}$. For example, if S is ultrametric, then the boundary of $B_r(x)$ is empty, even if the sphere is not.

Exercise 2.2.19. If A is any subset of a metric space, then ∂A is a closed set.

Suppose $T \subseteq S$. Then T is a metric space in its own right. If $A \subseteq T$, then also $A \subseteq S$. The boundary of A is a concept that depends on the complement of A, as well as on A itself, so it makes a difference whether the boundary is taken in T or in S. For example, $S = \mathbb{R}$ contains $T = [0, 1]$. If $A = [0, 1/2]$, then $\partial_T A = \{1/2\}$, but $\partial_S A = \{0, 1/2\}$.

Theorem 2.2.20. *Let S be a metric space, $A \subseteq S$, and $T \subseteq S$. Then $\partial_T(A \cap T) \subseteq \partial_S A$.*

Proof. Let $x \in \partial_T(A \cap T)$. For every $\varepsilon > 0$, there exist points $y \in A \cap T$ with $\varrho(y, x) < \varepsilon$ and points $z \in T \backslash (A \cap T)$ with $\varrho(z, x) < \varepsilon$. But points such as y are points of A, and points such as z are points of $S \backslash A$. So x is a boundary point of A in S.   □

Contraction Mapping

A point x is a **fixed point** of a function f iff $f(x) = x$. A function $f \colon S \to S$ is a **contraction** iff there is a constant $r < 1$ such that

$$\varrho(f(x), f(y)) \leq r \varrho(x, y)$$

for all $x, y \in S$. A contraction is easily seen to be continuous. There is a useful theorem on contraction mappings.

Theorem 2.2.21 (Contraction mapping theorem). *A contraction mapping f on a complete nonempty metric space S has a unique fixed point.*

---

* "Clopen" means "closed and open".

Proof. First, there is at most one fixed point. If x and y are both fixed points, then $\varrho(x, y) = \varrho(f(x), f(y)) \leq r\varrho(x, y)$. But $0 \leq r < 1$, so this is impossible if $\varrho(x, y) > 0$. Therefore $\varrho(x, y) = 0$, so $x = y$.

Now let x_0 be any point of S. (Recall that S is nonempty.) Then define recursively

$$x_{n+1} = f(x_n) \qquad \text{for } n \geq 0.$$

I claim that (x_n) is a Cauchy sequence. Write $a = \varrho(x_0, x_1)$. It follows by induction that $\varrho(x_{n+1}, x_n) \leq ar^n$. But then, if $m < n$, we have

$$\varrho(x_m, x_n) \leq \sum_{j=m}^{n-1} \varrho(x_{j+1}, x_j) \leq \sum_{j=m}^{n-1} ar^j$$
$$= \frac{ar^m - ar^n}{1 - r} = \frac{ar^m(1 - r^{n-m})}{1 - r}$$
$$\leq \frac{ar^m}{1 - r}.$$

Therefore, if $\varepsilon > 0$ is given, choose N large enough that $ar^N/(1 - r) < \varepsilon$. Then, for $n, m \geq N$, we have $\varrho(x_m, x_n) < \varepsilon$.

Now S is complete and (x_n) is a Cauchy sequence, so it converges. Let x be the limit. Now f is continuous, so from $x_n \to x$ follows also $f(x_n) \to f(x)$. But $f(x_n) = x_{n+1}$, so $f(x_n) \to x$. Therefore the two limits are equal, $x = f(x)$, so x is a fixed point. $\qquad\square$

The preceding theorem can be used to prove the existence of certain points in a complete metric space. But more than that is true: the proof of the theorem shows a way to "construct" the point in question. We record this consequence of the proof.

Corollary 2.2.22. *Let f be a contraction mapping on a complete metric space S. If x_0 is any point of S, and*

$$x_{n+1} = f(x_n) \qquad \text{for } n \geq 0,$$

then the sequence x_n converges to the fixed point of f.

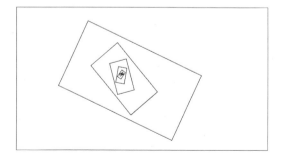

Fig. 2.2.21. Illustration for Corollary 2.2.22.

A function $f\colon S \to T$ is a **Lipschitz** function iff there is a constant B with

$$\varrho(f(x), f(y)) \leq B\,\varrho(x, y) \qquad \text{for all } x, y \in S.$$

Exercise 2.2.23. Suppose f is a Lipschitz function. Does it follow that f is continuous?

Let A be a nonempty set in a metric space S. Then for $x \in S$, the distance $\operatorname{dist}(\{x\}, A)$ satisfies the Lipschitz condition

$$\big|\operatorname{dist}(\{x\}, A) - \operatorname{dist}(\{y\}, A)\big| \leq \varrho(x, y).$$

We will usually write $\operatorname{dist}(x, A)$ for $\operatorname{dist}(\{x\}, A)$.

A function $f\colon S \to T$ is **inverse Lipschitz** iff there is a constant $A > 0$ with

$$\varrho(f(x), f(y)) \geq A\,\varrho(x, y) \qquad \text{for all } x, y \in S.$$

Exercise 2.2.24. Suppose f is an inverse Lipschitz function. Does it follow that f is continuous?

A function $f\colon S \to T$ is a **lipeomorphism** iff it is both Lipschitz and inverse Lipschitz. That is, there are positive constants A, B such that

$$A\,\varrho(x, y) \leq \varrho(f(x), f(y)) \leq B\,\varrho(x, y).$$

Such a function f is also called a **metric equivalence**.

Exercise 2.2.25. Suppose f is a lipeomorphism. Does it follow that f is a homeomorphism?

Exercise 2.2.26. Find an example of a sequence F_n of closed sets in $\mathbb{R}$ such that $\bigcup_{n \in \mathbb{N}} F_n$ is not closed.

Exercise 2.2.27. Suppose that, for each $n \in \mathbb{Z}$, we have a closed set $F_n \subseteq [n, n+1]$. Then show that $\bigcup_{n \in \mathbb{Z}} F_n$ is closed.

Separation

An important consideration for topological dimension will be "separation" of sets. Here is our first theorem of this type.

Theorem 2.2.28. *Let A and B be disjoint closed subsets of a metric space S. Then there exist disjoint open sets U and V in S with $U \supseteq A$ and $V \supseteq B$.*

Proof. Write $U = \{\, x \in S : \operatorname{dist}(x, A) < \operatorname{dist}(x, B)\,\}$. By the triangle inequality, we have

$$|\operatorname{dist}(x, A) - \operatorname{dist}(y, A)| \leq \varrho(x, y),$$

so that $\operatorname{dist}(x, A)$ is a continuous function of x. Similarly, $\operatorname{dist}(x, B)$ is a continuous function of x. It follows that U is an open set. Since A is closed, we have $\operatorname{dist}(x, A) = 0$ if and only if $x \in A$. So if $x \in A$, we have $\operatorname{dist}(x, A) = 0 < \operatorname{dist}(x, B)$; this shows that $A \subseteq U$. Let $V = \{\, x \in S : \operatorname{dist}(x, A) > \operatorname{dist}(x, B)\,\}$. As before, V is open and $B \subseteq V$. Clearly $U \cap V = \varnothing$. $\qquad\square$

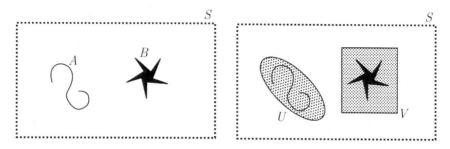

Fig. 2.2.28. Illustration for Theorem 2.2.28

This result can be rephrased in another form:

Corollary 2.2.29. *Suppose F is closed and U is open. If $F \subseteq U$, then there is an open set V with $F \subseteq V \subseteq \overline{V} \subseteq U$.*

Exercise 2.2.30. The same idea of proof can be used to prove some variants of the theorem:

(1) Let A and B be subsets of a metric space S. Suppose $\overline{A} \cap B = \varnothing = A \cap \overline{B}$. Then there exist disjoint open sets U and V in S with $U \supseteq A$ and $V \supseteq B$.

(2) Let A and B be disjoint closed subsets of a metric space S. Then there exist open sets U and V in S with $U \supseteq A$, $V \supseteq B$, and $\overline{U} \cap \overline{V} = \varnothing$. Rephrase this in the manner of Corollary 2.2.29.

2.3 Separable and Compact Spaces

Sometimes it is best to deal not with the most general metric space, but a more specialized class of metric spaces. This section deals with two important special classes of metric spaces.

Separable

A family $\mathcal{U}$ of subsets of S is said to **cover** a set A iff A is contained in the union of the family $\mathcal{U}$. A family which covers a set is known as **a cover** of the set. A cover consisting of a finite number of sets is called a **finite cover**. A cover consisting of a countable number of sets is called a **countable cover**. An **open cover** of a set A is a cover of A consisting only of open sets. If $\mathcal{U}$ is a cover of A, then a **subcover** is a subfamily of $\mathcal{U}$ that still covers A.

Theorem 2.3.1. *Let S be a metric space. The following are equivalent:*

(1) *There is a countable set D dense in S. [S is a **separable** space.]*

(2) *There is a countable base for the open sets of S. [the **second axiom of countability**.]*

(3) *Every open cover of S has a countable subcover.* [the ***Lindelöf
property.***]

Proof. (1) $\implies$ (2). Suppose S admits a countable dense set D. Let $\mathcal{B} = \{ B_{1/n}(a) : a \in D, n \in \mathbb{N} \}$. Then $\mathcal{B}$ is a countable family of open sets. I claim it is a base of the open sets of S. Let U be any open set of S, and let $x \in U$. Then there is $\varepsilon > 0$ so that $B_{\varepsilon}(x) \subseteq U$. Choose n so that $2/n < \varepsilon$. Since D is dense in S, we know that $x \in \overline{D}$. So there is a point $a \in B_{1/n}(x) \cap D$. Then $B_{1/n}(a) \in \mathcal{B}$, and we have $x \in B_{1/n}(a) \subseteq B_{2/n}(x) \subseteq U$. Thus $\mathcal{B}$ is a countable base for the open sets of S.

(2) $\implies$ (3). Suppose there is a countable base $\mathcal{B}$ for the open sets of S. Let $\mathcal{U}$ be an open cover of S. For each point $x \in S$, choose a set $U_x \in \mathcal{U}$ with $x \in U_x$. Then choose a basic set $D_x \in \mathcal{B}$ with $x \in D_x \subseteq U_x$. Now $\{ D_x : x \in S \}$ is a subfamily of $\mathcal{B}$, so it is countable. So it has the form

$$\{ D_x : x \in S \} = \{ D_{x_n} : n \in \mathbb{N} \}.$$

Now write $\mathcal{V} = \{ U_{x_n} : n \in \mathbb{N} \}$. This is a countable subfamily of $\mathcal{U}$. If $x \in S$, then $D_x = D_{x_n}$ for some n, and therefore $x \in D_{x_n} \subseteq U_{x_n}$. So $\mathcal{V}$ is a countable subcover.

(3) $\implies$ (1). Suppose that S has the property of Lindelöf. For $n \in \mathbb{N}$, the collection

$$\mathcal{B}_n = \{ B_{1/n}(x) : x \in S \}$$

is an open cover of S. Therefore it has a countable subcover, say

$$\mathcal{A}_n = \{ B_{1/n}(y) : y \in Y_n \},$$

for a countable set Y_n. Let

$$D = \bigcup_{n \in \mathbb{N}} Y_n.$$

Then D is countable, since it is a countable union of countable sets. If $x \in S$ is any point, and $\varepsilon > 0$, choose n with $1/n < \varepsilon$. Since $\mathcal{A}_n$ is a cover of S, there is $y \in Y_n \subseteq D$ with $x \in B_{1/n}(y)$. Therefore $y \in B_{1/n}(x) \subseteq B_{\varepsilon}(x)$. This shows that D is dense in S. $\qquad \square$

A metric space S will be called ***separable*** iff it has one (and therefore all) of the properties of Theorem 2.3.1. A subset of a metric space will be called ***separable*** iff it is a separable metric space when considered to be a metric space in its own right. Separable metric spaces have many useful properties. By (2), any subset of a separable metric space is separable. By (1), a union of countably many separable sets is separable. All of the examples of metric spaces in this book are separable:

Exercise 2.3.2. The set

$$\{ (x_1, x_2, \cdots, x_d) \in \mathbb{R}^d : \text{all } x_j \text{ are rational numbers} \}$$

is countable and dense in Euclidean space $\mathbb{R}^d$.

Exercise 2.3.3. Consider the space $E^{(\omega)}$ of infinite strings from the alphabet $\{0, 1\}$ under the metric $\varrho_{1/2}$. The set

$$\left\{ [\alpha] : \alpha \in E^{(*)} \right\}$$

is a countable base for the open sets. In fact, every open ball is one of the sets $[\alpha]$.

Compact

We will begin by considering compactness of a closed interval $[a, b]$ in $\mathbb{R}$.

Theorem 2.3.4 (The Bolzano–Weierstrass theorem). *Let $a < b$ be real numbers. If (x_n) is any sequence in the interval $[a, b]$, then (x_n) has at least one cluster point.*

Proof. We will define recursively a sequence (I_k) of closed intervals, such that each I_k contains x_n for infinitely many n. Let $I_0 = [1, b]$. Then I_0 contains x_n for all n. If I_k has been defined, say $I_k = [a_k, b_k]$, then we will consider the midpoint $c_k = (a_k + b_k)/2$. Since I_k contains x_n for infinitely many n, either the left half $[a_k, c_k]$ or the right half $[c_k, b_k]$ also contains x_n for infinitely many n. Let I_{k+1} be a half that contains x_n for infinitely many n. Now also note from the definition that the length of I_{k+1} is half the length of I_k, so the length of I_k is $2^{-k}(b - a)$. This converges to 0. Also note that $I_{k+1} \subseteq I_k$. This means that $a_m \in I_k$ for all $m \geq k$. So (a_k) is a Cauchy sequence, and hence a convergent sequence. Let x be its limit. The interval I_k is closed, so the limit x is in I_k. I claim that x is a cluster point of (x_n). If $\varepsilon > 0$, choose k so that the length of I_k is less than ε. Now x_n is in I_k for infinitely many n, so if N is given, then there is $n \geq N$ with $x_n \in I_k$. Also $x \in I_k$, so $|x_n - x| < \varepsilon$. Thus (x_n) has a cluster point. $\qquad\square$

A metric space S is called ***sequentially compact*** iff every sequence in S has at least one cluster point (in S).

Let $r > 0$. A subset A of a metric space S is an ***r-net*** for S iff every point of S is within distance at most r of some element of A. For example, the countable set $\{ rn : n \in \mathbb{Z} \}$ is an r-net in $\mathbb{R}$.

Proposition 2.3.5. *Let S be a sequentially compact metric space, and let $r > 0$. Then S has a finite r-net.*

Proof. Suppose S has no finite r-net.

We will define a sequence (x_n) recursively, with $\varrho(x_n, x_m) > r$ for all $m \neq n$. First, $S \neq \varnothing$ (since $\varnothing$ is a finite r-net in $\varnothing$). So we may choose $x_1 \in S$. Now, assume $x_1, x_2, \cdots, x_n$ have been chosen. Since $\{x_1, x_2, \cdots, x_n\}$ is not an r-net, there exists a point (call it x_{n+1}) such that $\varrho(x_j, x_{n+1}) > r$ for $1 \leq j \leq n$. This completes the definition of the sequence (x_n).

Now I claim that this sequence (x_n) has no cluster point. If x were a cluster point, then the ball $B_{r/2}(x)$ would contain at least two of the points x_n, which is impossible since they have distance exceeding r. Therefore S is not sequentially compact. $\square$

Corollary 2.3.6. *A sequentially compact metric space is separable.*

Proof. Suppose S is sequentially compact. For each n let D_n be a finite $1/n$-net for S. Then $D = \bigcup_{n \in \mathbb{N}} D_n$ is a countable set dense in S. $\square$

Proposition 2.3.7. *Let $a < b$ be real numbers. If A is any infinite subset of the interval $[a, b]$, then A has at least one accumulation point.*

Proof. If A is an infinite set, we may choose a sequence $x_n \in A$ of distinct elements. By Theorem 2.3.4, (x_n) has a cluster point x. If $\varepsilon > 0$, then $x_n \in B_\varepsilon(x)$ for infinitely many n, so $x_n \in B_\varepsilon(x)$ for some $x_n \neq x$. This shows that x is an accumulation point of A. $\square$

A metric space S is called **countably compact** iff every infinite subset of S has at least one accumulation point (in S). Let $\mathcal{F}$ be a family of subsets of a set S. We say that $\mathcal{F}$ has the **finite intersection property** iff any intersection of finitely many sets from $\mathcal{F}$ is nonempty.

Theorem 2.3.8 (The Heine–Borel theorem). *Let $a < b$ be real numbers. Let $\mathcal{F}$ be a family of closed subsets of the interval $[a, b]$. If $\mathcal{F}$ has the finite intersection property, then the intersection*

$$\bigcap_{F \in \mathcal{F}} F$$

of the entire family is not empty.

Proof. First, by Exercise 2.3.2, the line $\mathbb{R}$ is separable; therefore also $[a, b]$ is separable. Suppose (for purposes of contradiction), that

$$\bigcap_{F \in \mathcal{F}} F = \varnothing. \tag{1}$$

That means that $\{ [a, b] \setminus F : F \in \mathcal{F} \}$ is an open cover of $[a, b]$. So by the Lindelöf property (Theorem 2.3.1), there is a countable subcover. That means that there is a countable number of the sets of $\mathcal{F}$ with empty intersection. Say

$$\bigcap_{n \in \mathbb{N}} F_n = \varnothing, \tag{2}$$

where $F_n \in \mathcal{F}$. Now, for each n, the finite intersection $F_1 \cap F_2 \cap \cdots \cap F_n$ is nonempty; choose an element x_n. By sequentialcompactness (Theorem 2.3.4), the sequence x_n has a cluster point, say x. Since F_n is closed and $x_m \in F_n$ for all $m \geq n$, we have $x \in F_n$. This is true for all n, so

$$x \in \bigcap_{n \in \mathbb{N}} F_n,$$

which contradicts (2). This contradiction arose from assuming (1). Therefore

$$\bigcap_{F \in \mathcal{F}} F \neq \varnothing. \qquad \square$$

A metric space S is (temporarily) called **bicompact** iff every family of closed sets with the finite intersection property has nonempty intersection. As in the proof above, simply taking complements shows that this is equivalent to saying that every open cover of S has a finite subcover.

Theorem 2.3.9. *Let S be a metric space. The following are equivalent:*

(1) *S is sequentially compact,*
(2) *S is countably compact,*
(3) *S is bicompact.*

Proof. (3) $\implies$ (2). Suppose S is not countably compact. Then there is an infinite subset A of S with no accumulation points. For each point $x \in S$, choose an open ball B_x such that B_x contains no points of A (except possibly x itself). Then $\mathcal{U} = \{ B_x : x \in S \}$ is an open cover of S. Any finite subcollection of $\mathcal{U}$ contains only finitely many points of A, so $\mathcal{U}$ does not admit a finite subcover. So S is not bicompact.

(2) $\implies$ (1). Suppose S is countably compact. Let (x_n) be any sequence in S. If there is a point x with $x_n = x$ for infinitely many n, then that x is a cluster point of the sequence (x_n). On the other hand, if there is no such point, then the set $A = \{ x_n : n \in \mathbb{N} \}$ of values is an infinite set; so A has an accumulation point, which is easily seen to be a cluster point of the sequence (x_n). So in all cases (x_n) has a cluster point. Thus S is sequentially compact.

(1) $\implies$ (3). Suppose S is sequentially compact. Then by Corollary 2.3.6, S is separable. The rest of the proof follows the proof of Theorem 2.3.8 word for word. $\qquad \square$

A metric space S will be called **compact** iff it has one (and therefore all) of the properties of Theorem 2.3.9. A subset of a metric space will be called **compact** iff it is a compact metric space when considered to be a metric space in its own right. One of the most useful ways to prove that a set is compact is the following:

Proposition 2.3.10. *A closed subset of a compact space is compact.*

Proof. Suppose S is compact and $T \subseteq S$ is closed. Let (x_n) be a sequence in T. Now by the compactness of S, there is $x \in S$ which is the limit of a subsequence (x_{k_i}) of (x_n). But T is closed and $x_{k_i} \in T$ for all i, so the limit x is also in T. Thus (x_n) has a cluster point in T. This shows that T is compact.   □

Proposition 2.3.11. *Let $A \subseteq \mathbb{R}^d$. Then A is compact if and only if A is closed and bounded.*

Proof. First, suppose A is closed and bounded. Then A is a subset of a large cube,

$$C = \{ x = (x_1, x_2, \cdots, x_d) : -a \leq x_j \leq a \text{ for all } j \}.$$

By Proposition 2.3.10, if I show that C is compact, it will follow that A is compact. Now let (y_n) be a sequence in C. For notation, write

$$y_n = (y_{n1}, y_{n2}, \cdots, y_{nd}).$$

Now the sequence of first coordinates $(y_{n1})_{n \in \mathbb{N}}$ is a sequence in $[-a, a]$, which is compact. So there is a subsequence that converges, so there is an infinite set $N_1 = \{ n_1 < n_2 < \cdots \}$ and

$$\lim_{n \in N_1} y_{n1} = z_1.$$

Next, the sequence of second coordinates $(y_{n2})_{n \in N_1}$ is a sequence in $[-a, a]$, which is compact. So there is a subsequence that converges, say $N_2 \subseteq N_1$ and

$$\lim_{n \in N_2} y_{n2} = z_2.$$

Similarly, we get $N_3 \supseteq \cdots \supseteq N_d$, with

$$\lim_{n \in N_j} y_{nj} = z_j.$$

Finally, the subsequence $(y_n)_{n \in N_d}$ has all coordinates convergent, and its limit is $z = (z_1, z_2, \cdots, z_d) \in C$. This proves that C is compact.

Conversely, suppose that A is compact. If A is unbounded, then

$$\{ B_n(0) \cap A : n \in \mathbb{N} \}$$

is an open cover of A with no finite subcover. If A is not closed, there is an accumulation point x of A that is not in A. So there is a sequence (x_n) in A converging to x. This sequence has no cluster point in A.   □

Exercise 2.3.12. The metric space $E^{(\omega)}$ of all infinite strings constructed using the alphabet $E = \{0, 1\}$ is compact under the metric $\varrho_{1/2}$.

Of course compactness is a topological property: If S and T are homeomorphic, and S is compact, then T is compact.

Exercise 2.3.13. A compact subset of a metric space is closed.

Exercise 2.3.14. The union of finitely many compact sets is compact.

Image and Inverse Image

If $f\colon S \to T$ is a continuous function, and $A \subseteq S$, some properties of the set A are related to properties of the image set

$$f[A] = \{\, f(x) : x \in A \,\}.$$

Theorem 2.3.15. *Let $f\colon S \to T$ be continuous. Let $A \subseteq S$ be compact. Then $f[A]$ is compact.*

Proof. Let (y_n) be a sequence in $f[A]$. Then there exist points $x_n \in A$ with $f(x_n) = y_n$. By the compactness of A, there is a subsequence (x_{k_i}) that converges to some point $x \in A$. But since f in continuous, this implies that $y_{k_i} = f(x_{k_i}) \to f(x)$. Now $f(x) \in f[A]$, so (y_n) has a cluster point in $f[A]$. This shows that $f[A]$ is compact. $\square$

Corollary 2.3.16. *Let S be a compact metric space, and let the function $f\colon S \to \mathbb{R}$ be continuous. Then f is bounded; that is: there is $B \in \mathbb{R}$ such that $|f(x)| \le B$ for all $x \in S$.*

Proof. By Theorem 2.3.15, $f[S]$ is a compact subset of $\mathbb{R}$. By Proposition 2.3.11, it is a bounded set. $\square$

Let $f\colon S \to T$ be continuous. Properties of a set $B \subseteq T$ may be related to properties of the inverse image

$$f^{-1}[B] = \{\, x \in S : f(x) \in B \,\}.$$

Proposition 2.3.17. *Let $f\colon S \to T$ be continuous. Consider a set $B \subseteq T$. Then:*

(1) *If B is open, then $f^{-1}[B]$ is open.*
(2) *If B is closed, then $f^{-1}[B]$ is closed.*

Proof. (1) follows from Theorem 2.2.6. (2) follows from (1) by taking complements. $\square$

Recall that $\operatorname{dist}(A, B) = 0$ can happen if $A \neq B$. This happens, for example, if $A \cap B \neq \varnothing$. Or even if A and B have an accumulation point in common. But it can happen in other ways too:

Exercise 2.3.18. Give an example of two disjoint nonempty closed sets with $\operatorname{dist}(A, B) = 0$.

If one of the sets is compact, then we do get positive distance.

Theorem 2.3.19. *If A is closed, B is compact, and $A \cap B = \varnothing$, then $\operatorname{dist}(A, B) > 0$.*

Proof. Suppose $\operatorname{dist}(A, B) = 0$. Then there exist points $x_n \in A$ and $y_n \in B$ with $\varrho(x_n, y_n) < 1/n$. Now B is compact, so (by replacing the sequences with subsequences) we may assume that (y_n) converges. Say $y_n \to y \in B$. Then $x_n \to y$, also. But A is closed, so $y \in A$. Therefore $A \cap B \neq \varnothing$. $\square$

For $A \subseteq S$ and $x \in S$, recall that we write $\operatorname{dist}(x, A) = \operatorname{dist}(\{x\}, A)$. Now $\{x\}$ is a compact set, so we have $x \in \overline{A}$ if and only if $\operatorname{dist}(x, A) = 0$.

Uniform Continuity

Let $f\colon S \to T$ be a function. Let us recall a definition: f is ***continuous*** iff, for every $x \in S$ and every $\varepsilon > 0$, there exists $\delta > 0$ so that $f[B_\delta(x)] \subseteq B_\varepsilon(f(x))$. Compare that to this definition: f is ***uniformly continuous*** iff, for every $\varepsilon > 0$, there exists $\delta > 0$ so that for every $x \in S$, we have $f[B_\delta(x)] \subseteq B_\varepsilon(f(x))$. What is the difference: in the first case, δ is allowed to depend not only on ε, but also on x; but in the second case, while δ still depends on ε, it does not depend on x.

Clearly every uniformly continuous function is continuous. But the converse may fail:

Exercise 2.3.20. Let $f\colon \mathbb{R} \to \mathbb{R}$ be defined by $f(x) = x^2$. Then f is continuous, but not uniformly continuous.

However, when S is compact, then the two kinds of continuity are the same:

Theorem 2.3.21. *Let S be a compact metric space, let T be a metric space, and let $f\colon S \to T$ be a function. If f is continuous, then f is uniformly continuous.*

Proof. Let $\varepsilon > 0$ be given. Then by continuity, for every $x \in S$, there is a positive number $\delta(x)$ such that $f[B_{\delta(x)}(x)] \subseteq B_{\varepsilon/2}(f(x))$. Now the collection

$$\{ B_{\delta(x)/2}(x) : x \in S \}$$

is an open cover of S. By compactness, there is a finite subcover, say

$$B_{\delta(x_1)/2}(x_1), B_{\delta(x_2)/2}(x_2), \cdots, B_{\delta(x_n)/2}(x_n).$$

Let $\delta = \min\{\delta(x_1)/2, \delta(x_2)/2, \cdots, \delta(x_n)/2\}$. I claim that for this δ, and any $x \in S$, we have $f[B_\delta(x)] \subseteq B_\varepsilon(f(x))$. Indeed, suppose $y \in B_\delta(x)$. For some i, we have $y \in B_{\delta(x_i)/2}(x_i)$. Therefore $\varrho(y, x_i) < \delta(x_i)/2$, so $\varrho(f(y), f(x_i)) < \varepsilon/2$. But also $\varrho(x, x_i) \leq \varrho(x, y) + \varrho(y, x_i) < \delta(x_i)$, so $\varrho(f(x), f(x_i)) < \varepsilon/2$. So we have $\varrho(f(x), f(y)) < \varepsilon$. That is, we have $f[B_\delta(x)] \subseteq B_\varepsilon(f(x))$, so f is uniformly continuous. $\square$

Theorem 2.3.22. *Let S be a compact metric space, and let $\mathfrak{U}$ be an open cover of S. Then there is a positive number r such that for any set $A \subseteq S$ with $\operatorname{diam} A < r$, there is a set $U \in \mathfrak{U}$ with $A \subseteq U$.*

Proof. First, there is a finite subcover, say $\{U_1, U_2, \cdots, U_n\}$. Suppose the assertion is false. Then for every $k \in \mathbb{N}$, there is a set A_k of diameter at most $1/k$ that is not contained in any U_i. So there are points $x_{ik} \in A_k \setminus U_i$, $(1 \leq i \leq n, k \in \mathbb{N})$. By taking subsequences, we may assume that $\lim_k x_{ik} = x_i$ exists for each i. But the points x_i have distance 0 from each other, so they are all equal. This point x_i does not belong to any set U_i, contradicting the fact that the sets U_i cover S. $\square$

The largest number r with the property of the Theorem is known as the **Lebesgue number** of the cover $\mathcal{U}$.

Number Systems

Recall the situation from Sect. 1.6. Let b be a complex number, $|b| > 1$, and let D be a finite set of complex numbers, including 0. We are interested in representing complex numbers in the number system they define.

Write F for the set of "fractions"; that is numbers of the form

$$\sum_{j=1}^{\infty} a_j b^{-j}, \qquad a_j \in D.$$

Proposition 2.3.23. *The set F of fractions is a compact set.*

Proof. Let $A \subseteq F$ be infinite. For each digit $d \in D$, write

$$A(d) = \left\{ x \in A : x = \sum_{j=1}^{\infty} a_j b^{-j}, a_1 = d \right\}$$

Then we have $A = \bigcup_{d \in D} A(d)$; since A is infinite, at least one of the sets $A(d)$ is also infinite. Choose $d_1 \in D$ so that $A(d_1)$ is infinite. Then write

$$A(d_1, d) = \left\{ x \in A : x = \sum_{j=1}^{\infty} a_j b^{-j}, a_1 = d_1, a_2 = d \right\}.$$

There is $d_2 \in D$ so that $A(d_1, d_2)$ is infinite. We may continue in this way: we obtain a sequence (d_j) of digits so that $A(d_1, d_2, \cdots, d_k)$ is infinite for all k. Then the number

$$\sum_{j=1}^{\infty} d_j b^{-j}$$

is an accumulation point of the set A. This shows that F is compact.   □

2.4 Uniform Convergence

In Chap. 1 I had several occasions to mention "convergence" of a sequence of sets. There are two ways that will be used to make that idea precise. The first way, in this section, is "uniform convergence" of functions. The second way, in Sect. 2.5, is the metric of Hausdorff. In both cases, the most natural setting is that of an appropriate metric space. This is one of the reasons that we have been considering abstract metric spaces, rather than just subsets of Euclidean space.

Uniform Convergence in General

Let S and T be two metric spaces. We will be considering functions $f \colon S \to T$.

Let f_n be a sequence of functions from S to T, and let f be another function from S to T. The sequence f_n **converges uniformly** (on S) to the function f iff for every $\varepsilon > 0$, there exists $N \in \mathbb{N}$ so that for all $x \in S$ and all $n \geq N$, we have $\varrho(f_n(x), f(x)) < \varepsilon$.

This definition may be rephrased as convergence for a metric. We will say that two functions f and g are within **uniform distance** r of each other iff

$$\varrho(f(x), g(x)) \leq r \qquad \text{for all } x \in S.$$

Then we write $\varrho_{\mathrm{u}}(f, g) \leq r$. More precisely, $\varrho_{\mathrm{u}}(f, g)$ is the smallest number r that works:

$$\varrho_{\mathrm{u}}(f, g) = \sup \{ \varrho(f(x), g(x)) : x \in S \}.$$

Uniform convergence makes sense even if the functions are not continuous. Here is the most important property of uniform convergence:

Theorem 2.4.1. *Suppose* $f_n \colon S \to T$ *is a sequence of functions from the metric space* S *to the metric space* T. *Suppose* f_n *converges uniformly to a function* f. *If all functions* f_n *are continuous, then* f *is continuous.*

Proof. Let $x \in S$ be given. I must show that f is continuous at x. So let a positive number ε be given. I must find a corresponding value δ. Now f_n converges to f uniformly, and $\varepsilon/3$ is a positive number, so there exists $N \in \mathbb{N}$ such that for all $n \geq N$, and all $y \in S$, we have $\varrho(f(y), f_n(y)) < \varepsilon/3$. Once N is known, we can use the fact that the single function f_N is continuous at x to conclude that there exists $\delta > 0$ so that $\varrho(y, x) < \delta$ implies $\varrho(f_N(x), f_N(y)) < \varepsilon/3$. But then, for any y with $\varrho(x, y) < \delta$, we have

$$\varrho(f(x), f(y)) \leq \varrho(f(x), f_N(x)) + \varrho(f_N(x), f_N(y)) + \varrho(f_N(y), f(y))$$
$$< \frac{\varepsilon}{3} + \frac{\varepsilon}{3} + \frac{\varepsilon}{3} = \varepsilon.$$

This shows that f is continuous at x. □

If the set of functions we deal with is properly restricted, then ϱ_{u} will be a metric. This is useful, since then all of the work we have done on metric spaces will be relevant. If S and T are metric spaces, we will write $\mathcal{C}(S, T)$ for the set of all continuous functions from S to T.

Theorem 2.4.2. *Let* S *be a compact metric space, and let* T *be a metric space. Then* ϱ_{u} *is a metric on* $\mathcal{C}(S, T)$.

Proof. First, $\varrho_{\mathrm{u}}(f, g) \geq 0$ and $\varrho_{\mathrm{u}}(f, g) = \varrho_{\mathrm{u}}(g, f)$ are clear. Since S is compact, we have $\varrho_{\mathrm{u}}(f, g) < \infty$, as in Corollary 2.3.16. If $f = g$, then $\varrho_{\mathrm{u}}(f, g) = 0$. If $\varrho_{\mathrm{u}}(f, g) = 0$, then $0 \leq \varrho(f(x), g(x)) \leq 0$ for all $x \in S$, so $f(x) = g(x)$ for all

$x \in S$, which means $f = g$. Finally, I must prove the triangle inequality. For any $x \in S$, we have

$$\varrho\big(f(x), h(x)\big) \le \varrho\big(f(x), g(x)\big) + \varrho\big(g(x), h(x)\big) \le \varrho_u(f, g) + \varrho_u(g, h).$$

Therefore f and h are within uniform distance $\varrho_u(f, g) + \varrho_u(g, h)$ of each other, so $\varrho_u(f, h) \le \varrho_u(f, g) + \varrho_u(g, h)$. □

The metric ϱ_u will be called the **uniform metric**.

It makes sense to ask about the metric properties of the metric space $\mathcal{C}(S, T)$. Completeness will be a very useful property.

Theorem 2.4.3. *Suppose S is a compact metric space, and T is a complete metric space. Then the metric space $\mathcal{C}(S, T)$ is complete.*

Proof. Let $(f_n)_{n \in \mathbb{N}}$ be a Cauchy sequence in $\mathcal{C}(S, T)$. Let $x \in S$. Then we have

$$\varrho\big(f_n(x), f_m(x)\big) \le \varrho_u(f_n, f_m),$$

so $(f_n(x))_{n \in \mathbb{N}}$ is a Cauchy sequence in T. Since T is complete, this Cauchy sequence converges. Call its limit $f(x)$. This construction is valid for each $x \in S$, so it defines a function $f: S \to T$. I must show that f_n converges uniformly to f. Let $\varepsilon > 0$ be given. There is $N \in \mathbb{N}$ so that $\varrho_u(f_n, f_m) < \varepsilon/2$ for all $n, m \ge N$. For any $x \in S$, there is $N_x \in \mathbb{N}$ so that $\varrho\big(f_n(x), f(x)\big) \le \varepsilon/2$ for $n \ge N_x$. We may assume $N_x \ge N$. Therefore, for any $n \ge N$, and any x, we have

$$\varrho\big(f_n(x), f(x)\big) \le \varrho\big(f_n(x), f_{N_x}(x)\big) + \varrho\big(f_{N_x}(x), f(x)\big) < \varepsilon.$$

Therefore $\varrho_u(f_n, f) \le \varepsilon$. This proves that f_n converges uniformly to f. By Theorem 2.4.1, the limit f is continuous, $f \in \mathcal{C}(S, T)$. So the space $\mathcal{C}(S, T)$ is complete under the uniform metric ϱ_u. □

Exercise 2.4.4. Under what conditions on S and T is the metric space $\mathcal{C}(S, T)$ ultrametric?

Exercise 2.4.5. Under what conditions on S and T is the metric space $\mathcal{C}(S, T)$ compact?

Continuous Curves

Suppose the metric space T is a convex subset of a Euclidean space $\mathbb{R}^k$, and S is a Euclidean space $\mathbb{R}^d$. A function $f: T \to S$ is **affine** iff it satisfies

$$f(tx + (1 - t)y) = tf(x) + (1 - t)f(y)$$

whenever $x, y \in T$ and $0 \le t \le 1$.

Proposition 2.4.6. *An affine map f of an interval $[u, v]$ into $\mathbb{R}$ must be of the form $f(x) = mx + b$, for some $m, b \in \mathbb{R}$.*

Proof. If $u = v$, then of course $f(x) = f(u)$ for all $x \in [u, v]$. So suppose $u < v$. Now if $x \in [u, v]$, then

$$x = \frac{v - x}{v - u} u + \frac{x - u}{v - u} v,$$

which is of the form $tu + (1 - t)v$ with $0 \le t \le 1$. So

$$f(x) = \frac{v - x}{v - u} f(u) + \frac{x - u}{v - u} f(v)$$
$$= \frac{f(v) - f(u)}{v - u} x + \frac{vf(u) - uf(v)}{v - u},$$

which has the required form. □

Exercise 2.4.7. An affine map f of an interval $[u, v]$ into $\mathbb{R}^d$ must be of the form $f(x) = xm + b$, for some $m, b \in \mathbb{R}^d$.

A ***continuous curve*** in a metric space S is a continuous function $f: [0, 1] \to S$. (Illustration in Fig. 2.4.8.) Sometimes we will use the term to refer to the range $f\big[[0, 1]\big]$ of such a function.

A continuous curve $f: [0, 1] \to \mathbb{R}^d$ is ***piecewise affine*** iff there exist "division points" $0 = a_0 < a_1 < \cdots < a_n = 1$, such that f is affine on each of the subintervals $[a_{j-1}, a_j]$. The range of such a function is called a ***polygonal curve***.

Consider the dragon construction of the Koch curve (p. 18). There is a sequence (P_k) of polygonal curves. This sequence is supposed to "converge" to the Koch curve P. Since we do not have $P_{k+1} \subseteq P_k$, it is not reasonable to take the intersection. Here is one way to produce a reasonable limit.

The curve P_k consists of 4^k line segments. Divide the interval $[0, 1]$ into 4^k subintervals of equal length. Define a function $g_k: [0, 1] \to \mathbb{R}^2$, so that it

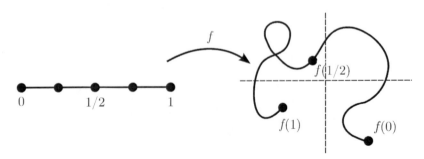

Fig. 2.4.8. Continuous curve

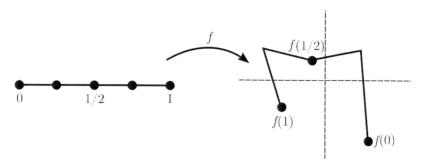

Fig. 2.4.9. Piecewise affine curve

is affine on each of these subintervals, and continuous on all of $[0, 1]$, so that the subintervals are mapped to the line segments of P_k. (To be explicit, start with $g_k(0)$ at the left end of P_k.) Thus we have $g_k[[0,1]] = P_k$.

Proposition 2.4.10. *The sequence* (g_k), *in the dragon construction of the Koch curve, converges uniformly.*

Proof. First note that every point of P_1 is within distance 1 of every point of P_0. So certainly $\varrho_u(g_0, g_1) \leq 1$. Now we estimate $\varrho_u(g_k, g_{k+1})$. Let $t \in [0, 1]$. Then t is in one of the 4^k subintervals J on which g_k is affine. Its image $g_k[J]$ is a line segment of length 3^{-k}. When g_{k+1} is constructed, the interval J is subdivided into 4 equal intervals, and a miniature copy of g_1 is used there (properly translated, rotated, and dilated by 3^{-k}). So $\varrho(g_k(t), g_{k+1}(t)) \leq 3^{-k}\varrho(g_0, g_1)$. Therefore $\varrho_u(g_k, g_{k+1}) \leq 3^{-k}$.

But then for $m > k$ we have

$$\varrho_u(g_k, g_m) \leq \sum_{j=k}^{m-1} \varrho_u(g_j, g_{j+1}) \leq \sum_{j=k}^{m-1} 3^{-j} < 3^{-k+1}/2.$$

This shows that (g_k) is a Cauchy sequence for the uniform metric, so it converges uniformly by Theorem 2.4.3.   □

Write $g(t) = \lim_{k \to \infty} g_k(t)$ for all $t \in [0, 1]$. By Theorem 2.4.1, the function g is continuous. The limit set, the Koch curve, is then $P = g[[0, 1]]$. This also explains the use of the word "curve".

Exercise 2.4.11. Provide similar description for Heighway's dragon. Be sure to prove the uniform convergence that is required.

Now that we have a rigorous definition for the "convergence" that is involved in the definitions, we may try to prove some of the properties of the dragon curves.

Is there a uniform convergence construction for the limit of the sequence (L_k) on p. 5? Not in the same sense as for the dragons. The limit set (which is presumably the Cantor dust) *is not* a continuous curve. (We will prove this later, Proposition 3.1.6.) But there is a similar sort of uniform convergence construction. We must replace the interval $[0, 1]$ with a different parameter space.

Consider the space $E^{(\omega)}$ of infinite strings from the alphabet $\{0, 1\}$. Define functions $g_k \colon E^{(\omega)} \to \mathbb{R}$ as follows. Begin with $g_0(\sigma) = 0$ for all σ. Then $g_0\big[E^{(\omega)}\big] = \{0\} = L_0$. Next, define $g_1(0\sigma) = 0$ and $g_1(1\sigma) = 2/3$. Then $g_1\big[E^{(\omega)}\big] = \{0, 2/3\} = L_1$. The function g_1 is continuous since the sets $[0]$ and $[1]$ are open sets. We may continue recursively: If g_k has been defined, then

$$g_{k+1}(0\sigma) = \frac{g_k(\sigma)}{3}$$

$$g_{k+1}(1\sigma) = \frac{g_k(\sigma) + 2}{3}.$$

The remainder is left to the reader:

Exercise 2.4.14. These functions $g_k \colon E^{(\omega)} \to \mathbb{R}$ are continuous and satisfy $g_k\big[E^{(\omega)}\big] = L_k$. The sequence (g_k) converges uniformly.

Exercise 2.4.15. Show that the limit function $g = \lim g_k$ is the addressing function for the Cantor dust described on p. 14. So $g\big[E^{(\omega)}\big]$ is exactly the triadic Cantor dust.

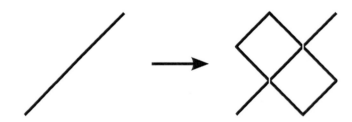

Fig. 2.4.12. Construction for Peano curve

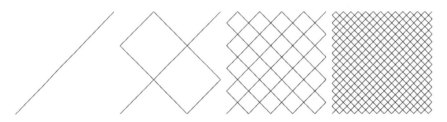

Fig. 2.4.13. Peano polygons

Space-Filling Curves

A continuous curve $f\colon [0,1] \to \mathbb{R}^d$, where $d \geq 2$ is called a ***space-filling curve*** iff $f\big[[0,1]\big]$ contains a ball $B_r(x)$. This possibility was realized by Peano, so it might be called a Peano curve. (We will see later that Heighway's dragon is a space-filling curve, Proposition 2.5.9.) Here is an easy example of a Peano curve. It can be done as a dragon curve. To go from one stage to the next, each line segment is replaced by nine segments with $1/3$ the length, as in Fig. 2.4.12. Figure 2.4.13 shows a few stages, as usual.

Exercise 2.4.16. Prove that the limit curve is space filling.

2.5 The Hausdorff Metric

Felix Hausdorff devised a way to describe convergence of sets. It is (for some purposes) better than the one discussed in the previous section, since it does not require finding an appropriate parameter space and parameterizations for the sets. It is simply a definition of a metric that applies to sets.

Convergence of Sets

Let S be a metric space. Let A and B be subsets of S. We say that A and B are within ***Hausdorff distance*** r of each other iff every point of A is within distance r of some point of B, and every point of B is within distance r of some point of A.

This idea can be made into a metric, called the ***Hausdorff metric***, D. If A is a set and $r > 0$, then the ***open r-neighborhood*** of A is

$$N_r(A) = \{\, y \colon \varrho(x,y) < r \text{ for some } x \in A \,\}.$$

The definition of the Hausdorff metric D:

$$D(A,B) = \inf \{\, r > 0 \colon A \subseteq N_r(B) \text{ and } B \subseteq N_r(A) \,\}.$$

By convention, $\inf \varnothing = \infty$.

This definition D does not define a metric, however. There are various problems. For example, in $\mathbb{R}$, what is the distance between $\{0\}$ and $[0,\infty)$? It is infinite. That is not allowed in the definition of metric. Therefore, we will restrict the use of D to bounded sets. What is the distance $D(\varnothing, \{0\})$? Again, infinite. So we will restrict the use of D to nonempty sets. What is the distance $D\big((0,1), [0,1]\big)$? Now the distance is 0, even though the two sets are not equal. Therefore we will restrict the use of D to closed sets. In fact for the purposes of this book, we will apply D only to nonempty compact sets.

If S is a metric space, we will write $\mathbb{H}(S)$ for the collection of all nonempty compact subsets of S. This is called the ***hyperspace*** for S.

Theorem 2.5.1. *Let S be a metric space. The Hausdorff function D is a metric on the set $\mathbb{H}(S)$.*

Proof. First, clearly $D(A, B) \geq 0$ and $D(A, B) = D(B, A)$. Since A and B are compact, they are bounded, so $D(A, B) < \infty$.

If $A = B$, then for every $\varepsilon > 0$ we have $A \subseteq N_\varepsilon(B)$; therefore $D(A, B) = 0$. Conversely, suppose $A, B \in \mathbb{H}(S)$ satisfy $D(A, B) = 0$. If $x \in A$, then for every $\varepsilon > 0$, we have $x \in N_\varepsilon(B)$, so $\mathrm{dist}(x, B) = 0$. Now B is compact, hence closed, so $x \in B$. This shows $A \subseteq B$. Similarly $B \subseteq A$, so $A = B$.

Finally we have the triangle inequality. Let $A, B, C \in \mathbb{H}(S)$. Let $\varepsilon > 0$. If $x \in A$, then there is $y \in B$ with $\varrho(x, y) < D(A, B) + \varepsilon$. Then there is $z \in C$ with $\varrho(y, z) < D(B, C) + \varepsilon$. This shows that A is contained in the $(D(A, B) + D(B, C) + 2\varepsilon)$-neighborhood of C. Similarly, C is contained in the $(D(A, B) + D(B, C) + 2\varepsilon)$-neighborhood of A. Therefore $D(A, C) \leq D(A, B) + D(B, C) + 2\varepsilon$. This is true for all $\varepsilon > 0$, so $D(A, C) \leq D(A, B) + D(B, C)$.   □

Here is one way to describe the limit.

Exercise 2.5.2. Let A_n be a sequence of nonempty compact subsets of S and let A be a nonempty compact subset of S. If A_n converges to A in the Hausdorff metric, then

$$A = \{\, x : \text{there is a sequence } (x_n) \text{ with } x_n \in A_n \text{ and } x_n \to x \,\}.$$

We may ask about the metric properties of the metric space $\mathbb{H}(S)$. The most important one will be completeness.

Theorem 2.5.3. *Suppose S is a complete metric space. Then the space $\mathbb{H}(S)$ is complete.*

Proof. Suppose (A_n) is a Cauchy sequence in $\mathbb{H}(S)$. I must show that A_n converges. Let $A = \{\, x : \text{there is a sequence } (x_k) \text{ with } x_k \in A_k \text{ and } x_k \to x \,\}$. I must show that $D(A_n, A) \to 0$ and A is nonempty and compact.

Let $\varepsilon > 0$ be given. Then there is $N \in \mathbb{N}$ so that $n, m \geq N$ implies $D(A_n, A_m) < \varepsilon/2$. Let $n \geq N$. I claim that $D(A_n, A) \leq \varepsilon$.

If $x \in A$, then there is a sequence (x_k) with $x_k \in A_k$ and $x_k \to x$. So, for large enough k, we have $\varrho(x_k, x) < \varepsilon/2$. Thus, if $k \geq N$, then (since $D(A_k, A_n) < \varepsilon/2$) there is $y \in A_n$ with $\varrho(x_k, y) < \varepsilon/2$, and we have $\varrho(y, x) \leq \varrho(y, x_k) + \varrho(x_k, x) < \varepsilon$. This shows that $A \subseteq N_\varepsilon(A_n)$.

Now suppose $y \in A_n$. Choose integers $k_1 < k_2 < \cdots$ so that $k_1 = n$ and $D(A_{k_j}, A_m) < 2^{-j}\varepsilon$ for all $m \geq k_j$. Then define a sequence (y_k) with $y_k \in A_k$ as follows: For $k < n$, choose $y_k \in A_k$ arbitrarily. Choose $y_n = y$. If y_{k_j} has been chosen, and $k_j < k \leq k_{j+1}$, choose $y_k \in A_k$ with $\varrho(y_{k_j}, y_k) < 2^{-j}\varepsilon$. Then y_k is a Cauchy sequence, so it converges. Let x be its limit. So $x \in A$. We have $\varrho(y, x) = \lim_k \varrho(y, y_k) < \varepsilon$. So $y \in N_\varepsilon(A)$. This shows that $A_n \subseteq N_\varepsilon(A)$. Note that, taking $\varepsilon = 1$ in this argument, I have also proved that $A \neq \varnothing$.

So we have $D(A, A_n) \leq \varepsilon$. This concludes the proof that (A_n) converges to A.

Next I show that A is "totally bounded": that is, for every $\varepsilon > 0$, there is a finite ε-net in A. Choose n so that $D(A_n, A) < \varepsilon/3$. By Proposition 2.3.5, there is a finite $(\varepsilon/3)$-net for A_n, say $\{y_1, y_2, \cdots, y_m\}$. Now for each y_i, there is $x_i \in A$ with $\varrho(x_i, y_i) < \varepsilon/3$. The finite set $\{x_1, x_2, \cdots, x_m\}$ is an ε-net for A.

Now I will show that A is a closed subset of S. Let x belong to the closure $\overline{A}$ of A. Then there exists a sequence (y_n) in A with $\varrho(x, y_n) < 2^{-n}$. For each n there is a point $z_n \in A_n$ with $\varrho(z_n, y_n) < D(A_n, A) + 2^{-n}$. Now

$$\varrho(z_n, x) \leq \varrho(z_n, y_n) + \varrho(y_n, x) < D(A_n, A) + 2^{-n} + 2^{-n}.$$

This converges to 0, so $z_n \to x$. Thus $x \in A$. This shows that A is closed.

Finally, to show that A is compact, I will show that it is countably compact. Let F be an infinite subset of A. There is a finite $(1/2)$-net B for A, so each element of F is within distance $1/2$ of some element of B. Now F is infinite and B is finite, so there is an element of B within distance $1/2$ of infinitely many elements of F. Let $F_1 \subseteq F$ be that infinite subset. The points of F_1 are all within distance 1 of each other; that is, $\operatorname{diam} F_1 \leq 1$. In the same way, there is an infinite set $F_2 \subseteq F_1$ with $\operatorname{diam} F_2 \leq 1/2$; and so on. There are infinite sets F_j with $\operatorname{diam} F_j \leq 2^{-j}$ and $F_{j+1} \subseteq F_j$ for all j. Now if x_j is chosen from F_j, we have $\varrho(x_j, x_k) \leq 2^{-j}$ if $j < k$, so (x_j) is a Cauchy sequence. Since S is complete, (x_j) converges, say $x_j \to x$. Since A is closed, $x \in A$. But then x is a cluster point of the set F. Therefore A is compact.   □

Exercise 2.5.4. Under what conditions on S is $\mathbb{H}(S)$ compact?

Exercise 2.5.5. Under what conditions on S is $\mathbb{H}(S)$ ultrametric?

Convergence in the Examples

Whenever we have used the idea of "convergence" for a sequence of sets in a metric space, we have been talking about nonempty compact sets. The Hausdorff metric is the proper way to interpret it. In fact, it agrees with the other interpretations that have been used.

Proposition 2.5.6. *Let A_n be a sequence of nonempty compact sets, and suppose they decrease: $A_1 \supseteq A_2 \supseteq \cdots$. Then A_n converges to the intersection $A = \bigcap_{n \in \mathbb{N}} A_n$ in the Hausdorff metric.*

Proof. Let $\varepsilon > 0$ be given. Now $A \subseteq A_n$, so $A \subseteq N_\varepsilon(A_n)$. For the other direction, note that the ε neighborhood of A,

$$N_\varepsilon(A) = \{ y : \varrho(x, y) < \varepsilon \text{ for some } x \in A \},$$

is an open set. The family $\{N_\varepsilon(A)\} \cup \{ S \setminus A_n : n \in \mathbb{N} \}$ is an open cover of A_1. Since A_1 is compact, there is a finite subcover. This means that, for some $N \in \mathbb{N}$, we have $(S \setminus A_n) \cup N_\varepsilon(A) \supseteq A_1$ for all $n \geq N$. Therefore we have $A_n \subseteq N_\varepsilon(A)$. So $D(A, A_n) \leq \varepsilon$ for all $n \geq N$. This shows that $A_n \to A$.   □

Proposition 2.5.7. *Suppose S is a compact metric space, T is a metric space and the sequence $f_n \colon S \to T$ converges uniformly to f. Then the image sets $f_n[S]$ converge to $f[S]$ according to the Hausdorff metric on $\mathbb{H}(T)$.*

Proof. This follows from the inequality $D\big(f[S], f_n[S]\big) \le \varrho_{\mathrm{u}}(f, f_n).$ □

Exercise 2.5.8. Let A_n be a sequence of nonempty compact subsets of S. If A is a cluster point of the sequence (A_n), then A is contained in the set of all points $x \in S$ for which there exists a sequence (x_n) such that $x_n \in A_n$ and x is a cluster point of (x_n).

Heighway's Dragon Tiles the Plane

Let us apply some of the properties of set convergence to study the Highway dragon fractal (p. 20).

Proposition 2.5.9. *The Heighway dragon is a space-filling curve.*

Proof. The information and notation used here are from the proof of Proposition 1.5.7. Suppose, at some stage of the construction, the polygon P_n contains all of the sides of a square S of the lattice L_n together with all of the sides of the four adjoining squares, as in Fig. 2.5.9(a). Then the polygon P_{n+2}, shown in Fig. 2.5.9(b), contains all four sides of all of the squares contained in S, together with all of the sides of all their adjoining squares.

   Now the polygons P_7 and P_8 each contain such a square S. By induction, all sides of all of the subsquares of S are contained in the polygons P_k, for $k \ge 7$. By Exercise 2.5.2, the limit P contains all points of the interior of the square S.   □

   We claim, in fact, that P tiles the plane. By this we mean: $\mathbb{R}^2$ is the union of countably many sets, each congruent to P, and any two of them intersect at most in their boundaries.

   Start with the vertical lines $x = k$ with integer k and horizontal lines $y = j$ with integer j. They subdivide the plane into squares of side 1. Imagine these

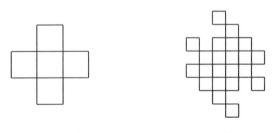

Fig. 2.5.9. (a)   (b)

squares classified in the checkerboard pattern as "black" and "white" so that squares that share an edge have opposite colors. For our tiling we begin with the edges of this square subdivision. Each edge is a line-segment of length 1. Starting with the edge, the process described for producing the dragon can be carried out. We need a "direction" for each edge. Let's say the vertical edges have the black square on their left, and the horizontal edges have the black square on their right. Then each edge of this square lattice occurs once. When we carry out one step in the process, replacing each edge by two shorter ones, we will see (as in the proof of proof of Proposition 1.5.7) that the result will be a new square lattice in the plane, with segments $1/\sqrt{2}$ times the length, but still every segment in the lattice occurs exactly once. In the limit, each edge from the original square lattice gives rise to one congruent copy of Heighway's dragon.

See a few stages in Plate 16. One particular edge is shown in black at the center, and a few of its neighboring edges are shown in in other colors. We know (Proposition 1.5.5) that the approximating polygons for any given dragon remain in a bounded region of the plane. So: given any point (x, y) in the plane, the squares surrounding it in the stages of the process can only come from a finite number of original edges. (In the plate, only a finite number of colors can be involved.) So there is at least one color for which a sequence in that color converges to the point (x, y). So (x, y) belongs to at least one of the dragons. It could happen that a point (x, y) belongs to two or more dragons: these will be boundary points for two or more dragons. Plate 17 shows nine dragons of the tiling that result from the nine edges of Plate 16.

2.6 Metrics for Strings

In Sect. 2.1, we defined a metric $\varrho_{1/2}$ for the space $E^{(\omega)}$ of infinite strings from the two-letter alphabet $\{0, 1\}$. There are other, equally good, metrics for the same space.

Metrics for 01-Strings

Let r be a real number satisfying $0 < r < 1$. A metric ϱ_r is defined on $E^{(\omega)}$ in the same way as the metric $\varrho_{1/2}$: if $\sigma = \alpha\sigma'$, $\tau = \alpha\tau'$, where the first character of σ' is different than the first character of τ', and if $k = |\alpha|$ is the length of α, then

$$\varrho_r(\sigma, \tau) = r^k.$$

Exercise 2.6.1. (1) ϱ_r is a metric on $E^{(\omega)}$.
(2) The basic set $[\alpha]$ has diameter $r^{|\alpha|}$, for all $\alpha \in E^{(*)}$.
(3) The space $(E^{(\omega)}, \varrho_r)$ is complete, compact, and separable.

Proposition 2.6.2. *The metric spaces constructed from $E^{(\omega)}$ using the different metrics ϱ_r are all homeomorphic to each other.*

Proof. Let $0 < r, s < 1$. If $h\colon E^{(\omega)} \to E^{(\omega)}$ is the identity function $h(\sigma) = \sigma$, then we will show that h is a homeomorphism from $\left(E^{(\omega)}, \varrho_r\right)$ to $\left(E^{(\omega)}, \varrho_s\right)$. It is enough to show that h is continuous, since applying this result with r and s interchanged will then show that h^{-1} is continuous.

Given $\varepsilon > 0$, choose k so that $s^k < \varepsilon$, then choose $\delta = r^k$. So if $\sigma, \tau \in E^{(\omega)}$ with $\varrho_r(\sigma, \tau) < \delta$, then $\varrho_r(\sigma, \tau) < r^k$, so σ and τ have at least their first k letters in common. But then $\varrho_s(\sigma, \tau) \leq s^k < \varepsilon$. This shows that h is continuous. (In fact, uniformly continuous.)   □

Since all of these metrics are homeomorphic, changing from one metric to another will make no difference for topological properties, such as compactness or separability, or such as the topological dimension (Chap. 3). So we will normally use the metric $\varrho_{1/2}$ when dealing with topological properties. But changing from one metric to another does make a difference for metric properties, such as fractal dimension (Chap. 6). Here is a simple example where only one of the metrics ϱ_r is appropriate: the addressing function for the Cantor dust (see p. 14) is a lipeomorphism.

Proposition 2.6.3. *Let $h\colon E^{(\omega)} \to \mathbb{R}$ be the addressing function for the Cantor dust, satisfying*

$$h(0\sigma) = \frac{h(\sigma)}{3}$$

$$h(1\sigma) = \frac{h(\sigma) + 2}{3}.$$

Then we have

$$\frac{1}{3}\varrho_{1/3}(\sigma, \tau) \leq \left|h(\sigma) - h(\tau)\right| \leq \varrho_{1/3}(\sigma, \tau).$$

Proof. Let σ, τ be given. Let α be their longest common prefix:

$$\sigma = \alpha\sigma',$$
$$\tau = \alpha\tau',$$

where the first letters of σ' and τ' are different. I will prove the inequalities by induction on the length $|\alpha|$. If $|\alpha| = 0$, that is α is the empty string Λ, then $\varrho_{1/3}(\sigma, \tau) = 1$; the two real numbers $h(\sigma), h(\tau)$ both lie in $[0, 1]$, so $\left|h(\sigma) - h(\tau)\right| \leq 1$; and σ, τ begin with different letters, so one of $h(\sigma), h(\tau)$ is in $[0, 1/3]$ and the other is in $[2/3, 1]$, and thus $\left|h(\sigma) - h(\tau)\right| \geq 1/3$. So the result is true when the length $|\alpha| = 0$.

Suppose the result is known when $|\alpha| = k$, and consider the case $|\alpha| = k+1$. Then either $\alpha = 0\beta$ or $\alpha = 1\beta$ for some β with $|\beta| = k$. We will take the case $\alpha = 1\beta$; the other case is similar. By the induction hypothesis, applied to $\beta\sigma'$ and $\beta\tau'$,

$$\frac{1}{3}\varrho_{1/3}(\beta\sigma', \beta\tau') \leq \left|h(\beta\sigma') - h(\beta\tau')\right| \leq \varrho_{1/3}(\beta\sigma', \beta\tau').$$

But
$$\varrho_{1/3}(\sigma, \tau) = \frac{1}{3}\,\varrho_{1/3}(\beta\sigma', \beta\tau'),$$

and
$$\begin{aligned}
\left|h(\sigma) - h(\tau)\right| &= \left|h(\alpha\sigma') - h(\alpha\tau')\right| \\
&= \left|h(1\beta\sigma') - h(1\beta\tau')\right| \\
&= \left|\frac{h(\beta\sigma') + 2}{3} - \frac{h(\beta\tau') + 2}{3}\right| \\
&= \frac{1}{3}\left|h(\beta\sigma') - h(\beta\tau')\right|.
\end{aligned}$$

So the inequalities for this case follow.   □

Exercise 2.6.4. If $r \neq 1/3$, then the addressing function h is not a lipeomorphism from $(E^{(\omega)}, \varrho_r)$ to $\mathbb{R}$.

The range $h\big[\{0,1\}^{(\omega)}\big]$ of the function h in Proposition 2.6.3 is, of course, exactly the triadic Cantor dust C. The proposition shows that the space $\{0,1\}^{(\omega)}$ of infinite strings is homeomorphic to the Cantor dust C.

There are other (useful) ways to define a metric on the space $E^{(\omega)}$. Now let us assign a positive real number w_α to each node α of the binary tree $E^{(*)}$. If they satisfy the right conditions, then we will be able to define a metric ϱ on $E^{(\omega)}$ so that the basic set $[\alpha]$ will have diameter exactly w_α.

Proposition 2.6.5. *Let a family w_α of positive real numbers be given, one for each node α in the infinite binary tree $E^{(*)}$. Suppose:*

$$w_\alpha > w_\beta \qquad \text{if } \alpha < \beta$$
$$\lim_{n \to \infty} w_{\sigma \restriction n} = 0 \qquad \text{for } \sigma \in E^{(\omega)}.$$

Then there is an ultrametric ϱ on $E^{(\omega)}$ such that $\operatorname{diam}[\alpha] = w_\alpha$ *for all* α.

Proof. Define ϱ as follows. If $\sigma = \tau$, then $\varrho(\sigma, \tau) = 0$. If $\sigma \neq \tau$, then $\varrho(\sigma, \tau) = w_\alpha$, where α is the longest common prefix of σ and τ.

Clearly $\varrho(\sigma, \tau) \geq 0$. If $\sigma \neq \tau$, then $\varrho(\sigma, \tau) = w_k > 0$. The equation $\varrho(\sigma, \tau) = \varrho(\tau, \sigma)$ is also easy.

So I must verify the ultra-triangle inequality,

$$\varrho(\sigma, \tau) \leq \max\{\varrho(\sigma, \theta), \varrho(\theta, \tau)\}.$$

If two (or three) of the strings σ, θ, τ are equal, then this is trivial. So suppose they are all different. Let α be the longest common prefix of σ and θ; let β be the longest common prefix of θ and τ; let γ be the longest common prefix of σ and τ. Now α and β are both prefixes of θ, so one of them is a prefix of the other. If $\alpha \leq \beta$, then α is a prefix of both σ and τ, so $\alpha \leq \gamma$, and therefore

$$\varrho(\sigma, \tau) = w_\gamma \leq w_\alpha = \varrho(\sigma, \theta) \leq \max\{\varrho(\sigma, \theta), \varrho(\theta, \tau)\}.$$

The case $\beta \leq \alpha$ is similar.

Now I must show $\mathrm{diam}[\alpha] = w_\alpha$. If $\sigma, \tau \in [\alpha]$, then the longest common prefix β of σ and τ is $\geq \alpha$, so $\varrho(\sigma, \tau) = w_\beta \leq w_\alpha$. Therefore $\mathrm{diam}[\alpha] \leq w_\alpha$. Choose any $\sigma \in E^{(\omega)}$; then $\alpha 0 \sigma, \alpha 1 \sigma \in [\alpha]$ and $\varrho(\alpha 0 \sigma, \alpha 1 \sigma) = w_\alpha$. Therefore $\mathrm{diam}[\alpha] \geq w_\alpha$. □

Exercise 2.6.6. Let ϱ be the metric defined on $E^{(\omega)}$ using diameters w_α, where

$$w_\alpha > w_\beta \qquad \text{if } \alpha < \beta$$

$$\lim_{n \to \infty} w_{\sigma \restriction n} = 0 \qquad \text{for } \sigma \in E^{(\omega)}.$$

Then

(1) $(E^{(\omega)}, \varrho)$ is homeomorphic to $(E^{(\omega)}, \varrho_{1/2})$.
(2) The countable set $\{ [\alpha] : \alpha \in E^{(*)} \}$ of "basic open sets" is equal to the set $\{ B_r(\sigma) : \sigma \in E^{(\omega)}, r > 0 \}$ of all open balls, and to the set of all closed balls.

Proposition 2.6.7. *Let A be any subset of $E^{(\omega)}$ with at least two elements. Then there is a basic open set $[\alpha]$ such that $A \subseteq [\alpha]$ and $\mathrm{diam}\, A = \mathrm{diam}([\alpha])$.*

Proof. Let α be the longest common prefix for the set A (as in Proposition 1.3.2). The string α is finite (possibly empty) since A has at least two elements. Clearly $A \subseteq [\alpha]$, and therefore $\mathrm{diam}\, A \leq \mathrm{diam}[\alpha]$. Now choose $\sigma \in A$. If k is the length $|\alpha|$, then of course $\sigma \restriction k = \alpha$. Since $\sigma \restriction (k+1)$ is not a common prefix for A, there is $\tau \in A$ with $\tau \restriction (k+1) \neq \sigma \restriction (k+1)$. That means that α is the longest common prefix for the pair $\{\sigma, \tau\}$, and therefore $\varrho(\sigma, \tau) = w_\alpha$. Therefore $\mathrm{diam}\, A \geq w_\alpha = \mathrm{diam}[\alpha]$. □

Other String Spaces

It is probably clear to the reader by now that what has been done for strings from the alphabet $\{0, 1\}$ depends very little on that particular choice. Most of it could equally well be done for any (finite) alphabet.

Let E be any finite set; we assume that E has at least two elements. We call the elements of E "letters", and we call E an "alphabet".

Exercise 2.6.8. Formulate definitions of the following: string from the alphabet E; $E^{(n)}$; length $|\alpha|$ of a string α; $E^{(*)}$; $E^{(\omega)}$; $[\alpha]$. Formulate variants of the following results for this setting: 1.3.2, 2.1.8, 2.6.1, 2.6.2, 2.6.5, 2.6.6, 2.6.7.

The "continuous" function is useful, but perhaps unfamiliar, on these string spaces. Roughly speaking, continuity of a function f means that any finite amount of information about $f(\sigma)$ is determined by only a finite amount of information about σ. Here is a more precise statement:

Exercise 2.6.9. Let E_1 and E_2 be two finite sets. Let $\sigma \in E_1^{(\omega)}$ and let $f\colon E_1^{(\omega)} \to E_2^{(\omega)}$ be a function. Then the following are equivalent:

(1) f is continuous at σ;

(2) for every integer n there exists m so that for $\tau \in E_1^{(\omega)}$, if we have $\tau{\restriction}m = \sigma{\restriction}m$ then we have $f(\tau){\restriction}n = f(\sigma){\restriction}n$.

Path Spaces of Graphs

In Chaps. 4 and 7 we will require a further generalization of the string spaces $E^{(\omega)}$.

Figure 2.6.10 shows examples of certain "graphs"   that we will be using. There is a finite set V of **vertices** or **nodes**, and there are **edges**. Each edge goes from one vertex to another (or possibly back to the same one). The direction is important, so this is a **directed graph**. There may be more than one edge connecting a given pair of nodes, so this is a **multigraph**. This informal description (together with the pictures) is probably enough to tell you what a directed multigraph is. But we will need a more mathematically sound definition.

A **directed multigraph** consists of two (finite) sets V and E, and two functions $i\colon E \to V$ and $t\colon E \to V$. The elements of V are called **vertices** or **nodes**; the elements of E are called **edges** or **arrows**. For an edge e, we call $i(e)$ the **initial vertex** of e, and we call $t(e)$ the **terminal vertex** of e. We will often write E_{uv} for the set of all edges e with $i(e) = u$ and $t(e) = v$.

There is an example pictured in Fig, 2.6.11. In this case, $V = \{\mathsf{S}, \mathsf{T}\}$ and $E = \{\mathsf{a}, \mathsf{b}, \mathsf{c}, \mathsf{d}, \mathsf{e}, \mathsf{f}\}$. We have, for example, $i(\mathsf{c}) = \mathsf{S}$, $t(\mathsf{c}) = \mathsf{T}$, $i(\mathsf{f}) = t(\mathsf{f}) = \mathsf{T}$.

A **path** in a directed multigraph is a sequence of edges, taken in some order, so that the terminal vertex of one edge is the initial vertex of the next

Fig. 2.6.10. Directed Multigraphs

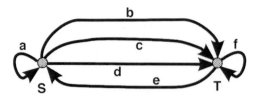

Fig. 2.6.11. An Example

edge. A path will often be identified with a string made up of the labels of the edges. Here are some examples of paths in the example: c or bedfe or aaaa. The **initial vertex** of a path is, by definition, the initial vertex of the first edge in the path. Similarly, the **terminal vertex** of a path is the terminal vertex of the last edge in the path. We extend the functions i and t accordingly. For example, if $\alpha = $ bedfe, then $i(\alpha) = $ S and $t(\alpha) = $ S.

We will write $E_{uv}^{(*)}$ for the set of all paths with initial vertex u and terminal vertex v. We may say that such a path **goes from** u **to** v, or that it **connects** u to v. The number of edges in a path is its **length**, written $|\alpha|$. So in the example $\alpha = $ bedfe, we have $|\alpha| = 5$. We will write $E_{uv}^{(n)}$ for the set of all paths from u to v of length n; and $E_u^{(n)}$ for the set of all paths of length n with initial vertex u; and $E^{(n)}$ for the set of all paths of length n. The empty set conventions will work out best if we say (by convention) that for each $u \in V$, the set $E_{uu}^{(0)}$ has one element, the empty path from u to itself, written Λ_u. Of course, we may identify E with $E^{(1)}$ and E_{uv} with $E_{uv}^{(1)}$. Note that the conventions have been set so that if the strings α, β represent paths, and the terminal vertex of α is equal to the initial vertex of β, then the concatenated string $\alpha\beta$ represents a path, as well.

The case that we have considered before, where all strings from a given alphabet E are allowed, corresponds to a graph with only one vertex, and the alphabet E as edge set.

A path α with $i(\alpha) = t(\alpha)$ is called a **cycle**. A cycle of length 1 is a **loop**. A directed multigraph is **strongly connected** iff, for each pair u, v of vertices, there is a path from u to v.

Let (V, E, i, t) be a directed multigraph. We will consider the set $E^{(*)}$ of all paths in the graph. This naturally has the structure of a "tree": If α is a path, then the children of α are the paths αe, for edges e with $i(e) = t(\alpha)$. Actually this is not a tree: it is a finite disjoint union of trees, one tree $E_v^{(*)}$ corresponding to each node v of the graph. A disjoint union of trees is sometimes called a **forest**. So we will call this the **path forest** of the graph.

What about infinite paths, corresponding to the infinite strings that we have considered before? An infinite string σ corresponds to an infinite path if the terminal vertex for each edge matches the initial vertex for the next edge. We write $E^{(\omega)}$ for the set of all infinite paths for the graph (V, E, i, t). If

$v \in V$ is a vertex, then we write $E_v^{(\omega)}$ for the set of all infinite paths starting at v. (It usually does not make sense to assign a terminal vertex to an infinite path.) There will be one of these path spaces $E_v^{(\omega)}$ for each vertex of the graph (V, E, i, t), or one for each tree $E_v^{(*)}$ in the forest $E^{(*)}$ of finite paths. If $\alpha \in E^{(*)}$, then we write $[\alpha] = \left\{ \sigma \in E^{(\omega)} : \alpha \leq \sigma \right\}$, that is, the set of all infinite paths that begin with the finite path α.

Metrics may be defined on the path spaces $E_v^{(\omega)}$ in much the same way as has already been done. Some notice must be taken of the possibility that some nodes in $E^{(*)}$ have no children, or only one child. Now if α has no children, then $[\alpha] = \varnothing$, so its diameter must be 0. If α has only one child β, then $[\alpha] = [\beta]$, so $\mathrm{diam}[\alpha]$ must be equal to $\mathrm{diam}[\beta]$. These cases will not occur if, in the graph (V, E, i, t), each node has at least two edges leaving it.

Let (V, E, i, t) be a directed multigraph. Let a family w_α of positive real numbers be given, one for each node α in the path forest $E^{(*)}$. We want to define distances $\varrho(\sigma, \tau)$. In fact, only distances between infinite strings with the same initial vertex will be needed. So we can think of defining several (disjoint) metric spaces $E_v^{(\omega)}$, one for each node $v \in V$. We will normally write simply ϱ for all the metrics involved.

Suppose the family w_α satisfies

$$w_\alpha > w_\beta \qquad \text{if } \alpha < \beta$$
$$\lim_{n \to \infty} w_{\sigma \restriction n} = 0 \qquad \text{for } \sigma \in E^{(\omega)}.$$

The definition for the metrics is as before. If $\sigma, \tau \in E_v^{(\omega)}$, and $\sigma \neq \tau$, then the two strings have a longest common prefix, since they have at least the prefix Λ_v in common. So define $\varrho(\sigma, \tau) = w_\alpha$, where α is the longest common prefix of σ and τ. This defines ultrametrics ϱ on the spaces $E_v^{(\omega)}$ such that $\mathrm{diam}[\alpha] = w_\alpha$ for "most" α. For which α might the equation fail?

Exercise 2.6.12. Adapt the following results to this setting: 2.6.6, 2.6.7

2.7 *Remarks

It is probably more conventional in fractal geometry to limit consideration to subsets of Euclidean space. But metric spaces are the proper setting for the theory of topological dimension, and perhaps even for the theory of Hausdorff dimension. We will need uniform convergence of functions, and the Hausdorff metric for sets; even if the primary interest is Euclidean space, these lead naturally to other metric spaces.

On the other hand, I have limited the discussion to metric spaces, rather than more general topological spaces. Certainly Hausdorff dimension belongs in a metric space. Topological dimension can be done in much greater generality, but I will leave that to the student who wants to go beyond this book. In fact, whenever it is convenient, the discussion will be restricted to separable metric spaces, or even to compact metric spaces.

Additional material along the lines of this chapter can be found in texts called "general topology" or "point-set topology", such as [38] or [41].

According to the definition in this chapter, d-dimensional Euclidean space is the set $\mathbb{R}^d$ of d-tuples of real numbers. For our purposes, this is reasonable. But in some other contexts, one would say that $\mathbb{R}^d$ is really more properly considered to be d-dimensional Euclidean space *together with* a distinguished coordinate system. This sort of distinction emphasizes the difference between two camps called "synthetic geometry" and "analytic geometry".

The reader aware of such things may have noticed that the axiom of choice is freely used in this chapter. There is no better way. If we want to do metric topology using sequences, then at least countable choice must be used.

The spaces $E^{(\omega)}$ that have been called here "string models" or "path models" are more commonly known as "shift models" and "subshift models". However, the word "shift" involved refers to the left shift on these spaces, which is never* mentioned in this book, so its use seemed inappropriate.

For the Hausdorff metric on sets in a metric space see for example [33, §28]. Felix Hausdorff's mathematical writings are few in number, but immensely influential.

Comments on Exercises

Do not read this until you have tried to do the exercise yourself!

Exercise 2.1.2: $x \le y \le z$ or $x \ge y \ge z$. That is, y is between x and z.

Exercise 2.1.7: $|x + y| = |x| + |y|$ if and only if one of x, y is a nonnegative scalar multiple of the other.

Exercise 2.1.14: An "if and only if" proof is done in two parts, an "if" part and an "only if" part. Two sets are equal when each is a subset of the other.

Exercise 2.2.10: This can be done from the ε, δ definition (good practice using the definition). Or it can be done using Theorem 2.2.9.

Exercise 2.2.15: Given a Cauchy sequence in $\mathbb{R}^3$, use the completeness of $\mathbb{R}$ three times; successively extract subsequences so that each of the three coordinates converges. Then show that the resulting subsequence converges.

Exercise 2.2.30: (1) $U = \{\, x : \operatorname{dist}(x, A) < \operatorname{dist}(x, B) \,\}$. (2)

$$U = \{\, x \in S : \operatorname{dist}(x, A) < (1/2)\operatorname{dist}(x, B) \,\}$$
$$V = \{\, x \in S : \operatorname{dist}(x, A) > 2\operatorname{dist}(x, B) \,\}.$$

Exercise 2.2.18: The sets $\mathbb{R}$ and $(0, 1)$ are homeomorphic.

Exercise 2.3.12: Show that $E^{(\omega)}$ is countably compact. If A is infinite, then at least one of the sets $A \cap [0]$, $A \cap [1]$ is infinite. Then at least one of the subsets $A \cap [\alpha]$, $\alpha \in E^{(2)}$, is infinite. Etc.

---

* With this exception

Exercise 2.3.18: In the plane, consider the two graphs $xy = 1$ and $xy = -1$.

Exercise 2.5.2: Suppose $A_n \to A$, and let

$$B = \{\, x : \text{there is a sequence } (x_n) \text{ with } x_n \in A_n \text{ and } x_n \to x \,\}.$$

First, $A \subseteq B$: Since any element $x \in A$ is within distance $D(A, A_n) + 1/n$ of some element x_n of A_n, we have $x_n \to x$. For $B \subseteq A$: Since any element of B is within distance ε of elements of A_n for large enough n, and those elements, in turn, are within distance ε of elements of A. But A is closed, so any point with distance 0 from A belongs to A.

I have, with difficulty, been prevented
by my friends from labeling [the book]:
What Every Young Analyst Should Know.
—J. L. Kelley, *General Topology*

3

Topological Dimension

The sets of elementary geometry have associated with them a ***dimension***. Points have dimension 0. Curves have dimension 1. Surfaces have dimension 2. Solids have dimension 3. When we leave elementary geometry, there is the possibility of considering point sets not falling into any of these clear-cut groups. Mathematicians have defined a general notion of dimension to help out in the study of such sets. In fact, they have done it in several different ways; one way of defining dimension may be useful for one purpose but not for another purpose.

We will discuss more than one of these definitions. The definitions that will be considered fall generally into two broad classes: topological dimension and fractal dimension. The first one is a topological dimension known as the "covering dimension". Keep in mind that these are ***topological*** dimensions also in a more technical sense: if two spaces S and T are homeomorphic, then their dimensions are equal.

3.1 Zero-Dimensional Spaces

Figure 3.1.1 suggests that if a set should be considered "1-dimensional", then it can be covered by small open sets that intersect only 2 at a time. (That is: any 3 of the sets have empty intersection; or, each point belongs to at most 2 of the sets.) A set is considered "0-dimensional" if it can be covered by small open sets that are disjoint. A set is "2-dimensional" if it can be covered by small open sets that intersect only 3 at a time. The usual "brick" packing shown in Fig. 3.1.2 covers the plane by closed rectangular sets that intersect at most 3 at a time—if the rectangles are enlarged slightly to open sets, we can still arrange that they intersect at most 3 at a time.

This idea is reasonable for compact metric spaces (see Theorem 3.2.2), but needs a bit of fine-tuning for non-compact spaces.

In this section we will discuss zero-dimensional spaces. In Sect. 3.2 we will go on to higher dimensions.

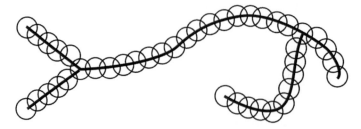

Fig. 3.1.1. A 1-dimensional space and a covering by disks

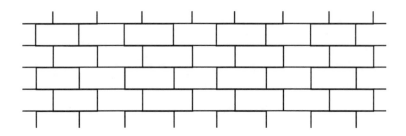

Fig. 3.1.2. A cover of the plane

Definition

If $\mathcal{A}$ and $\mathcal{B}$ are two collections of sets, we say that $\mathcal{B}$ is **subordinate** to $\mathcal{A}$ iff every $B \in \mathcal{B}$ there is $A \in \mathcal{A}$ with $B \subseteq A$. Let S be a metric space, and let $\mathcal{A}$ be an open cover of S. A **refinement** of $\mathcal{A}$ is an open cover $\mathcal{B}$ of S that is subordinate to $\mathcal{A}$. We also say say $\mathcal{B}$ **refines** $\mathcal{A}$. For example, a subcover of $\mathcal{A}$ is a refinement of $\mathcal{A}$.

Exercise 3.1.3. Let S be a metric space. Then S is compact if and only if every open cover has a finite refinement.

Let S be a metric space. A set A in a metric space S will be called **clopen** iff it is both a closed set and an open set. In particular, $\varnothing$ and S are clopen. If A, B are clopen, then so are $A \cap B$, $A \cup B$, and $A \setminus B$. (The collection of all clopen subsets of S is an algebra of sets in the sense to be defined on p. 147.)

A **clopen partition** of S is an open cover of S consisting of disjoint clopen sets. The space S is **zero-dimensional** iff every finite open cover of S has a finite refinement that is a clopen partition.

Example: If S is finite, then S is zero-dimensional. To see this: Each singleton $\{x\}$ is an open set, therefore a clopen set. And any open cover is refined by the open cover made up of all these singletons, which is a clopen partition of S.

Another good example is the Cantor dust C; see p. 2 for the definition and notations C, C_n. Let $\mathcal{A}$ be an open cover of C. For each $x \in C$, there is an open set $A \in \mathcal{A}$ with $x \in A$, and therefore a positive r with $B_r(x) \cap C \subseteq A$.

But then there is an integer n with $3^{-n} < r$, so the interval I of C_n that contains x has length 3^{-n} and $I \cap C \subseteq A$. The complement in C of I is a finite union of closed intervals, so $I \cap C$ is clopen in C. For each $x \in C$ choose such an interval I_x with $I_x \cap C$ clopen in C and $I_x \subseteq A$ for some $A \in \mathcal{A}$. Thus

$$\mathcal{A}_1 = \{ I_x : x \in C \}$$

is an open cover of C. Since C is compact, there is a finite subcover of $\mathcal{A}_1$, say $\mathcal{A}_2 = \{I_1, I_2, \cdots, I_k\}$. Now if we write $J_1 = I_1$, $J_2 = I_2 \setminus J_1$, $J_3 = I_3 \setminus J_2$, and so on, we get a finite cover $\mathcal{A}_3 = \{J_1, J_2, \cdots, J_k\}$ of C by sets clopen in C, which is still a refinement of $\mathcal{A}$. The sets $J_1, J_2, \cdots$ are disjoint, so the cover $\mathcal{A}_3$ is a clopen partition of C.

The next example to consider is the line $\mathbb{R}$. It is not zero-dimensional. To prove this, I will need to show that there are not very many clopen sets in $\mathbb{R}$.

Theorem 3.1.4. *The only clopen sets in the space $\mathbb{R}$ are $\varnothing$ and $\mathbb{R}$. Therefore, $\mathbb{R}$ is not zero-dimensional.*

Proof. Let $A \subseteq \mathbb{R}$, and suppose $A \neq \varnothing$ and $A \neq \mathbb{R}$. I must show that A is not a clopen set, or, equivalently, that A has a boundary point. We will define recursively two sequences, (x_n) and (y_n). First, we may choose a point $x_0 \in A$ since $A \neq \varnothing$. Also, we may choose a point $y_0 \notin A$ since $A \neq \mathbb{R}$. After x_n and y_n have been defined, with $x_n \in A$ and $y_n \notin A$, we want to define x_{n+1} and y_{n+1}. Consider the midpoint $z_n = (x_n + y_n)/2$. If $z_n \in A$, then define $x_{n+1} = z_n$, $y_{n+1} = y_n$; and if $z_n \notin A$, then define $x_{n+1} = x_n$, $y_{n+1} = z_n$. So in any case, we get $x_{n+1} \in A$ and $y_{n+1} \notin A$, with $|x_{n+1} - y_{n+1}| = |x_n - y_n|/2$. So by induction $|x_n - y_n| = |x_0 - y_0|/2^n$. Thus $|x_n - y_n| \to 0$ as $n \to \infty$. Also, $|x_{n+1} - x_n| \leq |x_n - y_n| = |x_0 - y_0|/2^n$, so (x_n) is a Cauchy sequence. Let $x = \lim_n x_n$. Because $|x_n - y_n| \to 0$, we have also $y_n \to x$. Therefore x is a boundary point of A. So A is not a clopen set.

Now the two-element cover $(-\infty, 1) \cup (-1, \infty)$ of $\mathbb{R}$ has no clopen refinement. So $\mathbb{R}$ is not zero-dimensional. □

It is worth pointing out the following, which has the same proof.

Corollary 3.1.5. *Let $a < b$. The only clopen subsets of the space $[a, b]$ are $\varnothing$ and $[a, b]$. Therefore, $[a, b]$ is not zero-dimensional.*

Here is a result promised in Chap. 2, p. 70:

Proposition 3.1.6. *Let C be the Cantor dust. There is no continuous function from the interval $[0, 1]$ onto C.*

Proof. Suppose $h : [0, 1] \to C$ is surjective. The set $M = [0, 1/3] \cap C$ is a clopen subset of C. Both M and its complement $C \setminus M$ are nonempty. So both the set $h^{-1}[M]$ and its complement $h^{-1}[C \setminus M]$ are nonempty. By Corollary 3.1.5, not both sets are open in $[0, 1]$. By Theorem 2.2.6, h is not continuous. □

Properties for Zero-Dimensionality

Here are some variants of the definition.

Theorem 3.1.7. *Let S be a metric space. The following are equivalent:*

(1) S *is zero-dimensional.*
(2) *If $\{U_1, U_2, \cdots, U_k\}$ is any finite open cover of S, then there exist sets $B_1 \subseteq U_1$, $B_2 \subseteq U_2, \cdots$, $B_k \subseteq U_k$, such that $\{B_1, B_2, \cdots, B_k\}$ is a clopen partition of S.*
(3) *If $\{U, V\}$ is an open cover of S, then there exist open sets $A \subseteq U$, and $B \subseteq V$ such that $A \cup B = S$ and $A \cap B = \varnothing$.*
(4) *If $\{U, V\}$ is an open cover of S, then there exist closed sets $A \subseteq U$, and $B \subseteq V$ such that $A \cup B = S$ and $A \cap B = \varnothing$.*

Proof. (2) $\Longrightarrow$ (3) and (2) $\Longrightarrow$ (1) are clear. (3) $\Longleftrightarrow$ (4) follows because in both cases the sets A, B are clopen.

(1) $\Longrightarrow$ (2). Suppose S is zero-dimensional. The open cover $\{U_1, U_2, \cdots, U_k\}$ is refined by a clopen partition $\mathcal{W}$. For each $W \in \mathcal{W}$ there is at least one i such that $W \subseteq U_i$; choose one of them, and call it $i(W)$. Now for each i, let

$$B_i = \bigcup \{ W \in \mathcal{W} : i(W) = i \} .$$

Then the sets B_i are open, and $\bigcup_i B_i = \bigcup_{W \in \mathcal{W}} W = S$. The complement of B_j is

$$\bigcup \{ W \in \mathcal{W} : i(W) \neq j \} ,$$

which is open. So B_j is clopen. If $x \in S$, then (since $\mathcal{W}$ is a partition) x belongs to at most one of the sets W. But $x \in B_i$ only if $x \in W$ for some W with $i(W) = i$. So x belongs to at most one of the sets B_i. That is, the sets B_i are disjoint.

(3) $\Longrightarrow$ (2). Suppose S has the property (3). Let $\{U_1, U_2, \cdots, U_k\}$ be an open cover of S. If $k = 1$, then this cover is already a clopen partition. So suppose $k \geq 2$. Now write $U = U_1$, $V = \bigcup_{i=2}^{k} U_i$. Then these sets cover S, so by the hypothesis (3), there exist clopen sets $A \subseteq U$, $B \subseteq V$ with $A \cup B = S$ and $A \cap B = \varnothing$. Let $B_1 = A$ and $B_i = B \cap U_i$ for $i \geq 2$. Then $B_i \subseteq U_i$ for all i, we have $\bigcup_{i=1}^{k} B_i = S$, and $B_1 \cap B_2 = \varnothing$. Repeat this construction a finite number of times, once for each two-element subset of $\{1, 2, \cdots, k\}$, to arrange the same conclusion with *all* of the intersections of two of the sets empty. $\square$

Separation

Recall Exercise 2.2.30(2): If A and B are disjoint closed sets, then there open sets U and V with $U \supseteq A$, $V \supseteq B$, and $\overline{U} \cap \overline{V} = \varnothing$. Zero-dimensional spaces have a stronger separation property.

Proposition 3.1.8. *Let S be a metric space. Then the following are equivalent:*

(1) *S is zero-dimensional.*
(2) *For every pair A, B of disjoint closed sets, there is a clopen set U with complement $V = S \setminus U$ so that $U \supseteq A$ and $V \supseteq B$.*

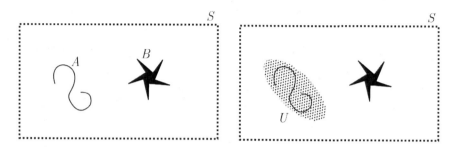

Fig. 3.1.8. Illustration for Theorem 3.1.8

Proof. Suppose S is zero-dimensional. Let A, B be disjoint closed sets. Then their complements $A' = S \setminus A$ and $B' = S \setminus B$ are open. Since $A \cap B = \varnothing$, the complements cover: $A' \cup B' = S$. So by Theorem 3.1.7(3), there are open sets U, V with $U \subseteq B'$, $V \subseteq A'$, $U \cup V = S$, $U \cap V = \varnothing$. So U, V are clopen, $U \supseteq A$ and $V \supseteq B$.

Conversely, suppose (2) holds. Let $\{U_1, U_2\}$ be an open cover of S. Then the complements $A = S \setminus U_1$, $B = S \setminus U_2$ are disjoint closed sets. Therefore there exists clopen set V with complement $U = S \setminus V$ so that $V \supseteq A$, $U \supseteq B$. So $V \subseteq U_2$ and $U \subseteq U_1$, and by Theorem 3.1.7(3), S is zero-dimensional.   □

Base for the Topology

Dimension zero is related to the existence of a clopen base for the topology.

Proposition 3.1.9. *Let S be a zero-dimensional metric space. Then there is a base for the topology of S consisting of clopen sets.*

Proof. Let $U \subseteq S$ be an open set and let $x_0 \in U$. The distance $r := \text{dist}(x_0, S \setminus U) > 0$. So S is covered by the two open sets U and $V = \{x \in S : \varrho(x, x_0) > r/2\}$. This cover is refined by a clopen partition $\mathcal{A}$. This shows there is a base for the topology of S consisting of clopen sets.   □

The converse in the compact case can be proved using the argument from the Cantor dust, above.

Proposition 3.1.10. *Let S be a nonempty compact metric space. Assume there is a base for the topology of S consisting of clopen sets. Then S is zero-dimensional.*

Proof. Let $\mathcal{B}$ be a base for the topology of S consisting of clopen sets. Let $\mathcal{A}$ be an open cover of S. For each $x \in S$, there is an open set $A \in \mathcal{A}$ with $x \in A$, and therefore there is some $B \in \mathcal{B}$ with $x \in B \subseteq A$. For each $x \in S$ choose such $B_x \in \mathcal{B}$. Then

$$\mathcal{A}_1 = \{\, B_x : x \in S \,\}$$

is an open cover of S. By compactness, there is a finite subcover, say $\mathcal{A}_2 = \{B_1, B_2, \cdots, B_k\}$. Now if we write $J_1 = B_1$, $J_2 = B_2 \setminus J_1$, $J_3 = B_3 \setminus J_2$, and so on, we get a finite cover $\mathcal{A}_3 = \{J_1, J_2, \cdots, J_k\}$ of C by clopen sets, which is still subordinate to $\mathcal{A}$. The sets $J_1, J_2, \cdots$ are disjoint, so $\mathcal{A}_3$ is a clopen partition of S. $\qquad\square$

In fact, the converse holds more generally for separable metric space.

Proposition 3.1.11. *Let S be a separable metric space. Then S is zero-dimensional if and only if there is a base for the topology of S consisting of clopen sets.*

Proof. One direction was done in Proposition 3.1.9. For the converse, assume there is a base $\mathcal{B}$ for the topology of S consisting of clopen sets. Let $\{U, V\}$ be an open cover of S. There is a collection $\mathcal{U}_1 \subseteq \mathcal{B}$ with $\bigcup \mathcal{U}_1 = U$. By the Lindelöf property (Theorem 2.3.1), there is a countable subcollection $\mathcal{U}_2 \subseteq \mathcal{U}_1$ with $\bigcup \mathcal{U}_2 = U$. Similarly there is a countable collection $\mathcal{V}_2 \subseteq \mathcal{B}$ with $\bigcup \mathcal{V}_2 = V$. Enumerate the union:

$$\mathcal{U}_2 \cup \mathcal{V}_2 = \{G_1, G_2, G_3, \cdots\}.$$

The sets G_m are clopen, and $\bigcup G_m = S$. Define $H_1 = G_1$, $H_2 = G_2 \setminus G_1$, and in general $H_m = G_m \setminus (G_1 \cup \cdots \cup G_{m-1})$. So the sets H_m are clopen, disjoint, and $\bigcup H_m = S$. Now let

$$E = \bigcup \{\, H_m : H_m \subseteq U \,\}, \qquad F = \bigcup \{\, H_m : H_m \not\subseteq U \,\}.$$

Then U, V are open, $U \cap V = \varnothing$, $U \cup V = S$, so U, V are clopen. Also $E \subseteq U$ and $F \subseteq V$. This shows that S is zero-dimensional. $\qquad\square$

Exercise 3.1.12. The set $\mathbb{Q}$ of rational numbers (with the usual metric) is zero-dimensional. The complementary set $\mathbb{R} \setminus \mathbb{Q}$ of irrationals is also zero-dimensional.

Exercise 3.1.13. Let K be a nonempty compact set in $\mathbb{R}^2$. Let $f_1, f_2, \cdots, f_n$ be contracting similarities of $\mathbb{R}^2$ to itself. Suppose K satisfies the self-referential equation

$$K = \bigcup_{i=1}^{n} f_i[K],$$

and that $f_i[K] \cap f_j[K] = \varnothing$ for $i \neq j$. Show K is zero-dimensional.

Sum Theorem

The real line is not zero-dimensional, but according to Exercise 3.1.12 it can be written as a union of two zero-dimensional sets. At least the union behaves well for closed sets.

Theorem 3.1.14. *Let S be a metric space. Let F_1 and F_2 be closed sets in S. If F_1 and F_2 are both zero-dimensional, then $F_1 \cup F_2$ is zero-dimensional.*

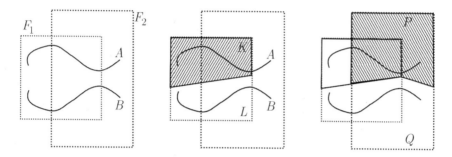

Fig. 3.1.14. Illustration for Theorem 3.1.14

Proof. Let F_1 and F_2 be zero-dimensional and closed. (Remark: Since F_1 is closed, a subset $E \subseteq F_1$ is closed in F_1 if and only if E is closed in S. The same may not be true for "open": a set $E \subseteq F_1$ open in F_1 need not be open in S.) We must show that $F_1 \cup F_2$ is also zero-dimensional. Let A, B be disjoint closed sets in $F_1 \cup F_2$. Now $A \cap F_1$ and $B \cap F_1$ are closed sets, so they are closed in F_1. They are disjoint. So, since F_1 is zero-dimensional, there exists a set K clopen in F_1 with complement $L = F_1 \setminus K$ such that $K \supseteq A \cap F_1$ and $L \supseteq B \cap F_1$. Now K and L are closed in S. The two sets $(K \cup A) \cap F_2$ and $(L \cup B) \cap F_2$ are disjoint and closed in F_2. So, since F_2 is zero-dimensional, there exists a set P clopen in F_2 with complement $Q = F_2 \setminus P$ such that $P \supseteq (K \cup A) \cap F_2$ and $Q \supseteq (L \cup B) \cap F_2$. Now let $U = K \cup P$, $V = L \cup Q$. So U and V are closed, disjoint, and $U \cup V = F_1 \cup F_2$. So U and V are clopen in $F_1 \cup F_2$. And $U \supseteq A$, $V \supseteq B$. So by Proposition 3.1.8, $F_1 \cup F_2$ is zero-dimensional. □

3.2 Covering Dimension

Let $n \geq -1$ be an integer. The **order** of a family $\mathcal{A}$ of sets is $\leq n$ iff any $n+2$ of the sets have empty intersection. If $n \geq 0$, then we say $\mathcal{A}$ has order n iff it has order $\leq n$ but does not have order $\leq n - 1$. For example, a family $\mathcal{A}$ of nonempty sets is disjoint if and only if its order is 0. A family of sets $\mathcal{A}$ has order -1 iff it is empty or is the singleton $\{\varnothing\}$. The set of disks in Fig. 3.1.1 has order 1. The set of rectangles in Fig. 3.1.2 has order 2.

The "covering dimension" will be defined now. The covering dimension of a metric space will be either an integer ≥ -1 or the extra symbol ∞, understood to be larger than all integers.

Let S be a metric space. Let $n \geq -1$ be an integer. We say that S has **covering dimension** $\leq n$ iff every finite open cover of S has a refinement with order $\leq n$. The covering dimension is n iff the covering dimension is $\leq n$ but not $\leq n-1$. We will write $\operatorname{Cov} S = n$ in that case. If the covering dimension is $\leq n$ for no integer n, then we say $\operatorname{Cov} S = \infty$. Covering dimension is also known as **Lebesgue dimension**.

Let us consider the simplest cases. If $\operatorname{Cov} S = -1$, then the open cover $\{S\}$ is refined by a cover of order -1, which is either $\varnothing$ or $\{\varnothing\}$; this is a cover only if $S = \varnothing$. So $\operatorname{Cov} S = -1$ if and only if $S = \varnothing$.

For nonempty S, the covering dimension is 0 if and only if S is zero-dimensional in the sense defined in Sect. 3.1.

Basics of Covering Dimension

Here are some simple variants of the definition of covering dimension.

Theorem 3.2.1. *Let S be a metric space, and let n be a nonnegative integer. The following are equivalent:*

(1) $\operatorname{Cov} S \leq n$
(2) *If $\{U_1, U_2, \cdots, U_k\}$ is any finite open cover of S, then there exist open sets $B_1 \subseteq U_1$, $B_2 \subseteq U_2, \cdots$, $B_k \subseteq U_k$, such that $\{B_1, B_2, \cdots, B_k\}$ is an open cover of S with order $\leq n$.*
(3) *If $\{U_1, U_2, \cdots, U_{n+2}\}$ is an open cover of S, then there exist open sets $B_1 \subseteq U_1$, $B_2 \subseteq U_2, \cdots$, $B_{n+2} \subseteq U_{n+2}$, such that $\bigcup_{i=1}^{n+2} B_i = S$ and $\bigcap_{i=1}^{n+2} B_i = \varnothing$.*
(4) *If $\{U_1, U_2, \cdots, U_{n+2}\}$ is an open cover of S, then there exist closed sets $F_1 \subseteq U_1$, $F_2 \subseteq U_2, \cdots$, $F_{n+2} \subseteq U_{n+2}$, such that $\bigcup_{i=1}^{n+2} F_i = S$ and $\bigcap_{i=1}^{n+2} F_i = \varnothing$.*

Proof. (2) $\Longrightarrow$ (3) and (2) $\Longrightarrow$ (1) are clear.

(1) $\Longrightarrow$ (2). Suppose $\operatorname{Cov} S \leq n$. The open cover $\{U_1, U_2, \cdots, U_k\}$ admits a finite refinement $\mathcal{W}$ of order $\leq n$. For each $W \in \mathcal{W}$ there is at least one i such that $W \subseteq U_i$; choose one of them, and call it $i(W)$. Now for each i, let

$$B_i = \bigcup \{ W \in \mathcal{W} : i(W) = i \}.$$

Then the sets B_i are open, and $\bigcup_i B_i = \bigcup_{W \in \mathcal{W}} W = S$. If $x \in S$, then (since $\mathcal{W}$ has order $\leq n$) it belongs to at most $n+1$ of the sets W. But $x \in B_i$ only if $x \in W$ for some W with $i(W) = i$. So x belongs to at most $n+1$ of the sets B_i.

(3) $\Longrightarrow$ (2). Suppose S has the property (3). Let $\{U_1, U_2, \cdots, U_k\}$ be an open cover of S. If $k \leq n + 1$, then this cover itself already has order $\leq n$. So suppose $k \geq n + 2$. Now write $W_1 = U_1$, $W_2 = U_2, \cdots$, $W_{n+1} = U_{n+1}$ and $W_{n+2} = \bigcup_{i=n+2}^{k} U_i$. Then these sets cover S, so by the hypothesis (3), there exist open sets $V_i \subseteq W_i$ with $\bigcup_{i=1}^{n+2} V_i = S$ and $\bigcap_{i=1}^{n+2} V_i = \varnothing$. Let $B_i = V_i$ for $i \leq n + 1$ and $B_i = V_{n+2} \cap U_i$ for $i \geq n + 2$. Then $B_i \subseteq U_i$ for all i, we have $\bigcup_{i=1}^{k} B_i = S$, and $\bigcap_{i=1}^{n+2} B_i = \varnothing$. Repeat this construction a finite number of times, once for each subset of $\{1, 2, \cdots, k\}$ with $n + 2$ elements, to arrange the same conclusion with *all* of the intersections of size $n + 2$ empty.

(3) $\Longrightarrow$ (4). There exist open sets $B_i \subseteq U_i$ with $\bigcup B_i = S$ and $\bigcap B_i = \varnothing$. Now $S \setminus B_1 \subseteq \bigcup_{i=2}^{n+2} B_i$, so by Corollary 2.2.29, there is an open set V_1 with $S \setminus B_1 \subseteq V_1 \subseteq \overline{V_1} \subseteq \bigcup_{i=2}^{n+2} B_i$. Let $F_1 = S \setminus V_1$. So we have $F_1 \subseteq B_1$ and $F \cup \bigcup_{i=2}^{n} B_i = S$. Next, there is an open set V_2 with $S \setminus B_2 \subseteq V_2 \subseteq \overline{V_2} \subseteq (S \setminus \overline{V_1}) \cup \bigcup_{i=2}^{n+2} B_i$. Let $F_2 = S \setminus V_2$. So we have $F_2 \subseteq B_2$ and $F_1 \cup F_2 \cup \bigcup_{i=3}^{n+2} B_i = S$. Continue in this way.

(4) $\Longrightarrow$ (3). There exist closed sets F_i as in (4). Now the closed set F_1 is a subset of the open set $U_1 \cap (S \setminus \bigcap_{i=2}^{n+2} F_i)$, so there is an open set B_1 with $F_2 \subseteq B_1 \subseteq \overline{B_1} \subseteq U_1 \cap (S \setminus \bigcap_{i=2}^{n+2} F_i)$. So $\overline{B_1} \subseteq U_1$ and $\overline{B_1} \cap \bigcap_{i=2}^{n+2} F_i = \varnothing$. Next, there is an open set B_2 with $F_2 \subseteq B_2 \subseteq \overline{B_2} \subseteq U_2 \cap \left(S \setminus (\overline{B_1} \cap \bigcap_{i=3}^{n+2} F_i)\right)$, so $\overline{B_2} \subseteq U_2$ and $\overline{B_1} \cap \overline{B_2} \cap \bigcap_{i=3}^{n+2} F_i = \varnothing$. Continue in this way.   □

Covering dimension for compact metric spaces may be defined in a simpler way. If $\mathcal{A}$ is a cover, then its **mesh** is $\sup_{A \in \mathcal{A}} \operatorname{diam} A$.

Theorem 3.2.2. *Let S be a compact metric space, and let $n \geq -1$ be an integer. Then $\operatorname{Cov} S \leq n$ if and only if for every $\varepsilon > 0$, there is an open cover of S with order $\leq n$ and mesh $\leq \varepsilon$.*

Proof. Suppose $\operatorname{Cov} S \leq n$. Let $\varepsilon > 0$. The collection $\mathcal{U}$ of all open sets of diameter $\leq \varepsilon$ is a cover of S. So it has a refinement $\mathcal{B}$ of order at most n. But $\mathcal{B}$ has mesh $\leq \varepsilon$.

Conversely, suppose that for every $\varepsilon > 0$, there is an open cover of S with order $\leq n$ and mesh $\leq \varepsilon$. Let $\mathcal{U}$ be any finite open cover of S. By Theorem 2.3.22, the Lebesgue number r of $\mathcal{U}$ is positive. Let $\mathcal{B}$ be an open cover of S with order at most n and mesh less than the minimum of r and ε. Then by the defining property of the Lebesgue number, $\mathcal{B}$ is a refinement of $\mathcal{U}$. So $\operatorname{Cov} S \leq n$.   □

Separation

The separation theorem for dimension 1 looks like this:

Proposition 3.2.3. *Let S be a metric space with $\operatorname{Cov} S \leq 1$. Let A, B, C be closed subsets of S with $A \cap B \cap C = \varnothing$. Then there exist open sets U, V, W such that $U \supseteq A$, $V \supseteq B$, $W \supseteq C$, $U \cup V \cup W = S$ and $U \cap V \cap W = \varnothing$.*

Proof. Sets A, B, C are closed, so the complements $A' = S \setminus A$, $B' = S \setminus B$, $C' = S \setminus C$ are open. These complements cover S. So by Theorem 3.2.1(4), there exist closed sets F, G, H with $F \subseteq A'$, $G \subseteq B'$, $H \subseteq C'$, $F \cup G \cup H = S$, and $F \cap G \cap H = \varnothing$. Define $U = S \setminus F$, $V = S \setminus G$, $W = S \setminus H$. Then U, V, W are open, $U \supseteq A$, $V \supseteq B$, $W \supseteq C$, $U \cup V \cup W = S$, and $U \cap V \cap W = \varnothing$. $\square$

Using Theorem 3.2.1(3) in place of Theorem 3.2.1(4) we obtain a variant:

Proposition 3.2.4. *Let S be a metric space with $\mathrm{Cov}\, S \leq 1$. Let A, B, C be closed subsets of S with $A \cap B \cap C = \varnothing$. Then there exist closed sets K, L, M such that $K \supseteq A$, $L \supseteq B$, $M \supseteq C$, $K \cup L \cup M = S$ and $K \cap L \cap M = \varnothing$.*

Exercise 3.2.5. Prove converses for Propositions 3.2.3 and 3.2.4.

Exercise 3.2.6. For general n, formulate the separation theorem corresponding to covering dimension n.

Here is the three-set separation for arbitrary metric space. Of course we cannot arrange that $U_1 \cup U_2 \cup U_3 = S$ in general.

Proposition 3.2.7. *Let S be a metric space. Let $A_1, A_2, A_3 \subseteq S$ be closed sets. Assume $A_1 \cap A_2 \cap A_3 = \varnothing$. Then there exist open sets $U_1, U_2, U_3 \subseteq S$ with $U_1 \supseteq A_1$, $U_2 \supseteq A_2$, $U_3 \supseteq A_3$, and $\overline{U_1} \cap \overline{U_2} \cap \overline{U_3} = \varnothing$.*

Proof. For any $x \in S$, define $d_1(x) = \mathrm{dist}(x, A_1)$, $d_2(x) = \mathrm{dist}(x, A_2)$, $d_3(x) = \mathrm{dist}(x, A_3)$. These are continuous functions of x and $d_1(x) + d_2(x) + d_3(x) > 0$ for all x. Define

$$U_1 = \{\, x \in S : d_1(x) < (1/4)(d_1(x) + d_2(x) + d_3(x)) \,\},$$
$$U_2 = \{\, x \in S : d_2(x) < (1/4)(d_1(x) + d_2(x) + d_3(x)) \,\},$$
$$U_3 = \{\, x \in S : d_3(x) < (1/4)(d_1(x) + d_2(x) + d_3(x)) \,\}.$$

Then U_j is open, $U_j \supseteq A_j$, and

$$\overline{U_j} \subseteq \{\, x \in S : d_j(x) \leq (1/4)(d_1(x) + d_2(x) + d_3(x)) \,\}$$

for $j = 1, 2, 3$. We see that the intersection $\overline{U_1} \cap \overline{U_2} \cap \overline{U_3}$ is empty by adding the three inequalities. $\square$

Exercise 3.2.8. Formulate the separation theorem corresponding to n closed sets in a general metric space.

Sum Theorem

The sum theorem for dimension 1 is similar to the sum theorem for dimension 0. First, the union of two closed sets.

Theorem 3.2.9. *Let S be a metric space, and let F_1, F_2 be closed subsets. If* $\operatorname{Cov} F_1 \le 1$ *and* $\operatorname{Cov} F_2 \le 1$, *then* $\operatorname{Cov}(F_1 \cup F_2) \le 1$.

Proof. Let A, B, C be closed sets in $F_1 \cup F_2$ with $A \cap B \cap C = \varnothing$. The three sets $A \cap F_1$, $B \cap F_1$, $C \cap F_1$ are closed in F_1, and their intersection is empty. So (Proposition 3.2.4) there exist closed sets $K, L, M \subseteq F_1$ so that $K \supseteq A \cap F_1$, $L \supseteq B \cap F_1$, $M \supseteq C \cap F_1$, $K \cap L \cap M = \varnothing$, and $K \cup L \cup M = F_1$. The three sets $(K \cup A) \cap F_2$, $(L \cup B) \cap F_2$, $(M \cup C) \cap F_2$ are closed in F_2, and their intersection is empty. So there exist closed sets $P, Q, R \subseteq F_2$ so that $P \supseteq (K \cup A) \cap F_2$, $Q \supseteq (L \cup B) \cap F_2$, $R \supseteq (M \cup C) \cap F_2$, $P \cap Q \cap R = \varnothing$ and $P \cup Q \cup R = F_2$. Define $E = K \cup P$, $F = L \cup Q$, $G = M \cup R$. Then E, F, G are closed in $F_1 \cup F_2$, $E \supseteq A$, $F \supseteq B$, $G \supseteq C$, $E \cap F \cap G = \varnothing$, and $E \cup F \cup G = F_1 \cup F_2$. Therefore $\operatorname{Cov}(F_1 \cup F_2) \le 1$. $\qquad\square$

We will see later (Theorem 3.4.11) that the union of two zero-dimensional sets has dimension at most 1, even if the sets are not closed.

Exercise 3.2.10. Let n be a positive integer, let S be a metric space, and let F_1, F_2 be closed subsets. If $\operatorname{Cov} F_1 \le n$ and $\operatorname{Cov} F_2 \le n$, then $\operatorname{Cov}(F_1 \cup F_2) \le n$.

Next, the union of a sequence of closed sets.

Theorem 3.2.11. *Let S be a metric space, and for each $i \in \mathbb{N}$, let $F_i \subseteq S$ be a closed set. If* $\operatorname{Cov} F_i \le 1$ *for all i, then* $\operatorname{Cov} \bigcup_{i \in \mathbb{N}} F_i \le 1$.

Proof. We may assume $S = \operatorname{Cov} \bigcup_{i \in \mathbb{N}} F_i$. (So "open" will mean open in this set.) Let A, B, C be closed sets in S. By Proposition 3.2.7, there exist open sets U_0, V_0, W_0 such that $U_0 \supseteq A$, $V_0 \supseteq B$, $W_0 \supseteq C$, and $\overline{U_0} \cap \overline{V_0} \cap \overline{W_0} = \varnothing$.

Beginning with these, we will recursively define open sets U_i, V_i, W_i such that

$$U_{i-1} \subseteq U_i, \quad V_{i-1} \subseteq V_i, \quad W_{i-1} \subseteq W_i,$$
$$\overline{U_i} \cap \overline{V_i} \cap \overline{W_i} = \varnothing, \quad F_i \subseteq U_i \cup V_i \cup W_i, \qquad (*)$$

where $F_0 = \varnothing$. Now $(*)$ is satisfied for $i = 0$. Fix $k > 0$ and assume $(*)$ is satisfied for all $i < k$. The three sets $\overline{U_{k-1}} \cap F_k$, $\overline{V_{k-1}} \cap F_k$, $\overline{W_{k-1}} \cap F_k$ are closed and have empty intersection. Since $\operatorname{Cov} F_k \le 1$, by Proposition 3.2.4 there are closed sets $K_k, L_k, M_k \subseteq F_k$ with $K_k \supseteq \overline{U_{k-1}} \cap F_k$, $L_k \supseteq \overline{V_{k-1}} \cap F_k$, $M_k \supseteq \overline{W_{k-1}} \cap F_k$, $K_k \cap L_k \cap M_k = \varnothing$, and $K_k \cup L_k \cup M_k = F_k$. Now the three sets $\overline{U_{k-1}} \cup K_k$, $\overline{V_{k-1}} \cup L_k$, $\overline{W_{k-1}} \cup M_k$ are closed and have empty intersection. So there are open sets U_k, V_k, W_k with $U_k \supseteq \overline{U_{k-1}} \cup K_k$, $V_k \supseteq \overline{V_{k-1}} \cup L_k$, $W_k \supseteq \overline{W_{k-1}} \cup M_k$, $\overline{U_k} \cap \overline{V_k} \cap \overline{W_k} = \varnothing$. These sets satisfy $(*)$ for $i = k$. This completes the recursive construction.

Now define

$$U = \bigcup_{i \in \mathbb{N}} U_i, \qquad V = \bigcup_{i \in \mathbb{N}} V_i, \qquad W = \bigcup_{i \in \mathbb{N}} W_i.$$

These sets are open, and $U \supseteq A$, $V \supseteq B$, $W \supseteq C$, and $U \cap V \cap W = \emptyset$. This completes the proof that $\operatorname{Cov} S \leq 1$.   □

For the following exercise, the argument is like the one above. But this may test your ingenuity in finding a good notation for the proof.

Exercise 3.2.12. Let n be a positive integer and let S be a metric space. For each $i \in \mathbb{N}$, let $F_i \subseteq S$ be a closed set. If $\operatorname{Cov} F_i \leq n$ for all i, then $\operatorname{Cov} \bigcup_{i \in \mathbb{N}} F_i \leq n$.

Subset Theorem

Theorem 3.2.13. *Let S be a metric space, and $T \subseteq S$ a subset. Then* $\operatorname{Cov} T \leq \operatorname{Cov} S$.

Proof. If $\operatorname{Cov} S = \infty$, there is nothing to prove, so assume $\operatorname{Cov} S = n$ for some integer n. We must show $\operatorname{Cov} T \leq n$. This will be done in three stages.

(1) First, assume T is closed. Let $\mathcal{A}$ be an open cover of T. Each element $A \in \mathcal{A}$ has the form $A = E \cap T$, where E is a set open in S. Also, the complement $S \setminus T$ is open in S. Therefore

$$\mathcal{A}_1 = \{ E \subseteq S : E \text{ is open in S, } E \cap T \in \mathcal{A} \} \cup \{ S \setminus T \}$$

is an open cover of S. So there is a refinement $\mathcal{A}_2$ of $\mathcal{A}_1$ that is an open cover of S of order $\leq n$. So

$$\mathcal{A}_3 = \{ E \cap T : E \in \mathcal{A}_2 \}$$

is subordinate to $\mathcal{A}$, an open cover of T, and of order $\leq n$. So $\operatorname{Cov} T \leq n$.

(2) Now assume T is open. An open set T can be written as a countable union of closed subsets, $T = \bigcup_{j \in \mathbb{N}} F_j$, where

$$F_j = \left\{ x \in S : \operatorname{dist}(x, S \setminus T) \geq \frac{1}{j} \right\}.$$

Now by (1), $\operatorname{Cov} F_j \leq n$ for all j. So by the sum theorem 3.2.12 we have $\operatorname{Cov} T \leq n$.

(3) Finally, consider the general subset T of S. Let $\{U_1, U_2, \cdots, U_{n+2}\}$ be an open cover of T. For each U_i, let V_i be a set open in S with $U_i = V_i \cap T$. The set $V = \bigcup_{i=1}^{n+2} V_i$ is open, so by (2), $\operatorname{Cov} V \leq n$. Now $\{V_1, V_2, \cdots, V_{n+2}\}$ is an open cover of V, so there exist open sets $W_i \subseteq V_i$ with $\bigcap_{i=1}^{n+1} W_i = \emptyset$ and $\bigcup_{i=1}^{n+2} W_i = V$. But then $W_i \cap T$ are open in T, and they have intersection $\emptyset$ and union T. This shows $\operatorname{Cov} T \leq n$, as required.   □

Examples

We saw in Theorem 3.1.4 that $\mathrm{Cov}\,\mathbb{R} \geq 1$. But in fact, as expected, the dimension is exactly 1.

Lemma 3.2.14. *Let $a < b$ be real. Then $\mathrm{Cov}[a, b] = 1$.*

Proof. We must show $\mathrm{Cov}[a, b] \leq 1$. Let $\varepsilon > 0$ be given. Let $n \in \mathbb{N}$ be so large that $1/n \leq \varepsilon/2$. Then

$$\left\{ \left(\frac{k-1}{n}, \frac{k+1}{n} \right) : k \in \mathbb{Z} \right\}$$

is an open cover of $[a, b]$ with mesh $\leq \varepsilon$ and order 1. □

Theorem 3.2.15. *The line $\mathbb{R}$ has covering dimension 1.*

Proof. The line $\mathbb{R}$ is the union of the closed subsets $[-n, n]$ for $n \in \mathbb{N}$. By Lemma 3.2.14, $\mathrm{Cov}[-n, n] = 1$ for all n. Therefore by the Sum Theorem 3.2.11, $\mathrm{Cov}\,\mathbb{R} \leq 1$. □

Exercise 3.2.16. $\mathrm{Cov}\,\mathbb{R}^2 \leq 2$. Figure 3.1.2 may be helpful.

The fact that $\mathrm{Cov}\,\mathbb{R}^2 \geq 2$ is more difficult to prove than $\mathrm{Cov}\,\mathbb{R} \geq 1$. This will be done in Sect. 3.3.

Exercise 3.2.17. $\mathrm{Cov}\,\mathbb{R}^3 \leq 3$.

The Sierpiński gasket is described on p. 8.

Proposition 3.2.18. *The Sierpiński gasket has covering dimension 1.*

Proof. I will use the notation S and S_k from p. 8. Each set S_k is made up of 3^k filled-in triangles with side 2^{-k}. For a given value of k, any two of these triangles either meet in a point, or else have distance at least $2^{-k}/(\sqrt{3}/2)$ between them. (The trema adjoining one side of the triangle is a triangle with side at least 2^{-k} and therefore altitude at least $2^{-k}/(\sqrt{3}/2)$.) So if $r > 0$ is smaller than $2^{-k}/\sqrt{3}$, then the r-neighborhoods of these triangles are open sets which form a cover of S with order 1 and mesh at most $2^{-k} + 2r$. Therefore $\mathrm{Cov}\,S \leq 1$. But S contains line segments, which have dimension 1, so $\mathrm{Cov}\,S \geq 1$ also. □

Exercise 3.2.19. Let d, n be positive integers. Let K be a nonempty compact set in $\mathbb{R}^d$. Let $f_1, f_2, \cdots, f_n$ be similarities of $\mathbb{R}^d$ to itself. Suppose K satisfies the self-referential equation

$$K = \bigcup_{i=1}^{n} f_i[K],$$

and that the set of images $\{f_1[K], f_2[K], \cdots, f_n[K]\}$ has order 1. Does it follow that $\mathrm{Cov}\,K \leq 1$?

Exercise 3.2.20. Compute the covering dimension for other sets: the Heighway dragon (p. 20); the Koch curve (p. 19); the McWorter pentigree (p. 24); the twindragon (p. 33); the Eisenstein fractions (p. 34); 120-degree dragon (p. 23).

Number Systems

Recall the situation from Section 1.6. Let b be a real number, $|b| > 1$, and let D be a finite set of real numbers, including 0. We are interested in representing real numbers in the number system they define.

Write W for the set of "whole numbers"; that is, numbers of the form

$$\sum_{j=0}^{M} a_j b^j. \tag{1}$$

Write F for the set of "fractions"; that is numbers of the form

$$\sum_{j=-\infty}^{-1} a_j b^j. \tag{2}$$

The set of all numbers represented by this system is the sum of one of each:

$$\sum_{j=-\infty}^{M} a_j b^j. \tag{3}$$

Exercise 1.6.3 may be considered a problem on topological dimension.

Proposition 3.2.21. *Let b be a real number with $|b| > 1$, and let D be a finite set of real numbers, including 0. Then either some real number has no expansion in the form (3) or some real number has more than one expansion in the form (3).*

Proof. Suppose (for purposes of contradiction) that every real number has a unique expansion in the form (3). We will deduce that $\mathbb{R}$ is zero-dimensional, a contradiction.

The string space $D^{(\omega)}$ is compact (Exercise 2.6.8), and the map sending the infinite string of digits to the corresponding number (2) is continuous, so F is compact (Theorem 2.3.15).

I claim that there is a minimum distance between elements of W. Suppose not: then choose $x_n, y_n \in W$ with $x_n \neq y_n$ and $|x_n - y_n| \to 0$. By subtracting the places where they agree, we may assume that x_n and y_n have only zeros in places higher than the highest power of b where they disagree. Let M_n be the highest power of b where the expansions of x_n and y_n disagree. Then $x'_n = x_n b^{-M_n - 1}$ and $y'_n = y_n b^{-M_n - 1}$ are elements of F, they differ in the first place to the right of the radix point, and still $|x'_n - y'_n| \to 0$. By taking subsequences, we may assume that the first place of x'_n is the same for all n, the first place of y'_n is the same for all n, (x'_n) converges, and (y'_n) converges. The limits $x = \lim x'_n$ and $y = \lim y'_n$ are equal, yet have different first places. This contradicts the uniqueness. So there is a minimum distance between elements of W.

Now I claim that the set F is clopen. If not, there is a boundary point x. Now F is closed, so $x \in F$. There is a sequence in $\mathbb{R} \setminus F$ that converges to x. So there exist $y_n \in F$ and $w_n \in W$ with $w_n \neq 0$ and $y_n + w_n \to x$. By compactness of F we may assume that (y_n) converges, say to $y \in F$. Then $w_n \to x - y$. Since there is a minimum distance between elements of W, this means that $w_n = x - y$ for large enough n. So we have $x = y + w_n$, contradicting uniqueness of representations.

Now consider an interval $[a, b]$ in the line. Let $\varepsilon > 0$ be given. Let $r = \varepsilon / \operatorname{diam} F$, so that rF has diameter ε. Then

$$\bigcup_{w \in W} (rF + rw)$$

is a cover of $[a, b]$ by disjoint clopen sets with mesh ε. Therefore by Theorem 3.2.2 $\operatorname{Cov}[a, b] = 0$. This contradicts Corollary 3.1.5. $\qquad \square$

3.3 *Two-Dimensional Euclidean Space

In this section we will prove that, as expected, the covering dimension of the plane $\mathbb{R}^2$ is 2. We will use this notation:

$$D = \left\{ x \in \mathbb{R}^2 : |x| \le 1 \right\}, \qquad \text{the unit disk;}$$
$$S = \left\{ x \in \mathbb{R}^2 : |x| = 1 \right\}, \qquad \text{the unit circle.}$$

Degree (mod 2)

The preliminaries to the discussion of the dimension of the plane involve a "homological" discussion of the degree of a map of the circle to itself, and the Brouwer fixed point theorem.

Theorem 3.3.1. *There is no continuous function* $f : D \to S$ *with* $f(x) = x$ *for all* $x \in S$.

Proof. If n is a positive integer, we define the **subdivision** $A(n)$ of the circle to be the set $\{a_0, a_1, \cdots, a_{n-1}\}$ of points on the circle T, starting with $a_0 = (1, 0)$, and continuing counterclockwise around the circle with equal spacing. So

---

* Optional section.

$$a_j = \left(\cos \frac{2\pi j}{n}, \sin \frac{2\pi j}{n} \right).$$

When n is large, consecutive points a_{j-1} and a_j are close together. Sometimes we may write $a_n = a_0$.

Given a continuous function $g \colon S \to S$, consider the image points $g(a_0)$, $g(a_1), \cdots, g(a_n)$. They may not occur in order around the boundary of the circle. Assuming that two consecutive points $g(a_{j-1})$ and $g(a_j)$ are close together (say, no more than $1/4$ of the circumference apart), let U_j be the shorter of the two arcs of the circle with endpoints $g(a_{j-1})$ and $g(a_j)$. If $g(a_{j-1}) = g(a_j)$, then U_j is just that single point. Now if $y \in S$ is not one of the points $g(a_j)$, we let $N(A(n), g, y)$ denote the number of the intervals U_j that y is in. Let $\widetilde{N}(A(n), g, y)$ be the residue* modulo 2 of $N(A(n), g, y)$.

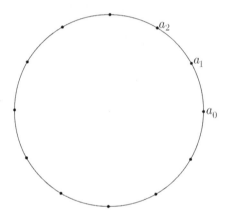

Fig. 3.3.1. Illustration for Theorem 3.3.1

I claim that the number $\widetilde{N}(A(n), g, y)$ is independent of the choice of y, as long as it is not one of the points $g(a_j)$. Indeed, think of moving y around the circle. As long as we do not cross one of the points $g(a_j)$, the number $N(A(n), g, y)$ remains unchanged. When we do cross one of the points $g(a_j)$, the number $N(A(n), g, y)$ may remain unchanged (if the two arcs U_j and U_{j+1} are on opposite sides of $g(a_j)$) or it may increase or decrease by 2 (if the two arcs U_j and U_{j+1} are on the same side of $g(a_j)$). It any case, the parity $\widetilde{N}(A(n), g, y)$ remains unchanged.

Therefore we will write $\widetilde{N}(A(n), g)$ for the value of all $\widetilde{N}(A(n), g, y)$.

Now let $f \colon D \to S$ be continuous, and suppose that $f(x) = x$ for all $x \in S$. For $0 \le r \le 1$, let

$$g_r(x) = f(rx) \qquad \text{for } x \in T.$$

---

* There are only two possible values: 0 represents "even" and 1 represents "odd".

Since D is compact, the function f is uniformly continuous. Therefore there is $\delta > 0$ so that $|x - y| < \delta$ implies $|f(x) - f(y)| < 1$. If n is large enough, consecutive points a_j of the subdivision $A(n)$ are within distance δ of each other. For this value of n, all of the functions g_r satisfy the assumption required above that consecutive points $g_r(a_{j-1})$, $g_r(a_j)$ have distance less than $1/4$ of the circumference of the circle. Fix this value of n.

Consider the numbers $N(A(n), g_r)$. The function g_1 is the identity on S, so clearly $N(A(n), g_1) = 1$. The function g_0 maps everything to the single point $f(0)$, so $N(A(n), g_0) = 0$. Now I claim that both of the sets $\{\, r \in [0, 1] : \widetilde{N}(A(n), g_r) = 0 \,\}$ and $\{\, r \in [0, 1] : \widetilde{N}(A(n), g_r) = 1 \,\}$ are open in $[0, 1]$. Fix some value $r_0 \in \{\, r \in [0, 1] : \widetilde{N}(A(n), g_r) = 0 \,\}$. Choose a point $y \in S$ not equal to any of the points $g_{r_0}(a_j)$. There is a minimum distance ε from y to the set $\{g_{r_0}(a_0), g_{r_0}(a_1), \cdots, g_{r_0}(a_{n-1})\}$. By the uniform continuity of f, there is $\delta > 0$ so that if $|r - r_0| < \delta$, then $|g_r(a_j) - g_{r_0}(a_j)| < \varepsilon$ for all j. This means that y lies in exactly the same arcs defined by g_r as arcs defined by g_{r_0}. So $N(A(n), g_r) = N(A(n), g_{r_0})$. Therefore $B_\delta(r_0)$ lies entirely in the set $\{\, r \in [0, 1] : \widetilde{N}(A(n), g_r) = 0 \,\}$. So it is an open set.

Thus $[0, 1]$ is the union of two disjoint nonempty subsets, both open in $[0, 1]$. This is impossible by Corollary 3.1.5. So the assumption that the function f exists is not tenable.   □

An easy consequence is Brouwer's fixed point theorem (for the disk). We will use a calculation, which is left to the reader. You can think of it as a test of your analytic geometry and algebra skills.

Exercise 3.3.2. Let (x, y) and (a, b) be distinct points in the disk D. Then the point (u, v) where the ray from (a, b) through (x, y) intersects the circle is given by $u = x + t(x - a)$ and $v = y + t(y - b)$, where

$$t = \frac{-(x - a)x - (y - b)y + \sqrt{(x - a)^2 + (y - b)^2 - (ay - bx)^2}}{(x - a)^2 + (y - b)^2}.$$

Theorem 3.3.3 (Brouwer's Fixed Point Theorem). If $g \colon D \to D$ is continuous, then there is a point $x \in D$ with $g(x) = x$.

Proof. Suppose there is no such fixed point. Define a function f as follows. If $x \in D$, then $g(x) \neq x$ is also in D; let $f(x)$ be the point where the ray from $g(x)$ through x intersects the circle S. By Exercise 3.3.2, we can see that the function f is continuous. Also, $f(x) = x$ if $x \in S$. This contradicts Theorem 3.3.1. So in fact g has a fixed point.   □

Topological Dimension of the Plane

Now I will show that the plane $\mathbb{R}^2$ has covering dimension 2. We already know (Exercise 3.2.16) that $\operatorname{Cov} \mathbb{R}^2 \leq 2$. To show $\operatorname{Cov} \mathbb{R}^2 \geq 2$ it is enough by

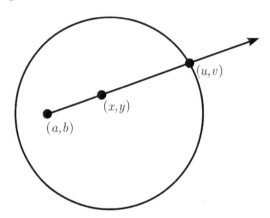

Fig. 3.3.2. Illustration for Exercise 3.3.2

Theorem 3.2.13 to show that $\operatorname{Cov} M \geq 2$ for a certain closed subset M of $\mathbb{R}^2$. The subset M will be a filled-in equilateral triangle, including the edges, the vertices, and the inside.

Theorem 3.3.4. *Let M be a filled-in equilateral triangle. Then $\operatorname{Cov} M \geq 2$.*

Proof. Assume, for purposes of contradiction, that $\operatorname{Cov} M \leq 1$. Since the triangle M is homeomorphic to the disk D, Brouwer's fixed point theorem also holds in M. Label the vertices of the triangle a, b, c, and the edges A, B, C in such a way that $A \cap B = \{c\}$, $B \cap C = \{a\}$, and $C \cap A = \{b\}$. The three subsets A, B, C are closed, and $A \cap B \cap C = \varnothing$. By Proposition 3.2.3, there exist open sets U, V, W with $U \supseteq A$, $V \supseteq B$, $W \supseteq C$, $U \cup V \cup W \supseteq M$, and $U \cap V \cap W = \varnothing$.

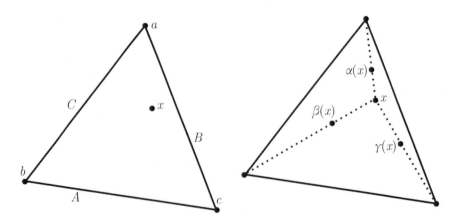

Fig. 3.3.4. Illustration for Theorem 3.3.4

Define a function $\alpha\colon M \to M$ as follows. The distance $\mathrm{dist}(x, S \setminus U)$ to the complement of U is a continuous function of the point x, because of the Lipschitz condition

$$\big|\,\mathrm{dist}(x, S \setminus U) - \mathrm{dist}(y, S \setminus U)\big| \le |x - y|.$$

Define

$$\alpha(x) = \begin{cases} x + \frac{\mathrm{dist}(x, S\setminus U)}{|a-x|}\,(a - x), & \text{if } x \neq a, \\ a, & \text{if } x = a. \end{cases}$$

The point a belongs to both B and C, and therefore to $V \cap W$; so $a \notin U$ and thus $\mathrm{dist}(x, S \setminus U) \le |a - x|$. So $\alpha(x)$ is a point on the line segment joining x and a. And therefore α is continuous even at the point a. But M is convex, so $\alpha(x) \in M$. If $x \in U$, then $\mathrm{dist}(x, S \setminus U) > 0$, so we have $\alpha(x) \neq x$. (Recall $a \notin U$.) On the other hand, if $x \notin U$, then $\alpha(x) = x$.

Define corresponding functions:

$$\beta(x) = \begin{cases} x + \frac{\mathrm{dist}(x, S\setminus V)}{|b-x|}\,(b - x), & \text{if } x \neq b, \\ b, & \text{if } x = b, \end{cases}$$

$$\gamma(x) = \begin{cases} x + \frac{\mathrm{dist}(x, S\setminus W)}{|c-x|}\,(c - x), & \text{if } x \neq c, \\ c, & \text{if } x = c. \end{cases}$$

These map M continuously into M. Also $x \in V$ if and only if $\beta(x) \neq x$; and $x \in W$ if and only if $\gamma(x) \neq x$. Now define

$$\varphi(x) = \frac{1}{3}\big(\alpha(x) + \beta(x) + \gamma(x)\big),$$

the centroid of the three points $\alpha(x), \beta(x), \gamma(x)$. This is a continuous function of X, and maps M into M because M is convex.

By the Brouwer fixed point theorem, φ has a fixed point in M, $\varphi(x_0) = x_0$. I claim that $x_0 \notin A$. Indeed, if $x_0 \in A \subseteq U$, then $\alpha(x_0)$ is not on the edge A, and thus the centroid $\varphi(x_0)$ is not in A, so $\varphi(x_0) \neq x_0$. Similarly, $x_0 \notin B$ and $x_0 \notin C$. So x_0 is strictly inside the triangle. Now consider which of the sets U, V, W the point x_0 belongs to. What if x_0 belongs to only one of the three sets? Say $x_0 \in U$, $x_0 \notin V$, $x_0 \notin W$. Then $\alpha(x_0) \neq x_0$ but $\beta(x_0) = x_0, \gamma(x_0) = x_0$, so $\varphi(x_0) \neq x_0$. What if x_0 belongs to exactly two of the sets U, V, W? Say $x_0 \in U, x_0 \in V$ but $x_0 \notin W$. Then the three points $\alpha(x_0), \beta(x_0), \gamma(x) = x_0$ are not collinear, so their centroid $\varphi(x_0)$ is not equal to a vertex x_0. The only remaining cases (x_0 belongs to all three of the sets U, V, W or to none of them) contradict the way we obtained the sets to start with. So we have a contradiction. This shows that $\mathrm{Cov}\, M \le 1$ is false. $\qquad\square$

Exercise 3.3.5. Assume known the 3-dimensional Brouwer fixed point theorem: a continuous function from the closed ball in $\mathbb{R}^3$ into itself has a fixed point. Use this to prove that the covering dimension of $\mathbb{R}^3$ is 3.

3.4 Inductive Dimension

Dimension zero was characterized by the existence of a clopen base for the topology. The (small) inductive dimension generalizes this.

Hermann Weyl explained dimension like this:* "We say that space is 3-dimensional because the walls of a prison are 2-dimensional."

If we have a point that we want to imprison, we can use a small cube as the prison. By making the cube small enough, when the point is forbidden to move through the faces of the cube, it can be confined to a very small region. The cube consists of 6 plane faces; we need to know that they are 2-dimensional. A point living in these faces (Flatland) can be imprisoned using a small circle. So saying that the faces of the cube are 2-dimensional requires knowing that a circle is 1-dimensional. A point living in the circle (Lineland) can be imprisoned using just 2 points as prison walls. So we need to know that a 2 point set is 0-dimensional. Finally, a point living in the 2 point set (Pointland) is already unable to move. So we need no prison walls at all. This will be the definition of a 0-dimensional set.

The Small Inductive Dimension

This is defined in an inductive manner. Each metric space S will be assigned a dimension, written $\operatorname{ind} S$, chosen from the set $\{-1, 0, 1, 2, 3, \cdots, \infty\}$, consisting of the integers ≥ -1 together with an extra symbol ∞. The empty metric space $\varnothing$ has $\operatorname{ind} \varnothing = -1$. If k is a nonnegative integer, then we say that $\operatorname{ind} S \leq k$ iff there is a base for the open sets of S consisting of sets U with† $\operatorname{ind} \partial U \leq k - 1$. We say $\operatorname{ind} S = k$ iff $\operatorname{ind} S \leq k$ but $\operatorname{ind} S \not\leq k - 1$. Finally, if $\operatorname{ind} S \leq k$ is false for all integers k, then we say $\operatorname{ind} S = \infty$. The small inductive dimension is also known as the **Urysohn–Menger dimension** or **weak inductive dimension**.

Note: by Proposition 3.1.11, if S is a separable metric space, then we have $\operatorname{Cov} S = 0$ if and only if $\operatorname{ind} S = 0$. This equivalence may fail for nonseparable spaces [55, Chap. 7, Sect. 4]. In fact, we will see later that if S is a compact metric space, then $\operatorname{Cov} S = \operatorname{ind} S$.

Exercise 3.4.1. Show $\operatorname{ind} \mathbb{R} = 1$.

Since the small inductive dimension is defined inductively, it is often possible to prove things about it by induction.

Theorem 3.4.2. *Topological dimension is a topological property: If S and T are homeomorphic, then $\operatorname{ind} S = \operatorname{ind} T$.*

---

* I have to quote from memory, because I cannot seem to find this quotation now.
† Recall the notation ∂U for the boundary of U.

Proof. The proof is by induction on $\text{ind}\,S$. If $\text{ind}\,S = -1$, then S is empty; since T is homeomorphic to S, it is also empty, so $\text{ind}\,T = -1$.

Suppose the theorem is known for spaces S with $\text{ind}\,S \leq k$, and consider a space S with $\text{ind}\,S = k + 1$. Let $h\colon S \to T$ be a homeomorphism. There is a base $\mathcal{B}$ for the open sets of S consisting of sets B with $\text{ind}\,\partial B \leq k$. Now $\{\, h[B] : B \in \mathcal{B} \,\}$ is a base for the open sets of T. If $B \in \mathcal{B}$, then $h[\partial B] = \partial h[B]$. The restriction of h to ∂B is a homeomorphism. By the induction hypothesis, $\text{ind}\,\partial h[B] = \text{ind}\,\partial B \leq k$. So we see that there is a base for the open sets of T consisting of sets with boundary of dimension $\leq k$. This shows that $\text{ind}\,T \leq k + 1$. But if $\text{ind}\,T \leq k$, then the induction hypothesis would show $\text{ind}\,S \leq k$, which is false. So $\text{ind}\,T = k + 1$. Therefore, by induction we see that if $\text{ind}\,S$ is an integer, then $\text{ind}\,S = \text{ind}\,T$.

If $\text{ind}\,S = \infty$, then $\text{ind}\,T = k$ is false for all integers k, so also $\text{ind}\,T = \infty$. So in all cases we have $\text{ind}\,S = \text{ind}\,T$.   □

Theorem 3.4.3. *Let S be a metric space, and let $T \subseteq S$. Then $\text{ind}\,T \leq \text{ind}\,S$.*

Proof. This is clear if $\text{ind}\,S = \infty$. So suppose $\text{ind}\,S < \infty$. The proof proceeds by induction on $\text{ind}\,S$.

If $\text{ind}\,S = -1$, then S is empty, so clearly $T \subseteq S$ is also empty, and hence $\text{ind}\,T = -1$.

Suppose the theorem is true for all pairs S, T with $T \subseteq S$ and $\text{ind}\,S \leq k$. Consider a pair $T \subseteq S$ with $\text{ind}\,S = k + 1$. I must show that there is a base for the open sets of T consisting of sets with boundary of dimension $\leq k$. So let $x \in T$, and let V be an open set in T with $x \in V$. I must find an open set U in T with $x \in U \subseteq V$ and $\text{ind}\,\partial_T U \leq k$. [Note that the boundary of a set in T may be different than the boundary of the same set in S; so the space is indicated as a subscript.] Now since V is open in T, there exists a set $\widetilde{V}$ open in S with $V = \widetilde{V} \cap T$. Since $\text{ind}\,S \leq k + 1$, and $x \in \widetilde{V}$, there is a set $\widetilde{U}$ open in S with $x \in \widetilde{U} \subseteq \widetilde{V}$ and $\text{ind}\,\partial_S \widetilde{U} \leq k$. Let $U = \widetilde{U} \cap T$. Then U is open in T, and $x \in U \subseteq V$. Now by Theorem 2.2.20, $\partial_T U \subseteq \partial_S \widetilde{U}$, so by the induction hypothesis, we have $\text{ind}\,\partial_T U \leq \text{ind}\,\partial_S \widetilde{U} \leq k$. Thus there is a base for the open sets of T consisting of sets U with $\text{ind}\,\partial_T U \leq k$. This means that $\text{ind}\,T \leq k + 1$.

Therefore, by induction, the theorem is true for all values of $\text{ind}\,S$.   □

The triadic Cantor dust C is homeomorphic to the space $\{0, 1\}^{(\omega)}$ of infinite strings based on the two-letter alphabet $\{0, 1\}$ (Proposition 2.6.3). Any metric space homeomorphic to these spaces may be called generically a "Cantor dust" or a "Cantor set". The Cantor dust is a ***universal zero-dimensional space*** in the following sense:

Theorem 3.4.4. *Let S be a nonempty separable metric space. Then $\text{ind}\,S = 0$ if and only if S is homeomorphic to a subset of the space $\{0, 1\}^{(\omega)}$.*

Proof. Suppose S is homeomorphic to $T \subseteq \{0,1\}^{(\omega)}$. By Theorem 3.4.2, ind $S = $ ind T. By Theorem 3.4.3, ind $T \leq $ ind$\{0,1\}^{(\omega)}$. But ind$\{0,1\}^{(\omega)} = 0$. Therefore ind $S \leq 0$. Since $S \neq \varnothing$, we have ind $S = 0$.

Conversely, suppose ind $S = 0$. There is a base $\mathcal{B}_1$ for the open sets of S consisting of clopen sets. By Theorem 2.3.1, there is a countable $\mathcal{B} \subseteq \mathcal{B}_1$, still a base for the open sets. Write $\mathcal{B} = \{U_1, U_2, \cdots\}$. (If $\mathcal{B}$ is finite, repeat the basic sets over and over.) For notation, we will use $U_i(1) = U_i$ and $U_i(0) = S \setminus U_i$; they are all clopen sets. If $\alpha \in \{0,1\}^{(k)}$, say $\alpha = e_1 e_2 \cdots e_k$, let $U(\alpha) = U_1(e_1) \cap U_2(e_2) \cap \cdots \cap U_k(e_k)$.

Define a map $h \colon S \to \{0,1\}^{(\omega)}$ as follows: given $x \in S$, the ith letter of $h(x)$ is 0 or 1 according as x belongs to $U_i(0)$ or $U_i(1)$. So this means $h(x) \in [\alpha]$ if and only if $x \in U(\alpha)$. Thus $h^{-1}\big[[\alpha]\big] = U(\alpha)$.

Now I claim h is one-to-one. Indeed, if $x \neq y$, then $S \setminus \{y\}$ is an open set containing x, so there is i with $x \in U_i \subseteq S \setminus \{y\}$, and therefore $h(x) \neq h(y)$. This shows that h is one-to-one. So the inverse function $h^{-1} \colon h[S] \to S$ exists. Now the sets $[\alpha]$ constitute a base for the open sets of $\{0,1\}^{(\omega)}$, and $h^{-1}\big[[\alpha]\big] = U(\alpha)$ is open for every α, so h is continuous. Similarly, the sets U_i constitute a base for the open sets of S, and

$$h[U_i] = \bigcup_{\alpha \in \{0,1\}^{(i-1)}} \big(h[S] \cap [\alpha 1]\big)$$

is open in $h[S]$ for every i. So h^{-1} is continuous. This means that h is a homeomorphism of S onto $h[S] \subseteq \{0,1\}^{(\omega)}$.   □

Exercise 3.4.5. Let S, T be metric spaces. Then $\mathrm{ind}(S \times T) \leq \mathrm{ind}\, S + \mathrm{ind}\, T$.

There are examples showing that strict inequality is possible [22, Example 1.5.17].

The Large Inductive Dimension

In Proposition 3.1.8, covering dimension zero was characterized in terms of a separation property. A generalization of this will be considered next.

Let A and B be disjoint subsets of a metric space S. We say that a set $L \subseteq S$ **separates** A and B iff there exist open sets U and V in S with $U \cap V = \varnothing$, $U \supseteq A$, $V \supseteq B$, and $L = S \setminus (U \cup V)$. (See Fig. 3.4.6.) So Proposition 3.1.8 says that in a zero-dimensional space S, any two disjoint closed sets are separated by the empty set. Note that a space S has ind $S \leq k$ if and only if a point $\{x\}$ and a closed set B not containing x can be separated by a set L with ind $L \leq k - 1$. Indeed, there is a basic open set U with $x \in U \subseteq S \setminus B$ and $L = \partial U$ separates $\{x\}$ and B.

The **large inductive dimension** is a topological dimension closely related to the small inductive dimension. Each metric space S will be assigned an element of the set $\{-1, 0, 1, 2, \cdots, \infty\}$, called the large inductive dimension of S, written Ind S. To begin, Ind $\varnothing = -1$. Next, if k is a nonnegative

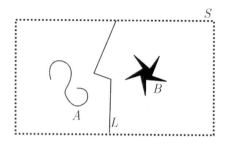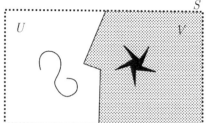

Fig. 3.4.6. L separates A and B

integer, we will say that $\operatorname{Ind} S \leq k$ iff any two disjoint closed sets in S can be separated by a set L with $\operatorname{Ind} L \leq k - 1$. We write $\operatorname{Ind} S = k$ iff $\operatorname{Ind} S \leq k$ but $\operatorname{Ind} S \not\leq k - 1$. We write $\operatorname{Ind} S = \infty$ iff $\operatorname{Ind} S \leq k$ is false for all integers k. The large inductive dimension is also called the **Čech dimension** or **strong inductive dimension.**

Exercise 3.4.7. Show $\operatorname{Ind} \mathbb{R} = 1$.

Exercise 3.4.8. Let S be a metric space. Then $\operatorname{ind} S \leq \operatorname{Ind} S$.

In general metric spaces, the large and small inductive dimension are not necessarily equal. But in separable spaces they are equal. This will be proved below (Corollary 3.4.17.)

Note that Proposition 3.1.8 now has the formulation: $\operatorname{Cov} S = 0$ if and only if $\operatorname{Ind} S = 0$. Proposition 3.1.11 says that if S is separable, then $\operatorname{Cov} S = 0$ if and only if $\operatorname{ind} S = 0$. So we conclude that all three definitions of "zero-dimensional" coincide for separable metric spaces.

Sum Theorems and Separation

Lemma 3.4.9. *Let S be a metric space, let A and B be disjoint closed sets, and let $T \subseteq S$. (a) Let U and V be open sets with $U \supseteq A$, $V \supseteq B$, and $\overline{U} \cap \overline{V} = \varnothing$. If $L' \subseteq T$ separates (in T) the sets $T \cap \overline{U}$ and $T \cap \overline{V}$, then there is a set $L \subseteq S$ that separates A and B in S with $L \cap T \subseteq L'$. (b) Suppose, in addition, that T is closed. Then, for any set $L' \subseteq T$ separating $T \cap A$ and $T \cap B$ in T, there is $L \subseteq S$ separating A and B in S with $L \cap T \subseteq L'$.*

Proof. (a) Since L' separates $T \cap \overline{U}$ and $T \cap \overline{V}$, we have $T \setminus L' = U' \cup V'$, with U' and V' open in T, $T \cap \overline{U} \subseteq U'$, and $T \cap \overline{V} \subseteq V'$. Now I claim that $A \cap \overline{V'} = \varnothing$. Indeed, $U \cap V' = T \cap U \cap V' \subseteq U' \cap V' = \varnothing$, and U is open, so $U \cap \overline{V'} = \varnothing$ and thus $A \cap \overline{V'} = \varnothing$. Similarly, $B \cap \overline{U'} = \varnothing$. Now U' and V' are disjoint and open in $U' \cup V'$, so $U' \cap \overline{V'} = \varnothing$ and $\overline{U'} \cap V' = \varnothing$. But then

$$(A \cup U') \cap \overline{(B \cup V')} = (A \cup U') \cap (\overline{B} \cup \overline{V'})$$
$$= (A \cap B) \cup (A \cap \overline{V'}) \cup (U' \cap B) \cup (U' \cap \overline{V'})$$
$$= \varnothing,$$

and similarly $\overline{(A \cup U')} \cap (B \cup V') = \varnothing$. So (by 2.2.30(1)) there exist disjoint open sets U'' and V'' with $A \cup U' \subseteq U''$ and $B \cup V' \subseteq V''$. Then let $L = S \setminus (U'' \cup V'')$. It separates A and B. Also

$$\begin{aligned} T \cap L &= T \setminus (U'' \cup V'') \\ &\subseteq T \setminus (U' \cup V') \\ &= L'. \end{aligned}$$

(b) Since L' separates $T \cap A$ and $T \cap B$ in T, there are disjoint sets U' and V', open in T, with $T \cap A \subseteq U'$, $T \cap B \subseteq V'$, and $T \setminus (U' \cup V') = L'$. Now the sets A and $(T \setminus U') \cup B$ are disjoint and closed, so there exists an open set U'' with

$$A \subseteq U'' \subseteq \overline{U''} \subseteq S \setminus ((T \setminus U') \cup B).$$

Similarly, B and $(T \setminus V') \cup \overline{U''}$ are disjoint and closed, so there exists an open set V'' with

$$B \subseteq V'' \subseteq \overline{V''} \subseteq S \setminus ((T \setminus V') \cup \overline{U''}).$$

So $A \subseteq U''$, $B \subseteq V''$, and $\overline{U''} \cap \overline{V''} = \varnothing$. Then apply part (a).   □

This enables us to prove a generalization of Proposition 3.1.8.

Corollary 3.4.10. *Let S be a separable metric space, let A and B be disjoint closed sets in S, and let $T \subseteq S$ be a subset with $\operatorname{ind} T = 0$. Then there is a set L that separates A and B in S such that $L \cap T = \varnothing$.*

Proof. Choose U and V open sets with $A \subseteq U$, $B \subseteq V$, and $\overline{U} \cap \overline{V} = \varnothing$. The sets $\overline{U} \cap T$ and $\overline{V} \cap T$ are disjoint and closed in T, which is zero-dimensional, so they can be separated by $L' = \varnothing$. Apply Lemma 3.4.9 to get L separating A and B with $L \cap T \subseteq L' = \varnothing$.   □

Here is a sum theorem for sets that are not necessarily closed.

Theorem 3.4.11. *Let S be a separable metric space, and let $A, B \subseteq S$. Then $\operatorname{ind}(A \cup B) \leq 1 + \operatorname{ind} A + \operatorname{ind} B$.*

Proof. If either $\operatorname{ind} A = \infty$ or $\operatorname{ind} B = \infty$, then the inequality is clear. So suppose they are both finite, say $\operatorname{ind} A = m$, $\operatorname{ind} B = n$. The proof proceeds by induction on the sum $m + n$.

For $n + m = -2$, both A and B are empty, so $A \cup B$ is also empty, and $\operatorname{ind}(A \cup B) = -1 = 1 + (-1) + (-1)$.

So assume that the result is known for smaller sums. Let $x \in A \cup B$. Then either $x \in A$ or $x \in B$. Suppose $x \in A$. Let V be open in $A \cup B$ with $x \in V$. Then the two sets $\{x\}$ and $A \setminus V$ are separated in A by a set L' with $\operatorname{ind} L' \leq m - 1$. So by Lemma 3.4.9, there is a set L that separates $\{x\}$ and $(A \cup B) \setminus V$ in $A \cup B$, and $L \cap A \subseteq L'$. Now $L = (L \cap A) \cup (L \cap B)$, so by the induction hypothesis, $\operatorname{ind} L \leq 1 + (m - 1) + n = m + n$. Therefore I have shown that $\operatorname{ind}(A \cup B) \leq m + n + 1$, as required.   □

Corollary 3.4.12. *The union of $n + 1$ zero-dimensional sets in a separable metric space has small inductive dimension $\leq n$.*

Next we prove the sum theorem for higher dimensions:

Theorem 3.4.13. *Let k be a nonnegative integer, and let S be a separable metric space. Let closed sets $T_n \subseteq S$ satisfy $\operatorname{ind} T_n \leq k$ for $n = 1, 2, \cdots$. Then $\operatorname{ind} \bigcup_{n \in \mathbb{N}} T_n \leq k$.*

Proof. Write $T = \bigcup_{n \in \mathbb{N}} T_n$. The result is clearly true for $k = \infty$. So suppose k is finite. The proof is by induction on k. The case $k = 0$ has been done already. Suppose $k \geq 1$, and assume the result is known for smaller values.

For each n, let $\mathcal{B}_n$ be a base for the open sets of T_n consisting of sets with boundary of dimension $< k$. By the Lindelöf property, we may assume that the bases $\mathcal{B}_n$ are countable. For all n and all $U \in \mathcal{B}_n$, we have $\operatorname{ind} \partial_{T_n} U \leq k-1$. By the induction hypothesis, the countable union

$$Y = \bigcup_{n \in \mathbb{N}} \ \bigcup_{U \in \mathcal{B}_n} (\partial_{T_n} U)$$

also has $\operatorname{ind} Y \leq k - 1$. But the space $Z_n = T_n \setminus Y$ has the family

$$\{ U \setminus Y : U \in \mathcal{B}_n \}$$

as a base for its open sets, and the sets $U \setminus Y$ are clopen in Z_n. So $\operatorname{ind} Z_n \leq 0$. Now consider the union

$$Z = \bigcup_{n \in \mathbb{N}} Z_n.$$

Each $Z_n = T_n \setminus Y = T_n \cap Z$ is closed in Z, so $\operatorname{ind} Z \leq 0$. Thus (by Theorem 3.4.11)

$$\operatorname{ind} T = \operatorname{ind}(Y \cup Z) \leq 1 + (k - 1) + 0 = k. \qquad \square$$

Examination of the proof yields the converse of Corollary 3.4.12:

Corollary 3.4.14. *Let S be a separable metric space with finite small inductive dimension k. Then S is the union of $k + 1$ zero-dimensional sets.*

Theorem 3.4.15. *Let S be a separable metric space, let A and B be disjoint closed subsets. Let $k \geq 0$ be an integer, and let $T \subseteq S$ have small inductive dimension k. Then there is a set L separating A and B with $\operatorname{ind}(T \cap L) \leq k-1$.*

Proof. The case $k = 0$ has already been done (Corollary 3.4.10). So suppose $k \geq 1$. By Corollary 3.4.14, we can write $T = Y \cup Z$, with $\operatorname{ind} Y = k - 1$ and $\operatorname{ind} Z = 0$. Now (again by Corollary 3.4.10) there is a set L separating A and B with $L \cap Z = \varnothing$. Then $L \cap T \subseteq Y$. So $\operatorname{ind}(L \cap T) \leq \operatorname{ind} Y \leq k - 1$. $\qquad \square$

Another consequence is this separation theorem, which generalizes Proposition 3.1.8 in another way:

Corollary 3.4.16. *Suppose S is a separable metric space with $\operatorname{ind} S \leq n-1$. Let $A_1, B_1, A_2, B_2, \cdots, A_n, B_n$ be $2n$ closed sets in S with $A_i \cap B_i = \varnothing$ for all i. Then there exist sets $L_1, L_2, \cdots, L_n$ such that L_i separates A_i and B_i for all i, and the intersection $\bigcap_{i=1}^{n} L_i = \varnothing$.*

Proof. First, there is a set L_1 that separates A_1 and B_1, such that $\operatorname{ind} L_1 \leq n-2$. Applying the theorem with $T = L_1$, we get a set L_2 that separates A_2 and B_2 with $\operatorname{ind}(L_1 \cap L_2) \leq n-3$. Continuing in this way, we get $L_1, L_2, \cdots, L_n$ with $\operatorname{ind}(L_1 \cap L_2 \cap \cdots \cap L_n) \leq -1$, so it is empty.   □

ind = Ind = Cov

It is true for general metric spaces S that $\operatorname{ind} S \leq \operatorname{Ind} S = \operatorname{Cov} S$, and for separable metric spaces S that $\operatorname{Cov} S = \operatorname{ind} S = \operatorname{Ind} S$. (See, for example, [35, Chap. V] or [22, Theorems 4.1.3, 4.1.5] or [22, Theorem 4.1.5]. An example with $\operatorname{ind} S = 0$, $\operatorname{Ind} S = \operatorname{Cov} S = 1$ is described in [55, Chap. 7, Sect. 4].) I will prove only this much: For separable S, $\operatorname{Cov} S \leq \operatorname{ind} S = \operatorname{Ind} S$; for compact S, $\operatorname{Cov} S = \operatorname{ind} S = \operatorname{Ind} S$.

Proposition 3.4.17. *Let S be a separable metric space. Then $\operatorname{Ind} S \leq \operatorname{ind} S$.*

Proof. This is clear if $\operatorname{ind} S = \infty$. So suppose $\operatorname{ind} S = k$ is finite. The case $k = 0$ is from Theorems 3.1.11 and 3.1.8. Suppose $k \geq 1$ and the result is known for smaller values of the dimension. Let A and B be disjoint closed sets in S. Then by Theorem 3.4.15, A and B can be separated by a set L with $\operatorname{ind} L \leq k-1$. But by the induction hypothesis, $\operatorname{Ind} L \leq k-1$. Therefore we have $\operatorname{Ind} S \leq k$.   □

Theorem 3.4.18. *Let S be a separable metric space. Then $\operatorname{Cov} S \leq \operatorname{ind} S$.*

Proof. The result is clearly true if $\operatorname{ind} S = \infty$. So suppose we have $\operatorname{ind} S < \infty$. Let $n = \operatorname{ind} S$. I must show $\operatorname{Cov} S \leq n$. Now by Corollary 3.4.14, there exist zero-dimensional sets $Y_1, Y_2, \cdots, Y_{n+1}$ with $S = \bigcup_{j=1}^{n+1} Y_j$. Let $\{U_1, U_2, \cdots, U_k\}$ be a finite open cover of S. Then for each j, the family $\{U_1 \cap Y_j, U_2 \cap Y_j, \cdots, U_k \cap Y_j\}$ is a finite open cover of Y_j. Now $\operatorname{ind} Y_j = 0$, so $\operatorname{Cov} Y_j = 0$, and therefore there exist disjoint open sets $B_{ij} \subseteq U_i$ with $\bigcup_{i=1}^{k} B_{ij} \supseteq Y_j$. Now the family

$$\mathcal{B} = \{\, B_{ij} : 1 \leq i \leq k, 1 \leq j \leq n+1 \,\}$$

covers S and refines $\{U_1, U_2, \cdots, U_k\}$. If we have any $n+2$ elements B_{ij} of $\mathcal{B}$, some two of them have the same second index j, so the intersection of the $n+2$ sets is empty. Therefore $\mathcal{B}$ has order $\leq n$. This completes the proof that $\operatorname{Cov} S \leq n$.   □

The second part of the equality (for compact spaces) of the covering dimension with the other topological dimensions will be proved in a way that also establishes another useful characterization of topological dimension for compact spaces. A function $h\colon W \to S$ is said to be ***at most n-to-one*** iff for each $x \in S$ there are at most n points $y \in W$ with $h(y) = x$.

Theorem 3.4.19. *Let S be a compact metric space. Let $n \geq 0$ be an integer. Then the following are equivalent:*

(1) $\operatorname{ind} S \leq n$.
(2) $\operatorname{Cov} S \leq n$.
(3) *There is a compact zero-dimensional metric space W and a continuous map h of W onto S that is at most $(n + 1)$-to-one.*

Proof. (1) $\Longrightarrow$ (2) is Theorem 3.4.18.

(2) $\Longrightarrow$ (3). Begin with the cover $\mathcal{U}_0 = \{S\}$, and (recursively) choose finite open covers $\mathcal{U}_k$ of order $\leq n$ such that the mesh of $\mathcal{U}_k$ is less than half of the Lebesgue number of $\mathcal{U}_{k-1}$. We can use this sequence of covers to define a tree: the nodes of the kth generation are the elements of $\mathcal{U}_k$; for each $U \in \mathcal{U}_k$, there is $V \in \mathcal{U}_{k-1}$ with $\overline{U} \subseteq V$; let V be the parent of U. (If there is more than one such set V, choose one.) If we label the edges of the tree (an edge goes from the parent to the child) by a countable set E, then the set of nodes of the tree is in one-to-one correspondence with the set $E_z^{(*)}$ of finite paths in the tree starting at the root z. Write U_α for the open set corresponding to the path $\alpha \in E_z^{(*)}$. Thus, the cover $\mathcal{U}_k$ is $\{\, U_\alpha : \alpha \in E^{(k)} \,\}$. Let $W = E_z^{(\omega)}$ be the set of infinite paths starting at the root. As usual, W is an ultrametric space under the metric $\varrho_{1/2}$, and therefore $\operatorname{ind} W = 0$. Also W is compact, since each node has only finitely many children.

Define the map $h\colon W \to S$ as follows: If $\sigma \in E_z^{(\omega)}$ is an infinite path, the sets $\overline{U_{\sigma \restriction k}}$ are compact, $\overline{U_{\sigma \restriction 0}} \supseteq \overline{U_{\sigma \restriction 1}} \supseteq \cdots$, and $\lim_{k \to \infty} \operatorname{diam} \overline{U_{\sigma \restriction k}} = 0$. The set

$$\bigcap_{k \in \mathbb{N}} \overline{U_{\sigma \restriction k}} = \bigcap_{k \in \mathbb{N}} U_{\sigma \restriction k}$$

is nonempty (by compactness) and has diameter 0, so it consists of a single point. Let that point be $h(\sigma)$. Now $h\big[[\sigma \restriction k]\big] \subseteq U_{\sigma \restriction k}$, so h is continuous. Since the families $\mathcal{U}_k$ cover S, we may deduce that h is surjective.

Finally, I must show that the map is at most $(n + 1)$-to-one. Suppose σ_1, $\sigma_2, \cdots, \sigma_{n+2} \in W$ are all different. Then there is a generation k so that

$$\sigma_1 {\restriction} k, \sigma_2 {\restriction} k, \cdots, \sigma_{n+2} {\restriction} k$$

are all different. But $\mathcal{U}_k$ has order at most n, so

$$\bigcap_{i=1}^{n+2} U_{\sigma_i \restriction k} = \varnothing.$$

Therefore not all of the points $h(\sigma_i)$ are equal.

(3) $\implies$ (1). The proof is by induction on n. The case $n = 0$ and the induction step for $n \geq 1$ begin in the same way, however.

Let $x \in S$ and let $\varepsilon > 0$. I must show that there is an open set $V \subseteq S$ such that $x \in V \subseteq B_\varepsilon(x)$ and $\operatorname{ind} \partial V \leq n - 1$. The set $D = h^{-1}[\{x\}]$ has at most $n + 1$ elements, say $D = \{z_1, z_2, \cdots, z_m\}$. The function h is continuous and W is zero-dimensional, so there exist clopen sets $U_i \subseteq W$ with $z_i \in U_i$ and $h[U_i] \subseteq B_\varepsilon(x)$. Let $F = \bigcup_{i=1}^m h[U_i]$. Now the sets U_i are clopen and W is compact, so the set U_i are compact, and thus F is compact and therefore closed. Let V be the interior of F, that is, the set of all the interior points of F. Then V is an open set.

First, I claim that $x \in V$, that is, x is an interior point of F. Suppose not. Then there is a sequence (x_k) in $S \setminus F$ with $x_k \to x$. Choose points $y_k \in W$ with $h(y_k) = x_k$. Taking a subsequence, we may assume that (y_k) converges. Its limit y satisfies $h(y) = x$, so in fact $y = z_i$ for some i. Now z_i is an interior point of U_i, so $y_k \in U_i$ for some k, and therefore $x_k = h(y_k) \in F$. This contradiction shows that x is an interior point of F.

Note that $\partial V \subseteq F \setminus V$. Now consider the subset

$$W_1 = h^{-1}[\partial V] \setminus (U_1 \cup U_2 \cup \cdots \cup U_m)$$

of W. It is closed, hence compact and zero-dimensional. I claim that h maps W_1 onto ∂V. Indeed, if $y \in \partial V$, then there is a sequence (y_k) in $S \setminus F$ with $y_k \to y$. Choose $w_k \in W$ with $h(w_k) = y_k$; by taking a subsequence, we may assume that $w_k \to w$ for some $w \in W$. Of course, $h(w) = y$. Now $w_k \notin U_1 \cup U_2 \cup \cdots \cup U_m$, so $w \notin U_1 \cup U_2 \cup \cdots \cup U_m$, and therefore $y \in W_1$.

Next I claim that $\operatorname{ind} \partial V \leq n - 1$. We must distinguish the cases $n = 0$ and $n \geq 1$. First, suppose $n = 0$. Then h is one-to-one. Now $\partial V \subseteq F$ and $W_1 \subseteq h^{-1}[F] \setminus U_1 = \varnothing$. Therefore $\partial V = \varnothing$, or $\operatorname{ind} \partial V = -1 = n - 1$. Next suppose $n \geq 1$ and the result is known for smaller values of n. If $y \in \partial V$, then since $\partial V \subseteq F$, the map h sends at least one point of $U_1 \cup U_2 \cup \cdots \cup U_m$ onto y, so at most n points of W_1 are sent onto y. Thus the restriction $h \colon W_1 \to \partial V$ satisfies the hypotheses of the theorem for $n - 1$. By the induction hypothesis, $\operatorname{ind} \partial V \leq n - 1$.

This completes the proof that $\operatorname{ind} S \leq n$. $\square$

In the corollary, the condition in (b) is called "σ-compact" and the condition in (c) is called "locally compact".

Corollary 3.4.20. (a) *If S is a compact metric space, then $\operatorname{Cov} S = \operatorname{ind} S = \operatorname{Ind} S$. (b) If S is a metric space and S is the union of countably many compact subsets, then $\operatorname{Cov} S = \operatorname{ind} S = \operatorname{Ind} S$. (c) If S is a separable metric space, and every point of S is in the interior of some compact subset, then $\operatorname{Cov} S = \operatorname{ind} S = \operatorname{Ind} S$.*

Proof. (a) $\operatorname{ind} S = \operatorname{Ind} S$ by Exercise 3.4.8 and Proposition 3.4.17; $\operatorname{ind} S = \operatorname{Cov} S$ by Theorem 3.4.19.

(b) Now $S = \bigcup_{n \in \mathbb{N}} F_n$ where F_n is compact. A compact metric space is separable, so S is separable. As before, $\operatorname{ind} S = \operatorname{Ind} S$ by Exercise 3.4.8 and Proposition 3.4.17; $\operatorname{Cov} S \leq \operatorname{ind} S$ by Theorem 3.4.18. So it remains to prove $\operatorname{Cov} S \geq \operatorname{ind} S$. There is nothing to do if $\operatorname{Cov} S = \infty$. So assume $\operatorname{Cov} S = k$ for some integer k. We will deduce that $\operatorname{ind} S \leq k$. Since $F_n \subseteq S$, we have $\operatorname{Cov} F_n \leq k$ by the subset theorem 3.2.13. So $\operatorname{ind} F_n \leq k$ by (a). Compact subsets are closed, so $\operatorname{ind} S \leq k$ by the sum theorem 3.4.13.

(c) The collection of all open sets V such that $\overline{V}$ is compact is an open cover of S. By the Lindelöf property, there is a countable subcover. So S satisfies (b). □

Although the result $\operatorname{Cov} S = \operatorname{ind} S = \operatorname{Ind} S$ is true for any separable metric space, we have not proved it here. There are cases not covered by this corollary. For example, it can be proved that the separable metric space $\mathbb{R} \backslash \mathbb{Q}$ of irrational numbers satisfies none of the conditions (a), (b), (c). But it does satisfy $\operatorname{ind} = \operatorname{Ind} = \operatorname{Cov} = 0$.

3.5 *Remarks

The classical reference on topological dimension is the book by W. Hurewicz and H. Wallman [35]. For this chapter, I have used that reference, as well as [22, 53, 54, 55]. Modern topological dimension theory deals with "topological spaces" and not just metric spaces. Topologists usually write "dim" for the covering dimension. But in this book, that is reserved for the Hausdorff dimension.

Warning: The definition of "order" for a family of sets is not the same in all references. In some texts it is adjusted by 1. The definition used in this book (and many others) has the advantage that dimension n corresponds to covers of order n. But the disadvantage that "family of order n" means that at the sets meet at most $n + 1$ at a time. The alternative definition (also used in many books) has the advantage that "family of order n" means that at the sets meet at most n at a time, but the disadvantage that dimension n corresponds to covers of order $n + 1$.

The small inductive dimension uses the catch-all symbol "∞" for spaces that do not have small inductive dimension otherwise specified by the definition. There is the possibility of a more refined classification of metric spaces of infinite dimension. The way in which the definition is formulated makes the use of transfinite ordinal numbers quite natural. If α is an ordinal number, and S is a metric space, then we say that $\operatorname{ind} S \leq \alpha$ iff there is a base $\mathcal{B}$ for the open sets of S such that $\operatorname{ind} \partial U < \alpha$ for $U \in \mathcal{B}$. *Questions:* Which ordinals α are the dimension of some metric space? Separable space? Compact space? Do we still need an extra symbol ∞, or does every metric space admit some ordinal as its dimension? What happens to the formula $\operatorname{ind}(A \cup B) \leq \operatorname{ind} A + \operatorname{ind} B + 1$ (when addition is not commutative)? References: [22, p. 50], [35, p. 83].

The proof in Sect. 3.3 that $\operatorname{Cov} \mathbb{R}^2 = 2$ is a small taste of the mathematical field known as **algebraic topology**. A ten-cent description of algebraic topology might say that an algebraic object is associated with a situation from topology, in such a way that useful information can be obtained about the topology from the algebra. In the proof given in Sect. 3.3, the algebraic object that is used is the integers modulo 2. This is not a very sophisticated algebraic object, but it is enough to distinguish between two kinds of functions from the circle to itself.

Hints for some exercises.

Exercise 3.1.12: The intervals $(a, b) \cap \mathbb{Q}$ with irrational a, b form a base for the topology of $\mathbb{Q}$.

Exercise 3.1.13: Show there is a positive minimum distance between the sets $f_i[K]$, and remember the proof that the Cantor set is zero-dimensional.

Exercise 3.2.17: Define a tiling of $\mathbb{R}^3$ by congruent rectangular solids that meet at most 4 at a time. Do it by "layers", where each layer looks like Fig 3.1.2, but adjacent layers are offset with respect to each other.

Exercise 3.2.20: Heighway dragon, 2; Koch curve, 1; the McWorter pentigree, 1; the twindragon, 2; the Eisenstein fractions, 2; 120-degree dragon, 2.

Exercise 3.4.1: Use the usual base of open intervals.

Exercise 3.4.5: $\partial(U \times V) = \left(\partial U \times \overline{V}\right) \cup \left(\overline{U} \times \partial V\right)$.

Exercise 3.4.7: If A, B are disjoint nonempty closed sets, then they are separated by $L = \{\, x \in \mathbb{R} : \operatorname{dist}(x, A) = \operatorname{dist}(x, B) \,\}$. Show L is zero-dimensional.

A line is breadthless length.
A surface is that which has length and breadth only.
A solid is that which has length, breadth, and depth.
—Euclid, *The Elements* (translation by T. L. Heath)

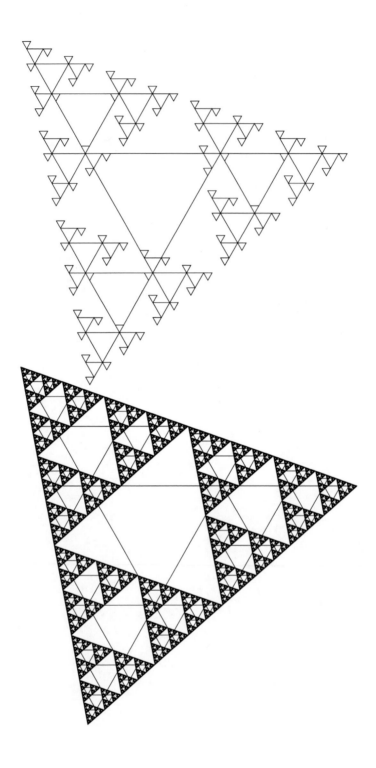

4

Self-Similarity

There are several variant notions of "dimension" that may be classified as fractal dimensions. Among the most widely used of these are the Hausdorff dimension, the packing dimension, and the box dimension. They will be considered in Chap. 6. We begin here with the **similarity dimension**, a fractal dimension that is easier to define (but not as useful).

At the same time, we will formally discuss **iterated function systems**. This is an efficient way of specifying many of the sets that we will be interested in. It has been publicized in recent years by Michael Barnsley; see [3], [4].

4.1 Ratio Lists

A **ratio list** is a finite list of positive numbers, $(r_1, r_2, \cdots, r_n)$. An **iterated function system** realizing a ratio list $(r_1, r_2, \cdots, r_n)$ in a metric space S is a list $(f_1, f_2, \cdots, f_n)$, where $f_i \colon S \to S$ is a similarity with ratio r_i.

A nonempty compact set $K \subseteq S$ is an **invariant set** or **attractor** for the iterated function system $(f_1, f_2, \cdots, f_n)$ iff $K = f_1[K] \cup f_2[K] \cup \cdots \cup f_n[K]$. The reason for the term "attractor" is in Corollary 4.1.4. We may sometimes say **set attractor** to distinguish it from other types of attractors that also occur.

Examples: The triadic Cantor dust (p. 6) is an invariant set for an iterated function system realizing the ratio list $(1/3, 1/3)$. The Sierpiński gasket (p. 9) is an invariant set for an iterated function system realizing the ratio list $(1/2, 1/2, 1/2)$.

The **sim-value** (or **similarity value**) of a ratio list $(r_1, r_2, \cdots, r_n)$ is the positive number s such that $r_1^s + r_2^s + \cdots + r_n^s = 1$.

Theorem 4.1.1. *Let $(r_1, r_2, \cdots, r_n)$ be a ratio list. Suppose each $r_i < 1$. Then there is a unique nonnegative number s satisfying $\sum_{i=1}^{n} r_i^s = 1$. The number s is 0 if and only if $n = 1$.*

Proof. Consider the function $\Phi\colon [0, \infty) \to [0, \infty)$ defined by

$$\Phi(s) = \sum_{i=1}^{n} r_i^s.$$

Then Φ is a continuous function, $\Phi(0) = n \geq 1$ and $\lim_{s \to \infty} \Phi(s) = 0 < 1$. Therefore, by the intermediate value theorem, there is at least one value s with $\Phi(s) = 1$. The derivative of Φ is

$$\sum_{i=1}^{n} r_i^s \log r_i.$$

This is < 0, so Φ is strictly decreasing. Therefore there is only one solution s to $\Phi(s) = 1$. If $n > 1$, then $\Phi(0) > 1$, so $s \neq 0$.   □

A ratio list $(r_1, r_2, \cdots, r_n)$ is called **contracting** (or **hyperbolic**) iff $r_i < 1$ for all i. Similarly an iterated function system may be called contracting or hyperbolic.

The number s is called the **similarity dimension** of a (nonempty compact) set K iff K satisfies a self-referential equation of the type

$$K = \bigcup_{i=1}^{n} f_i[K],$$

where $(f_1, f_2, \cdots, f_n)$ is a hyperbolic iterated function system of similarities whose ratio list has sim-value s.

It is, of course, conceivable that a given set admits two different decompositions, and therefore two different similarity dimensions. So the value s is more correctly considered to be a characteristic of an iterated function system (the sim-value) rather than of the set attractor K. We will see later that, under the right circumstances, the similarity dimension of a set K coincides with the Hausdorff dimension of K, which is uniquely determined by the set K.

Consider, for example, an interval $[a, b]$. It is the union of two smaller intervals, $[a, (a + b)/2]$ and $[(a + b)/2, b]$. Each of the parts is similar to the whole set $[a, b]$, with ratio $1/2$. The sim-value of the ratio list $(1/2, 1/2)$ is the solution s of the equation

$$2 \left(\frac{1}{2} \right)^s = 1,$$

so $s = 1$. The similarity dimension is 1.

Note, however, that we can also write $[0, 1] = [0, 2/3] \cup [1/3, 1]$, corresponding to ratio list $(2/3, 2/3)$, yielding a dimension larger than 1. So in order for the similarity dimension to be a characteristic of the set, we will need to limit overlap in some way. This is discussed in Sect. 6.4.

What do we want to do for a rectangle $[a, b] \times [c, d]$ in $\mathbb{R}^2$? It is the union of four rectangles with sides half the size. The ratio list is $(1/2, 1/2, 1/2, 1/2)$ and the similarity dimension is 2.

How about a closed ball $\overline{B}_r(a)$ in $\mathbb{R}^2$ (a "disk")? Some geometry shows that a disk is not the union of finitely many disks of smaller radius. This illustrates the main drawback of the similarity dimension: it is not defined for many sets, even very simple sets.

Now let us consider a more interesting example. The triadic Cantor dust is the invariant set for an iterated function system realizing ratio list $(1/3, 1/3)$ (p. 6). So the similarity dimension is the solution s of the equation

$$2\left(\frac{1}{3}\right)^s = 1,$$

so* $s = \log 2/\log 3$, or approximately 0.6309. This agrees with the assertion (p. 2) that the dimension should be less than 1.

Next take the Sierpiński gasket. The ratio list is $(1/2, 1/2, 1/2)$, so the similarity dimension is $\log 3/\log 2 \approx 1.585$. It was asserted (p. 9) that the dimension should be between 1 and 2.

Exercise 4.1.2. Calculate similarity dimensions for other sets: the Heighway dragon (p. 20); the Koch curve (p. 19); the McWorter pentigree (p. 24); the twindragon (p. 33); the Eisenstein fractions (p. 34); 120-degree dragon (p. 23).

In these examples, we begin with a set, and then try to find a corresponding iterated function system. It is often useful to do things the other way around: begin with an iterated function system, and use it to obtain a set.

Theorem 4.1.3. *Let S be a nonempty complete metric space, let $(r_1, \cdots, r_n)$ be a contracting ratio list, and let $(f_1, \cdots, f_n)$ be an iterated function system of similarities in S that realizes the ratio list. Then there is a unique nonempty compact invariant set for the iterated function system.*

Proof. Consider the hyperspace $\mathbb{H}(S)$ of nonempty compact subsets of S, with the Hausdorff metric D. Since S is complete, so is $\mathbb{H}(S)$, by Theorem 2.5.3. Define a function $F\colon \mathbb{H}(S) \to \mathbb{H}(S)$ as follows:

$$F(A) = \bigcup_{i=1}^{n} f_i[A].$$

The continuous image of a compact set is compact (Theorem 2.3.15), and the union of finitely many compact sets is compact (Exercise 2.3.14); thus if A is compact, then so is $F(A)$.

I claim that F is a contraction map. Let $r = \max\{r_1, r_2, \cdots, r_n\}$. Clearly $r < 1$. I will show that

$$D\big(F(A), F(B)\big) \leq rD(A, B).$$

---

* Since I am a mathematician, when I write "log" it refers to the natural logarithm. But in fact for the quotient of two logarithms, as we have here, it doesn't matter what the base is.

Let $q > D(A, B)$ be given. If x is any element of $F(A)$, then $x = f_i(x')$ for some i and some $x' \in A$. Since $q > D(A, B)$, there is a point $y' \in B$ with $\varrho(x', y') < q$. But then the point $y = f_i(y') \in F(B)$ satisfies $\varrho(x, y) = r_i \varrho(x', y') < rq$. This is true for all $x \in F(A)$, so $F(A)$ is contained in the rq-neighborhood of $F(B)$. Similarly, $F(B)$ is contained in the rq-neighborhood of $F(A)$. Therefore $D\big(F(A), F(B)\big) \le rq$. This is true for every $q > D(A, B)$, so we have $D\big(F(A), F(B)\big) \le r D(A, B)$.

Therefore we have a contraction map F defined on a complete metric space $\mathbb{H}(S)$. By the contraction mapping theorem (2.2.21), F has a unique fixed point. A fixed point of F is exactly the same thing as an invariant set for $(f_1, f_2, \cdots, f_n)$. $\qquad\square$

This proof, together with Corollary 2.2.22, provides a construction for the invariant set:

Corollary 4.1.4. *In the notation of Theorem 4.1.3, if A_0 is any nonempty compact set in S, and if*

$$A_{k+1} = \bigcup_{i=1}^{n} f_i[A_k]$$

for $k \ge 0$, then the sequence (A_k) converges in the Hausdorff metric to the invariant set of the iterated function system.

This illustrates the use of the term ***attractor***.

Exercise 4.1.5. Let $(r_1, r_2, \cdots, r_n)$ be a contracting ratio list. Suppose an iterated function system $(f_1, f_2, \cdots, f_n)$ consists not of similarities, but only of maps $f_i \colon S \to S$ satisfying

$$\varrho(f_i(x), f_i(y)) \le r_i \, \varrho(x, y).$$

Show that if S is complete, then there is a unique invariant set.

We have been talking about "iterated function systems." What does this have to do with "iteration"? Let (f_1, f_2, f_3) be the iterated function system associated with the Sierpiński gasket (p. 9). The iteration that we will be interested in involves starting with any point $a \in \mathbb{R}^2$, and then repeatedly applying these three functions, in any order.

Exercise 4.1.6. Let $k_1, k_2, k_3, \cdots$ be an infinite sequence in the set $\{1, 2, 3\}$. Let $a \in \mathbb{R}^2$ be a point. Let the sequence (x_n) be defined by

$$x_0 = a; \qquad x_n = f_{k_n}(x_{n-1}) \quad \text{for } n \ge 1.$$

Then: (1) Every cluster point of the sequence (x_n) belongs to the Sierpiński gasket S; (2) Every point of the Sierpiński gasket is a cluster point of such a sequence (x_n) for some choice of k_i; (3) There is a point a and choice sequence k_i so that S is exactly equal to the set of all cluster points of (x_n).

A more sophisticated version of (3) says that a "random" choice of (k_i) will (with probability one) have cluster set S. This fact has sometimes been used to produce a picture of the invariant set for an iterated function system on a computer. (See "the chaos game" in [3].)

Exercise 4.1.7. Let S be a complete metric space, let $(r_1, r_2, \cdots, r_n)$ be a contracting ratio list, and let $(f_1, f_2, \cdots, f_n)$ be a realization of the ratio list in S. State and prove the appropriate version of Exercise 4.1.6 for this situation.

Recall the situation from Sect. 1.6: Let b be a complex number, $|b| > 1$, and let $D = \{d_1, d_2, \cdots, d_k\}$ be a finite set of complex numbers, including 0. We are interested in representing complex numbers (or real numbers) in the number system they define.

Write F for the set of "fractions"; that is, numbers of the form

$$\sum_{j=-\infty}^{-1} a_j b^j.$$

Analysis of the set F is a "self-similar" set problem:

Exercise 4.1.8. The set F is nonempty and compact, and is the invariant set for an iterated function system of similarities. What is the sim-value of this iterated function system?

There are various ways that such a "number system" can be generalized. Consider the following way to define a set in $\mathbb{R}^d$. Let $b \in \mathbb{R}$ with $|b| > 1$ as before, but let D be a finite subset of $\mathbb{R}^d$, including 0. Then let F be the set of all vectors

$$\sum_{j=-\infty}^{-1} b^j a_j,$$

where $a_j \in D$. Is there an iterated function system to describe F?

The **Menger sponge** is a set in $\mathbb{R}^3$. (See Fig. 4.1.9.) Begin with a cube (filled in) of side 1. Subdivide it into 27 smaller cubes by trisecting the edges.

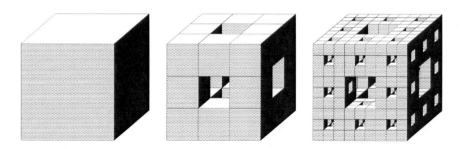

Fig. 4.1.9. Menger sponge

The trema to remove consists of the center cube and the 6 cubes in the centers of the faces. That means 20 cubes remain. (The boundaries of these 20 cubes must also remain, so that the set will be compact.) Continue in the same way with the small cubes.

Exercise 4.1.10. What is the topological dimension of the Menger sponge? What is the similarity dimension of the Menger sponge?

4.2 String Models

We will now consider "string models" in general. Given any contracting ratio list, there is a realization that is, in an appropriate sense, the "best" realization (Theorem 4.2.3). Any other realization of the same ratio list is an "image" of this one. One advantage will be seen in Chap. 6 when other fractal dimensions are computed.

Two instances of the string model have already been seen in Chap. 1, where the space of infinite strings in the alphabet $\{0, 1\}$ was used as a model for $[0, 1]$ and for the Cantor dust.

Example

Before we consider the general model, we will take one more example, the Sierpiński gasket. This time we will need to use an ***infinite ternary tree***. Each node has exactly three children; each node except the root has exactly one parent, and every node is a descendant of the root. To represent this using strings, we will need a three-letter alphabet. Any three letters will do; I will use this alphabet: $E = \{L, U, R\}$. (When I chose the letters I was looking at Fig. 1.2.1, and thinking of the words "left", "upper", "right".)

We will write (as before) $E^{(k)}$ for the set of all k-letter strings from this alphabet; and $E^{(*)}$ for the set of all finite strings; and $E^{(\omega)}$ for the set of all infinite strings. We may identify $E^{(*)}$ with the infinite ternary tree: the empty string Λ is the root, and if α is a string, then αL is the left child, αU is the middle child (perhaps also called the "upper child"), αR is the right child.

We want to define a map $h: E^{(\omega)} \to \mathbb{R}^2$ with range equal to the Sierpiński gasket S. The gasket itself is a union of three parts, the "left" part, the "upper" part, and the "right" part. There are three dilations of $\mathbb{R}^2$ corresponding to these three parts. They will be called now: f_L, f_U, f_R. The "addressing function" or "model map" $h: E^{(\omega)} \to \mathbb{R}^2$ should be defined so that $h[E^{(\omega)}] = S$. It should be continuous and satisfy $h(L\sigma) = f_L(h(\sigma))$, $h(U\sigma) = f_U(h(\sigma))$, $h(R\sigma) = f_R(h(\sigma))$ for all strings σ.

We will describe this addressing function h using base 2 expansions in the (u, v) coordinate system of Exercise 1.2.4. A string from the alphabet $\{L, U, R\}$ maps to a pair (u, v), according to the rules in the following table:

letter	digit of u	digit of v
L	0	0
U	0	1
R	1	0

For example,

$$h(\text{LRLUU}\cdots) = \big((0.01000\cdots)_2, (0.00011\cdots)_2\big).$$

By Exercise 1.2.4, the range of h is exactly the Sierpiński gasket S.

The metric to be used on $E^{(\omega)}$ will be called $\varrho_{1/2}$. If $\sigma, \tau \in E^{(\omega)}$, then

$$\varrho_{1/2}(\sigma, \tau) = \frac{1}{2^k}$$

where k is the length of the longest common prefix of σ and τ. With this definition, we have $\operatorname{diam}[\alpha] = (1/2)^{|\alpha|}$. (Recall the notation $|\alpha|$ for the length of a finite string α.)

Exercise 4.2.1. The addressing function h defined above satisfies the Lipschitz condition

$$|h(\sigma) - h(\tau)| \le \varrho_{1/2}(\sigma, \tau).$$

Exercise 4.2.2. Is the addressing function h inverse Lipschitz?

General Definition

Let $(r_1, r_2, \cdots, r_n)$ be a contracting ratio list. The model for this ratio list will be the space $E^{(\omega)}$ of infinite strings on an alphabet E with n letters. If no better choice suggests itself, the set $E = \{1, 2, \cdots, n\}$ may be used as the alphabet. Usually, however, we will choose letters that suggest the intended example. But there will be understood a one-to-one correspondence between the letters of E and the ratios in the ratio list. When the alphabet is known, we will often even use them to label the ratio list, so we may write $(r_e)_{e \in E}$ for $(r_1, r_2, \cdots, r_n)$.

For each letter $e \in E$, there is a corresponding function $\theta_e \colon E^{(\omega)} \to E^{(\omega)}$, called a **right shift**, defined by

$$\theta_e(\sigma) = e\sigma.$$

That is, insert the letter e at the beginning of the string. We will define a metric on $E^{(\omega)}$ so that the right shifts $(\theta_e)_{e \in E}$ form a realization of the given ratio list $(r_e)_{e \in E}$.

To define a metric on $E^{(\omega)}$, we will specify a "diameter" w_α for each node $\alpha \in E^{(*)}$ of the tree. This is done recursively:

$$w_\Lambda = 1,$$

$$w_{\alpha e} = w_\alpha \, r_e \qquad \text{for } \alpha \in E^{(*)} \text{ and } e \in E.$$

Alternatively, w_α is the product, over all the letters e that make up α, of the ratios r_e. In the Sierpiński gasket example above, for example, $w_{\mathsf{LRLU}} = r_{\mathsf{L}} r_{\mathsf{R}} r_{\mathsf{L}} r_{\mathsf{U}}$. The metric ϱ is defined from these diameters in the usual way. If there are at least two letters, then there are no exceptions to the formula $\mathrm{diam}[\alpha] = w_\alpha$.

Now it is easy to verify that the iterated function system $(\theta_e)_{e \in E}$ realizes the ratio list $(r_e)_{e \in E}$: Suppose $\sigma, \tau \in E^{(\omega)}$ have longest common prefix α. If $e \in E$, then the longest common prefix of $e\sigma$ and $e\tau$ is $e\alpha$. So

$$\varrho\big(\theta_e(\sigma), \theta_e(\tau)\big) = w_{e\alpha} = r_e w_\alpha = r_e \, \varrho(\sigma, \tau).$$

That is, θ_e is a similarity on $\big(E^{(\omega)}, \varrho\big)$ with ratio r_e.

The metric space $E^{(\omega)}$ is complete, so the right-shift realization $(\theta_e)_{e \in E}$ has a unique nonempty compact invariant set. In fact, that invariant set is the whole space $E^{(\omega)}$. The space $E^{(\omega)}$, together with the right shifts, will be called the **string model** of the ratio list $(r_e)_{e \in E}$.

Theorem 4.2.3 (String model theorem). *Let S be a nonempty complete metric space and let $(f_e)_{e \in E}$ be any iterated function system realizing the ratio list $(r_e)_{e \in E}$ in S. Assume that $r_e < 1$ for all e. Then there is a unique continuous function $h \colon E^{(\omega)} \to S$ such that*

$$h(e\sigma) = f_e\big(h(\sigma)\big)$$

for all $\sigma \in E^{(\omega)}$ and $e \in E$. The range $h\big[E^{(\omega)}\big]$ is the invariant set of the iterated function system $(f_e)_{e \in E}$.

Proof. We will use uniform convergence. We will define recursively a sequence (g_k) of continuous functions $g_k \colon E^{(\omega)} \to S$. Choose any point $a \in S$. Define $g_0(\sigma) = a$ for all σ. If g_k has been defined, then define g_{k+1} by:

$$g_{k+1}(e\sigma) = f_e\big(g_k(\sigma)\big)$$

for $e \in E$ and $\sigma \in E^{(\omega)}$. The function g_0 is clearly continuous. We can verify by induction that g_{k+1} is continuous, using the fact that each of the sets $[e]$ is open.

I claim that the sequence (g_k) converges uniformly. Let $r = \max_e r_e$, so that $r < 1$. Now $E^{(\omega)}$ is compact, so $\varrho_{\mathrm{u}}(g_1, g_0)$ is finite. We have

$$\varrho\big(g_{k+1}(e\sigma), g_k(e\sigma)\big) = \varrho\big(f_e(g_k(\sigma)), f_e(g_{k-1}(\sigma))\big)$$
$$\leq r_e \varrho\big(g_k(\sigma), g_{k-1}(\sigma)\big)$$
$$\leq r \varrho_{\mathrm{u}}(g_k, g_{k-1}).$$

Therefore $\varrho_u(g_{k+1}, g_k) \leq r\varrho_u(g_k, g_{k-1})$. So $\varrho_u(g_{k+1}, g_k) \leq r^k \varrho_u(g_1, g_0)$ by induction. By the triangle inequality, if $m \geq k$, then

$$\varrho_u(g_m, g_k) \leq \sum_{j=k}^{m-1} \varrho_u(g_{j+1}, g_j) \leq \sum_{j=k}^{\infty} r^j \varrho_u(g_1, g_0).$$

This is the tail of a convergent geometric series, so it approaches 0 as $k \to \infty$. Therefore the sequence (g_k) is a Cauchy sequence in $\mathcal{C}(E^{(\omega)}, S)$. So it converges uniformly. Write h for its limit.

Now we have by the definition of the sequence (g_k),

$$g_{k+1}\big[E^{(\omega)}\big] = \bigcup_{e \in E} f_e\Big[g_k\big[E^{(\omega)}\big]\Big].$$

The sequence of sets $\big(g_k[E^{(\omega)}]\big)$ converges to the invariant set by Corollary 4.1.4, and converges to $h\big[E^{(\omega)}\big]$ by Proposition 2.5.7. Therefore $h\big[E^{(\omega)}\big]$ is the invariant set.

For the uniqueness, suppose that $g\colon E^{(\omega)} \to S$ and $h\colon E^{(\omega)} \to S$ with $g(e\sigma) = f_e\big(g(\sigma)\big)$ and $h(e\sigma) = f_e\big(h(\sigma)\big)$. Now with $r = \max_e r_e$ as before, we have by the same calculation as above, $\varrho_u(g, h) \leq r\varrho_u(g, h)$. This is impossible unless $\varrho_u(g, h) = 0$. So $g = h$.   □

We call h the **addressing function**. When x and σ are related by $x = h(\sigma)$, we say that σ is an **address** of the point x.

Exercise 4.2.4. The addressing function h is Lipschitz when the string model has the metric described above.

Exercise 4.2.5. Let K be the attractor for the iterated function system $(f_e)_{e \in E}$, and let $h\colon E^{(\omega)} \to S$ be the addressing function. Assume the family $\{ f_e[K] : e \in E \}$ has order n (in the sense defined on p. 91). Show that the map h is at most $(n + 1)$-to-one. Conclude that ind $K \leq n$.

This can compute the topological dimension for most of the self-similar examples we have seen. But not all.

Exercise 4.2.6. Let K be the Menger sponge (p. 121). Compute its topological dimension.

4.3 Graph Self-Similarity

There is a generalization of self-similarity that provides a way to study a larger class of sets. A definitive formulation is due to Mauldin and Williams, but variants were used by others. Figs. 4.3.1 and 4.3.2 illustrate two examples. There will be a list of several nonempty compact sets to be constructed

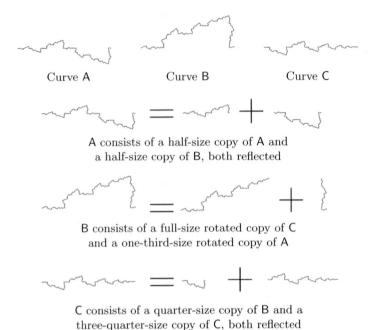

Curve A   Curve B   Curve C

A consists of a half-size copy of A and
a half-size copy of B, both reflected

B consists of a full-size rotated copy of C
and a one-third-size rotated copy of A

C consists of a quarter-size copy of B and a
three-quarter-size copy of C, both reflected

Fig. 4.3.1. The MW curves

simultaneously. Each of them is decomposed into parts obtained from sets in the list using certain similarities.

Li's lace fractal (p. 84) is made up of isosceles right trianglar blocks of two kinds, called P and Q. Each of these sets is made up of parts that are similar copies of the same sets P, Q. Sometimes an illustration (as in Fig 4.3.3) can be used to specify the information on how the similarities act. The transformations map the triangles P, Q to the smaller triangles that make them up. A nonsymmetric letter is used in the picture to show how the image should be oriented.

For each such construction, there corresponds in a natural way a directed multigraph, with a positive number associated with each edge. There is one

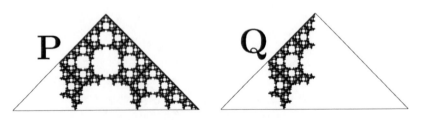

Fig. 4.3.2. Li's lace

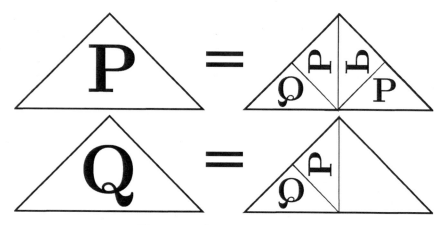

Fig. 4.3.3. Structure of Li's lace

node of the graph for each of the sets to be constructed. The edges from a node correspond to the subsets into which the corresponding set is decomposed, and the number associated with each edge corresponds to the ratio of the similarity. (Be careful to pay attention to which way the arrows go. It may seem more natural for some purposes to do it the other way around, but this is the direction chosen by Mauldin and Williams.) The structure involved, a directed multigraph (V, E, i, t) together with a function $r \colon E \to (0, \infty)$, will be called a **Mauldin–Williams graph.** For technical reasons we assume each vertex has at least one edge leaving it.

Suppose (V, E, i, t, r) is a Mauldin–Williams graph. An iterated function system realizing the graph is made up of metric spaces S_v, one for each vertex $v \in V$, and similarities f_e, one for each edge $e \in E$, such that $f_e \colon S_v \to S_u$ if $e \in E_{uv}$, and f_e has ratio $r(e)$. An **invariant set list** for such an iterated function system is a list of nonempty compact sets $K_v \subseteq S_v$, one for each node $v \in V$, such that

$$K_u = \bigcup_{\substack{v \in V \\ e \in E_{uv}}} f_e[K_v]$$

for all $u \in V$. (An "invariant set list" may sometimes be called an **invariant list** or an **attractor.**) When a graph (V, E, i, t, r) and an iterated function system (f_e) are related in this way, we may say that the iterated function system is **directed** by the graph, and call it a **graph-directed** iterated function system.

Compare this abstract definition to the two examples. The Mauldin–Williams graphs are shown in Figs. 4.3.4(a) and 4.3.4(b).

Each of the nonempty compact sets K_v satisfying such equations will be said to have **graph self-similarity.**

For this purpose, the ratio lists discussed above correspond to graphs with a single vertex, and one loop for each item in the ratio list.

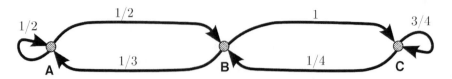

Fig. 4.3.4. (a) Graph for the MW curves

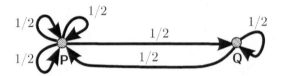

Fig. 4.3.4. (b) Graph for Li's lace

Existence of Invariant Set List

Just as in the case of ratio lists, if proper conditions are satisfied, then the invariant list of sets exists and is unique.

Theorem 4.3.5. *Let (V, E, i, t, r) be a Mauldin–Williams graph. Suppose $r(e) < 1$ for all $e \in E$. Let $(f_e)_{e \in E}$ realize the graph in nonempty complete metric spaces S_v. Then there is a unique list $(K_v)_{v \in V}$ of nonempty compact sets $(K_v \subseteq S_v)$ such that*

$$K_u = \bigcup_{\substack{v \in V \\ e \in E_{uv}}} f_e[K_v]$$

for all $u \in V$.

Proof. The proof is similar to the proof of Theorem 4.1.3. The metric spaces $\mathcal{K}(S_v)$ are complete. So the finite Cartesian product

$$\prod_{v \in V} \mathcal{K}(S_v)$$

is also complete with the metric given by the maximum of the coordinate metrics. Let us write $(A_v)_{v \in V}$ for a typical element of this product space. The function defined by

$$F\big((A_v)_{v \in V}\big) = \left(\bigcup_{\substack{v \in V \\ e \in E_{uv}}} f_e[A_v] \right)_{u \in V}$$

is a contraction mapping, and its unique fixed point is the invariant list required.   □

A Mauldin–Williams graph (V, E, i, t, r) will be called **strictly contracting** if the conditions $r(e) < 1$ are satisfied. You can note that the graph in Fig. 4.3.4 for the MW cirves has an edge with value 1, so it is not strictly contracting.

Exercise 4.3.6. Discuss the graph self-similarity of the boundary of Heighway's dragon.

Path Models

Let us consider the Mauldin–Williams graphs, and their realizations as iterated function systems of similarities. It will be useful to have a tree-type model for such a system. (Computing the Hausdorff dimension of the tree model will be much of the work in computing the Hausdorff dimension for sets with graph self-similarity.) The models will be analogous to the string models considered above. But (because of their construction) we will usually call them "path models".

Let (V, E, i, t, r) be a Mauldin–Williams graph. There is a "path forest" corresponding to it, as on p. 80. The definition of the edge-value function r can be extended to paths by defining: $r(\Lambda_u) = 1$ for empty paths, and $r(\alpha e) = r(\alpha) r(e)$ for any path α and any edge e with $t(\alpha) = i(e)$.

Given an iterated function system $(f_e)_{e \in E}$, acting on spaces S_v, we may extend the notation to paths: f_{Λ_u} is the identity function on S_u, and $f_{\alpha e}$ is the composite function $f_\alpha \circ f_e$, defined by $f_{\alpha e}(x) = f_\alpha(f_e(x))$. In cases when confusion will be minimal, we will save writing by using the edge e itself to stand for the function f_e; that is, we may write $e(x)$, where $x \in S_{t(e)}$, as an abbreviation of $f_e(x)$. Similarly, if α is a finite path, we may write $\alpha(x)$ for $f_\alpha(x)$.

Let (V, E, i, t, r) be a strictly contracting Mauldin–Williams graph. The spaces $E_v^{(\omega)}$ of infinite paths of the graph admit right shift maps as before: If $e \in E_{uv}$, then we define $\theta_e \colon E_v^{(\omega)} \to E_u^{(\omega)}$ by $\theta_e(\sigma) = e\sigma$. One way to define metrics on the spaces $E_v^{(\omega)}$ so that this family of maps realizes the Mauldin–Williams graph is as follows: If $\sigma, \tau \in E_v^{(\omega)}$, and α is their longest common prefix, then $\varrho(\sigma, \tau) = r(\alpha)$. [We will, however, use a different system of metrics in Sec. 6.6.]

Exercise 4.3.7. With the metrics ϱ defined, the right shift θ_e is a similarity with ratio $r(e)$.

Rescaling

Suppose (V, E, i, t, r) is a Mauldin–Williams graph. Let maps f_e realize it in spaces (S_v, ϱ_v). To **rescale** the realization, we replace each of the metrics ϱ_v by a constant multiple $\varrho'_v = a_v \varrho_v$ of itself. The sets S_v are unchanged, and the maps f_e are unchanged. Of course an invariant list for the original iterated function system is exactly the same thing as an invariant list for the rescaled iterated function system.

Under the new metrics, what happens to the contraction ratios of the maps f_e? If $e \in E_{uv}$, then

$$\varrho'_u\big(f_e(x), f_e(y)\big) = a_u \varrho_u\big(f_e(x), f_e(y)\big)$$
$$= a_u r(e) \varrho_v(x, y)$$
$$= \frac{a_u r(e)}{a_v}\, \varrho'_v(x, y).$$

Thus, with the new metrics, the maps f_e realize a Mauldin–Williams graph (V, E, i, t, r'), where

$$r'(e) = \frac{a_u}{a_v}\, r(e) \qquad \text{for } e \in E_{uv}.$$

The Mauldin–Williams graph (V, E, i, t, r') is called a **rescaling** of the graph (V, E, i, t, r). A Mauldin–Williams graph (V, E, i, t, r) will be called **contracting** iff it is a rescaling of a strictly contracting graph.

Exercise 4.3.8. Show that the graph in Fig. 4.3.4 is contracting.

Theorem 4.3.5 shows that a realization of a contracting graph in a complete space has a unique invariant list. That list can be constructed by the method analogous to Corollary 4.1.4.

Finally, we have an alleged criterion for contractivity.

Exercise 4.3.9. Prove or disprove: A Mauldin–Williams graph (V, E, i, t, r) is contracting if and only if $r(\alpha) < 1$ for all nonempty cycles $\alpha \in E^{(*)}$.

Similarity Dimension

A Mauldin–Williams graph has a "sim-value" associated with it, in the same way as a ratio list. The case of a Mauldin–Williams graph with 2 nodes will be discussed now. The definitions for graphs with more than 2 nodes are given in Sect. 6.6.

Let (V, E, i, t, r) be a Mauldin–Williams graph. Suppose $V = \{1, 2\}$. Write

$$A(s) = \sum_{e \in E_{11}} r(e)^s$$

$$B(s) = \sum_{e \in E_{12}} r(e)^s$$

$$C(s) = \sum_{e \in E_{21}} r(e)^s$$

$$D(s) = \sum_{e \in E_{22}} r(e)^s,$$

and let

$$\Phi(s) = \frac{A(s) + D(s) + \sqrt{(A(s) - D(s))^2 + 4B(s)C(s)}}{2}.$$

The **sim-value** of the graph is the solution s of the equation $\Phi(s) = 1$. (Unfortunately, we will not see the reason for such a complicated definition until later.)

As in the case of a ratio list, such a number s need not exist. But with reasonable additional conditions, the sim-value exists and is unique. Recall (p. 80) that a directed graph is **strongly connected** if there is a path from any node to any other node.

Proposition 4.3.10. *A strictly contracting, strongly connected Mauldin–Williams graph with 2 nodes has a unique sim-value.*

Proof. Since the graph is strictly contracting, we have $A(s) \to 0$, $B(s) \to 0$, $C(s) \to 0$, and $D(s) \to 0$ as $s \to \infty$, so $\Phi(s) \to 0$.

Since the graph is strongly connected, there is at least one edge from 1 to 2 and at least one edge from 2 to 1. So $B(0) \geq 1$ and $C(0) \geq 1$. Then $\Phi(0) \geq 1$. Equality $\Phi(0) = 1$ holds only if $A(0) = D(0) = 0$ and $B(0) = C(0) = 1$; that is, the two edges postulated are the only edges. In that case the sim-value is 0. In all other cases, $\Phi(0) > 1$. Thus: if the graph is strictly contracting and strongly connected, then there is a nonnegative solution s to the equation $\Phi(s) = 1$.

I claim that the solution is unique since $\Phi'(s) < 0$ for all s. The partial derivatives of Φ are

$$\frac{\partial \Phi}{\partial A} = \frac{1}{2} + \frac{1}{2} \frac{A - D}{\sqrt{(A - D)^2 + 4BC}},$$

$$\frac{\partial \Phi}{\partial B} = \frac{C}{\sqrt{(A - D)^2 + 4BC}},$$

$$\frac{\partial \Phi}{\partial C} = \frac{B}{\sqrt{(A - D)^2 + 4BC}},$$

$$\frac{\partial \Phi}{\partial D} = \frac{1}{2} + \frac{1}{2} \frac{D - A}{\sqrt{(A - D)^2 + 4BC}}.$$

Recall that $A(s) \geq 0$, $B(s) > 0$, $C(s) > 0$, and $D(s) \geq 0$. Then $|A - D| \leq \sqrt{(A - D)^2 + 4BC}$, so

$$\frac{\partial \Phi}{\partial A} \geq \frac{1}{2} - \frac{1}{2} = 0, \quad \frac{\partial \Phi}{\partial B} > 0, \quad \frac{\partial \Phi}{\partial C} > 0, \quad \frac{\partial \Phi}{\partial D} \geq \frac{1}{2} - \frac{1}{2} = 0.$$

The four derivatives satisfy $A'(s) \leq 0$, $B'(s) < 0$, $C'(s) < 0$, and $D'(s) \leq 0$. So we have $\Phi'(s) < 0$. $\qquad \square$

Suppose we have an invariant set list (K_v) directed by a contracting Mauldin–Williams graph. We say that the sim-value of the graph is the **similarity dimension** of the sets K_v. As before, this terminology may be inexact. But under the right conditions (discussed in Chap. 6) this sim-value equals the Hausdorff dimension of all the sets K_v.

Exercise 4.3.11. Compute the similarity dimension of the Li lace fractal. That is: Compute the sim-value of the graph in Fig. 4.3.4(b).

We will do examples later involving the Mauldin–Williams graphs in Figs. 4.3.12 and 4.3.13.

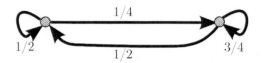

Fig. 4.3.12. A graph

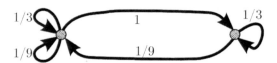

Fig. 4.3.13. A graph

Exercise 4.3.14. Compute the sim-value of the graph of Fig. 4.3.12.

Exercise 4.3.15. Compute the sim-value of the graph of Fig. 4.3.13.

Here is the way in which the sim-value of a graph will be used.

Proposition 4.3.16. *Let (V, E, i, t, r) be a Mauldin–Williams graph with $V = \{1, 2\}$. The number $s \geq 0$ is the sim-value of the graph if and only if there exist positive numbers x and y satisfying*

$$x = A(s)x + B(s)y$$
$$y = C(s)x + D(s)y.$$

Proof. Write $A = A(s)$, etc. They satisfy $A \geq 0$, $B > 0$, $C > 0$, and $D \geq 0$. I must show that $(A + D + \sqrt{(A - D)^2 + 4BC})/2 = 1$ if and only if there exist positive x and y with $Ax + By = x$, $Cx + Dy = y$.

First, suppose $(A + D + \sqrt{(A - D)^2 + 4BC})/2 = 1$. Then $A + D < 2$, so either $A < 1$ or $D < 1$. We will take the case $A < 1$; the other case is similar. Let $x = B$ and $y = 1 - A$. Then $x > 0$ and $y > 0$. The first equation is $Ax + By = AB + B(1 - A) = B = x$. Algebra applied to the equation

$(A + D + \sqrt{(A - D)^2 + 4BC})/2 = 1$ yields $BC - AD = 1 - A - D$. For the second equation, $Cx + Dy = CB + D(1 - A) = BC - AD + D = 1 - A = y$.

Conversely, suppose positive numbers x and y exist with $Ax + By = x$ and $Cx + Dy = y$. Since $y \neq 0$, we may solve:

$$\frac{B}{1 - A} = \frac{x}{y} = \frac{1 - D}{C}.$$

This means that $(1 - A)(1 - D) = BC$, so with a little algebra, $(A - D)^2 + 4BC = (2 - A - D)^2$. Now $B/(1 - A) = x/y > 0$, and $B > 0$, so $A < 1$. Similarly $D < 1$. So $2 - A - D > 0$, and therefore $\sqrt{(A - D)^2 + 4BC} = 2 - A - D$, or $(A + D + \sqrt{(A - D)^2 + 4BC})/2 = 1$.   □

Exercise 4.3.17. How is the sim-value of a two-node Mauldin–Williams graph affected by rescaling?

Exercise 4.3.18. State and prove a "path model theorem", analogous to Theorem 4.2.3, for graph self-similarity.

4.4 *Remarks

The idea of self-similarity is explored by P. A. P. Moran [51] and by John Hutchinson [36]. It will be useful in Chap. 6 in computation of Hausdorff dimensions. Theorem 4.3.5 on the existence of the invariant set is taken from [36].

The "sim-value" for an iterated function system or for an M-W graph was called the "dimension" in the first edition. But that can be confusing. So we have used a different term in the second edition. It is not actually a dimension of the graph itself, or of the iterated function system, but of a set assiciated with them.

Theorem 4.1.1 shows that the sim-value equation $\sum_{i=1}^{n} r_i^s = 1$ admits a unique nonnegative real solution s. But if we allow s to be a complex number, there are other solutions. These so-called "complex dimensions" and their analogs have been explored primarily by Michel Lapidus. See [42].

Graph self-similarity is a generalization of self-similarity. There are several sources for similar ideas; the one used as the model here is [48]. The example shown in Figs. 4.3.1 was concocted to illustrate the idea. Li's lace is from [43], Example (c), Sect. 3.3.

The Menger sponge is a universal 1-dimensional space. Karl Menger defined this space (and corresponding spaces in higher dimensions) for this purpose. The exact statement is like Theorem 3.4.4: Let S be a separable metric space. Then $\operatorname{Cov} S \leq 1$ if and only if S is homeomorphic to a subset of the Menger sponge. [8, p. 503*ff*].

Exercise 4.1.2: Approximate values of the similarity dimension: Heighway, 2; Koch, 1.26; McWorter, 1.86; twindragon, 2; Eisenstein, 2; 120-degree, 1.26.

Exercise 4.1.8: $\log k / \log |b|$.

Exercise 4.1.10: Topological dimension 1; similarity dimension approximately 2.73.

Exercise 4.3.14: 1.

Exercise 4.3.15: The dimension is $-\log x / \log 3$, where x is a solution of $x^3 - x^2 - 2x + 1 = 0$. So the dimension is approximately 0.737.

He had finally made some progress on the Barnsleyformer; because the trick wasn't in the morphogenesis after all, but in the fractal geometry.
—Michael F. Flynn, "Remembered Kisses", *ANALOG, December, 1988*

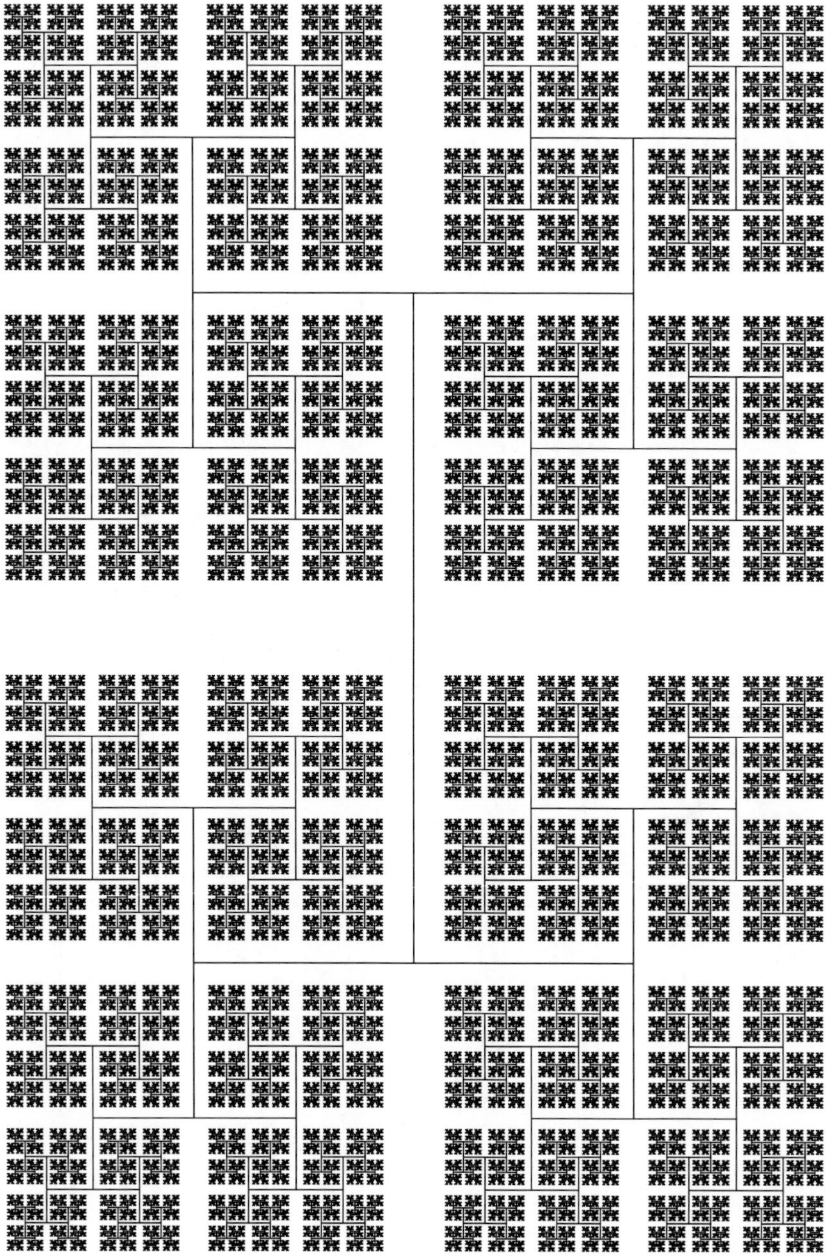

5

Measure Theory

This chapter contains the background from measure theory that is required to understand the Hausdorff dimension. It is true that the Hausdorff dimension can be defined in half a page without reference to measure theory, but when it is done that way there is no indication of the motivation for the definition.

Measure theory will also be indispensable in many of the proofs related to fractal dimension. It will simplify many of the proofs of lower bounds for Hausdorff dimension and of upper bounds for packing dimension. Instead of repeating a combinatorial calculation in each instance, we do the combinatorics once and for all in this chapter, and then repeatedly reap the benefits in Chap. 6.

Since measure theory (like metric topology) is a standard part of graduate mathematics curriculum today, most of the introductory remarks to Chap. 2 are also applicable here.

5.1 Lebesgue Measure

Certain calculations will be done with the symbols ∞ and $-\infty$. They are not real numbers, but they can be useful in connection with calculations involving real numbers. Most of the conventions are sensible when you think about them. Here are some examples:

(1) If $a \in \mathbb{R}$, then $-\infty < a < \infty$.
(2) If $a \in \mathbb{R}$, then $a + \infty = \infty$ and $a - \infty = -\infty$. Also $\infty + \infty = \infty$ and $-\infty - \infty = -\infty$. The combination $\infty - \infty$ is not defined.
(3) If $a \in \mathbb{R}$ is positive, then $a \times \infty = \infty$ and $a \times (-\infty) = -\infty$. If $a \in \mathbb{R}$ is negative, then $a \times \infty = -\infty$ and $a \times (-\infty) = \infty$. The combination $0 \times \infty$ is not defined. [However, we do understand that an infinite series $\sum_{n=1}^{\infty} a_n$, where every term $a_n = 0$, has sum 0.]

The **length** of one of the intervals

$$(a, b) \quad (a, b] \quad [a, b) \quad [a, b]$$

is $b-a$, where $a, b \in \mathbb{R}$ and $a < b$. The length of the degenerate interval $[a, a] = \{a\}$ is 0; the length of the empty set $\varnothing$ is 0. The length of an unbounded interval

$$(a, \infty) \quad [a, \infty) \quad (-\infty, b) \quad (-\infty, b] \quad (-\infty, \infty)$$

is ∞. This follows the conventions on calculation with ∞.

We will be interested in a substantial generalization of the notion of the "length" of a subset of $\mathbb{R}$. The lemma that makes it possible asserts that the length of a countable union of intervals cannot exceed the sum of the lengths of the parts.

Lemma 5.1.1. *Suppose the closed interval $[c, d]$ is covered by a countable family of open intervals:*

$$[c, d] \subseteq \bigcup_{i \in \mathbb{N}} (a_i, b_i).$$

Then

$$d - c < \sum_{i=1}^{\infty} (b_i - a_i).$$

Proof. First, since $[c, d]$ is a compact set, it is in fact covered by a finite number of the intervals:

$$[c, d] \subseteq \bigcup_{i=1}^{n} (a_i, b_i)$$

for some n. I will show that when this happens, the conclusion

$$d - c < \sum_{i=1}^{n} (b_i - a_i)$$

follows. The proof is by induction on n.

If $n = 1$, then $[c, d] \subseteq (a_1, b_1)$, so $a_1 < c$ and $d < b_1$. Thus $d - c < b_1 - a_1$, as required.

Now suppose $n \geq 2$, and the result is true for covers by at most $n - 1$ open intervals. Suppose

$$[c, d] \subseteq \bigcup_{i=1}^{n} (a_i, b_i).$$

If some interval (a_i, b_i) is disjoint from $[c, d]$, it may be omitted from the cover; then we have a cover by at most $n - 1$ sets, so we would be finished by the induction hypothesis. So assume $(a_i, b_i) \cap [c, d] \neq \varnothing$ for all i. Among all of the left endpoints a_i, there is one that is no larger than any of the others. By renumbering the intervals, let us assume that it is a_1. Since c is covered, we must have $a_1 < c$. Now if $b_1 > d$, we have $d - c < b_1 - a_1 \leq \sum_{i=1}^{n}(b_i - a_i)$, so we are finished. So suppose $b_1 \leq d$. Since (a_1, b_1) intersects $[c, d]$, we have $b_1 > c$. So $b_1 \in [c, d]$. At least one of the open intervals (a_i, b_i) covers the point b_1. By renumbering, we may assume it is (a_2, b_2). Finally, we have a cover of $[c, d]$ by $n - 1$ sets:

$$[c, d] \subseteq (a_1, b_2) \cup \bigcup_{i=3}^{n}(a_i, b_i).$$

So by the induction hypothesis,

$$d - c < (b_2 - a_1) + \sum_{i=3}^{n}(b_i - a_i)$$

$$\leq (b_2 - a_2) + (b_1 - a_1) + \sum_{i=3}^{n}(b_i - a_i)$$

as required. This completes the proof by induction.   □

A useful generalization of the notion of the length of a subset of $\mathbb{R}$ is the **Lebesgue measure** of the set. This will be defined in stages. We will use half-open intervals of the form $[a, b)$ in the definition. Intervals of other forms could be used instead, but these have been chosen because of this convenient property:

Lemma 5.1.2. *Let $a < b$ be real numbers, and $\varepsilon > 0$. Then $[a, b)$ can be written as a finite disjoint union*

$$[a, b) = \bigcup_{i=1}^{n}[a_i, b_i),$$

with $b - a = \sum_{i=1}^{n}(b_i - a_i)$ and $b_i - a_i \leq \varepsilon$ for all i.

Proof. Choose $n \in \mathbb{N}$ so large that $(b - a)/n \leq \varepsilon$. Let $b_i = a + i(b - a)/n$ for $0 \leq i \leq n$, and $a_i = b_{i-1}$.   □

Now let A be any subset of $\mathbb{R}$. The **Lebesgue outer measure** of A is obtained by covering A with countably many half-open intervals of total length as small as possible. In symbols,[*]

$$\overline{\mathcal{L}}(A) = \inf \sum_{j=1}^{\infty}(b_j - a_j)$$

where the infimum is over all countable families $\{[a_j, b_j) : j \in \mathbb{N}\}$ of half-open intervals with $A \subseteq \bigcup_{j \in \mathbb{N}}[a_j, b_j)$.

---

[*] In case you can't tell, the symbol $\mathcal{L}$ is supposed to be a fancy letter L, for "Lebesgue".

Lemma 5.1.3. *Let $A \subseteq \mathbb{R}$ and let $\varepsilon > 0$. Then*

$$\overline{\mathcal{L}}(A) = \inf \sum_{j=1}^{\infty} (b_j - a_j)$$

where the infimum is over all countable families $\{ [a_j, b_j) : j \in \mathbb{N} \}$ of half-open intervals with $A \subseteq \bigcup_{j \in \mathbb{N}} [a_j, b_j)$ and $b_j - a_j \leq \varepsilon$ for all j.

Proof. This follows from Lemma 5.1.2.   □

We must do some combinatorics on the line to see that the definition is not trivial.*

Theorem 5.1.4. *If A is an interval, then $\overline{\mathcal{L}}(A)$ is the length of A.*

Proof. Suppose $A = [a, b]$, where $a < b$ are real numbers. First, if $\varepsilon > 0$, then the singleton $\{[a, b + \varepsilon)\}$ covers the set A, so $\overline{\mathcal{L}}(A) \leq b - a + \varepsilon$. This is true for any $\varepsilon > 0$, so $\overline{\mathcal{L}}(A) \leq b - a$.

Now suppose $A \subseteq \bigcup_{j \in \mathbb{N}} [a_j, b_j)$. Let $\varepsilon > 0$, and write $a_j' = a_j - \varepsilon/2^j$. Then $A \subseteq \bigcup_{j \in \mathbb{N}} (a_j', b_j)$. By Lemma 5.1.1, $\sum_{j=1}^{\infty} (b_j - a_j') > b - a$. So we have $\sum_{j=1}^{\infty} (b_j - a_j) \geq \sum_{j=1}^{\infty} (b_j - a_j') - \varepsilon > b - a - \varepsilon$. This is true for any $\varepsilon > 0$, so $\sum_{j=1}^{\infty} (b_j - a_j) \geq b - a$. Therefore $\overline{\mathcal{L}}(A) \geq b - a$. So we have $\overline{\mathcal{L}}([a, b]) = b - a$.

Next consider $A = (a, b)$. Then $\overline{\mathcal{L}}(A) \leq \overline{\mathcal{L}}([a, b]) = b - a$ and on the other hand $\overline{\mathcal{L}}(A) \geq \overline{\mathcal{L}}([a + \varepsilon, b - \varepsilon]) = b - a - 2\varepsilon$ for any $\varepsilon > 0$. Similar arguments cover cases $[a, b)$ and $(a, b]$. If $A = [a, \infty)$, then $A \supseteq [a, a + t]$ for any $t > 0$, and therefore $\overline{\mathcal{L}}(A) \geq t$; this means that $\overline{\mathcal{L}}(A) = \infty$. Similar arguments cover the other cases of infinite length intervals.   □

Here are some of the basic properties of Lebesgue outer measure.

Theorem 5.1.5. (1) $\overline{\mathcal{L}}(\varnothing) = 0$;
(2) *if $A \subseteq B$, then $\overline{\mathcal{L}}(A) \leq \overline{\mathcal{L}}(B)$;*
(3) $\overline{\mathcal{L}} \left(\bigcup_{n \in \mathbb{N}} A_n \right) \leq \sum_{n=1}^{\infty} \overline{\mathcal{L}}(A_n)$.

Proof. For (1), note that $\varnothing \subseteq \bigcup_{i \in \mathbb{N}} [0, \varepsilon/2^i)$, so $\overline{\mathcal{L}}(\varnothing) \leq \varepsilon$. For (2), note that any cover of B is also a cover of A.

Now consider (3). If $\overline{\mathcal{L}}(A_n) = \infty$ for some n, then the inequality is clear. So suppose $\overline{\mathcal{L}}(A_n) < \infty$ for all n. Let $\varepsilon > 0$. For each n, choose a countable cover $\mathcal{D}_n$ of A_n by half-open intervals with

$$\sum_{D \in \mathcal{D}_n} \overline{\mathcal{L}}(D) \leq \overline{\mathcal{L}}(A_n) + 2^{-n}\varepsilon.$$

Now $\mathcal{D} = \bigcup_{n \in \mathbb{N}} \mathcal{D}_n$ is a countable cover of the union $\bigcup_{n \in \mathbb{N}} A_n$. Therefore

---

* I can easily write down the same definition for subsets of the rational numbers. But then every set turns out to have outer measure 0, so it is not a very useful definition.

$$\overline{\mathcal{L}}\left(\bigcup_{n\in\mathbb{N}} A_n\right) \le \sum_{D\in\mathcal{D}} \overline{\mathcal{L}}(D) \le \sum_{n=1}^{\infty} \sum_{D\in\mathcal{D}_n} \overline{\mathcal{L}}(D)$$

$$\le \sum_{n=1}^{\infty} \overline{\mathcal{L}}(A_n) + \sum_{n=1}^{\infty} 2^{-n}\varepsilon = \sum_{n=1}^{\infty} \overline{\mathcal{L}}(A_n) + \varepsilon.$$

Since ε was any positive number, we have

$$\overline{\mathcal{L}}\left(\bigcup_{n\in\mathbb{N}} A_n\right) \le \sum_{n=1}^{\infty} \overline{\mathcal{L}}(A_n). \qquad \square$$

In general, the inequality in part (3) is not equality, even for two disjoint sets. But we do have equality in some cases. The simplest case is the following:

Theorem 5.1.6. *Let* $A, B \subseteq \mathbb{R}$ *with* $\operatorname{dist}(A, B) > 0$. *Then* $\overline{\mathcal{L}}(A \cup B) = \overline{\mathcal{L}}(A) + \overline{\mathcal{L}}(B)$.

Proof. First, the inequality $\overline{\mathcal{L}}(A \cup B) \le \overline{\mathcal{L}}(A) + \overline{\mathcal{L}}(B)$ follows from part (3) of Theorem 5.1.5. Let $\varepsilon = \operatorname{dist}(A, B)/2$, and let $A \cup B \subseteq \bigcup_{j\in\mathbb{N}}[a_j, b_j)$, where $b_j - a_j \le \varepsilon$ for all j. Then each interval $[a_j, b_j)$ intersects at most one of the sets A and B. So the collection $\mathcal{D} = \{\, [a_j, b_j) : j \in \mathbb{N} \,\}$ can be written as the disjoint union of two collections, $\mathcal{D} = \mathcal{D}_1 \cup \mathcal{D}_2$, where $\mathcal{D}_1$ covers A and $\mathcal{D}_2$ covers B. Now $\overline{\mathcal{L}}(A) \le \sum_{D\in\mathcal{D}_1} \overline{\mathcal{L}}(D)$ and $\overline{\mathcal{L}}(B) \le \sum_{D\in\mathcal{D}_2} \overline{\mathcal{L}}(D)$, so

$$\overline{\mathcal{L}}(A) + \overline{\mathcal{L}}(B) \le \sum_{D\in\mathcal{D}_1} \overline{\mathcal{L}}(D) + \sum_{D\in\mathcal{D}_2} \overline{\mathcal{L}}(D) = \sum_{D\in\mathcal{D}} \overline{\mathcal{L}}(D) \le \sum_{j=1}^{\infty} b_j - a_j.$$

Therefore, by Lemma 5.1.3, we have $\overline{\mathcal{L}}(A) + \overline{\mathcal{L}}(B) \le \overline{\mathcal{L}}(A \cup B)$. $\qquad \square$

Corollary 5.1.7. *If* $A, B \subseteq \mathbb{R}$ *are disjoint and compact, then we have* $\overline{\mathcal{L}}(A) + \overline{\mathcal{L}}(B) = \overline{\mathcal{L}}(A \cup B)$.

Proof. Apply Theorems 2.3.19 and 5.1.6. $\qquad \square$

Theorem 5.1.8. *If* $A \subseteq \mathbb{R}$, *then*

$$\overline{\mathcal{L}}(A) = \inf\{\, \overline{\mathcal{L}}(U) : U \supseteq A, \ U \ open \,\}.$$

Proof. Certainly $\overline{\mathcal{L}}(A) \le \inf\{\, \overline{\mathcal{L}}(U) : U \supseteq A, \ U \ open \,\}$. So I must prove the opposite inequality. If $\overline{\mathcal{L}}(A) = \infty$, it is trivially true. So suppose $\overline{\mathcal{L}}(A) < \infty$. Let $\varepsilon > 0$. Then there exists a cover $\bigcup_{j\in\mathbb{N}}[a_j, b_j)$ of A with $\sum_{j=1}^{\infty}(b_j - a_j) \le \overline{\mathcal{L}}(A) + \varepsilon/2$. Now the set $U = \bigcup_{j\in\mathbb{N}}(a_j - \varepsilon/2^{j+1}, b_j)$ is open, $U \supseteq A$, and $\overline{\mathcal{L}}(U) \le \sum_{j=1}^{\infty}(b_j - a_j) + \varepsilon/2 \le \overline{\mathcal{L}}(A) + \varepsilon$. Therefore $\overline{\mathcal{L}}(A) + \varepsilon \ge \overline{\mathcal{L}}(U)$. This shows that $\overline{\mathcal{L}}(A) \ge \inf\{\, \overline{\mathcal{L}}(U) : U \supseteq A \,\}$. $\qquad \square$

The outer measure $\overline{\mathcal{L}}(A)$ of a set $A \subseteq \mathbb{R}$ is determined by approximating a set from the outside by open sets. There is a corresponding "inner measure", obtained by approximating a set from the inside. This time, however, we will use compact sets.

Let $A \subseteq \mathbb{R}$. The **Lebesgue inner measure** of the set A is

$$\underline{\mathcal{L}}(A) = \sup \left\{ \overline{\mathcal{L}}(K) :\ K \subseteq A, K \text{ compact} \right\}.$$

Again, we need an argument to see that the definition is interesting.

Theorem 5.1.9. *If A is an interval, then $\underline{\mathcal{L}}(A)$ is the length of A.*

Proof. We consider the case of an open interval $A = (a, b)$. Other kinds of intervals follow from this case as before.

If $K \subseteq A$ is compact, then K is covered by the single interval A, so that $\overline{\mathcal{L}}(K) \leq b - a$. Therefore $\underline{\mathcal{L}}(A) \leq b - a$. On the other hand, if $\varepsilon > 0$, then the set $[a + \varepsilon, b - \varepsilon]$ is compact, so $\underline{\mathcal{L}}(A) \geq \overline{\mathcal{L}}([a + \varepsilon, b - \varepsilon]) = b - a - 2\varepsilon$. This is true for any $\varepsilon > 0$, so $\underline{\mathcal{L}}(A) \geq b - a$. $\square$

Exercise 5.1.10. If $A \subseteq \mathbb{R}$ is any set, then $\underline{\mathcal{L}}(A) \leq \overline{\mathcal{L}}(A)$.

It is not possible to prove that $\underline{\mathcal{L}}(A) = \overline{\mathcal{L}}(A)$ in general. A set A is called **Lebesgue measurable**, roughly speaking, when this equation is true. Precisely: If $\overline{\mathcal{L}}(A) < \infty$, then A is Lebesgue measurable iff $\underline{\mathcal{L}}(A) = \overline{\mathcal{L}}(A)$. If $\overline{\mathcal{L}}(A) = \infty$, then A is Lebesgue measurable iff $A \cap [-n, n]$ is Lebesgue measurable for all $n \in \mathbb{N}$. If A is Lebesgue measurable, we will write $\mathcal{L}(A)$ for the common value of $\overline{\mathcal{L}}(A)$ and $\underline{\mathcal{L}}(A)$, and call it simply the **Lebesgue measure** of A. We will often say simply **measurable** when we mean Lebesgue measurable.

Theorem 5.1.11. *Let $A_1, A_2, \cdots$ be disjoint Lebesgue measurable sets. Then $\bigcup_n A_n$ is measurable, and $\mathcal{L}(\bigcup_n A_n) = \sum_n \mathcal{L}(A_n)$.*

Proof. It is enough to prove the theorem in the case that $\overline{\mathcal{L}}(\bigcup A_n) < \infty$, since the general case will then follow by applying this case to sets $A_n \cap [-m, m]$. We know by Theorem 5.1.5 that $\overline{\mathcal{L}}(\bigcup A_n) \leq \sum \mathcal{L}(A_n)$. Let $\varepsilon > 0$. For each n, choose a compact set $K_n \subseteq A_n$ with $\overline{\mathcal{L}}(K_n) \geq \underline{\mathcal{L}}(A_n) - \varepsilon/2^n$. Since A_n is measurable, $\overline{\mathcal{L}}(K_n) \geq \overline{\mathcal{L}}(A_n) - \varepsilon/2^n$. Now the sets K_n are disjoint, so by Corollary 5.1.7, the compact set $L_m = K_1 \cup K_2 \cup \cdots \cup K_m$ satisfies $\overline{\mathcal{L}}(L_m) = \overline{\mathcal{L}}(K_1) + \cdots + \overline{\mathcal{L}}(K_m)$. Therefore $\underline{\mathcal{L}}(\bigcup A_n) \geq \sum_{n=1}^{m} \overline{\mathcal{L}}(K_n)$. Now this is true for all m, so $\underline{\mathcal{L}}(\bigcup A_n) \geq \sum_{n=1}^{\infty} \overline{\mathcal{L}}(K_n) \geq \sum_{n=1}^{\infty} \mathcal{L}(A_n) - \varepsilon$. This is true for any positive ε, so we have $\underline{\mathcal{L}}(\bigcup_n A_n) \geq \sum \mathcal{L}(A_n)$.

So $\overline{\mathcal{L}}(\bigcup A_n) = \underline{\mathcal{L}}(\bigcup A_n)$, and therefore $\bigcup A_n$ is measurable and $\mathcal{L}(\bigcup A_n) = \sum \mathcal{L}(A_n)$. $\square$

Theorem 5.1.12. *Compact subsets, closed subsets, and open subsets of $\mathbb{R}$ are Lebesgue measurable.*

Proof. Let $K \subseteq \mathbb{R}$ be compact. It is bounded, so $K \subseteq [-n, n]$ for some n, and therefore $\overline{\mathcal{L}}(K) < \infty$. The compact set K is a subset of K, so $\underline{\mathcal{L}}(K) \geq \overline{\mathcal{L}}(K)$.

Let $F \subseteq \mathbb{R}$ be a closed set. Then for each $n \in \mathbb{N}$, the intersection $F \cap [-n, n]$ is compact, and therefore measurable. Thus F is measurable.

Let U be an open set. It is enough to do the case $\overline{\mathcal{L}}(U) < \infty$. For each $x \in U$, there is an open interval I with $x \in I \subseteq U$. By the Lindelöf property, U is the union of countably many of these intervals, say $U = \bigcup_{j \in \mathbb{N}} I_j$. Now each set $I_n \setminus \bigcup_{j=1}^{n-1} I_j$ is a finite union of intervals (open, closed, half-open) so that U is a disjoint union of countably many intervals. So U is measurable.   □

Theorem 5.1.13. *Let $A \subseteq \mathbb{R}$. Then A is measurable if and only if, for every $\varepsilon > 0$, there exist an open set U and a closed set F with $U \supseteq A \supseteq F$ and $\mathcal{L}(U \setminus F) < \varepsilon$.*

Proof. Suppose first that A is measurable. We consider first of all the case $\mathcal{L}(A) < \infty$. Then there exists an open set $U \supseteq A$ such that $\mathcal{L}(U) < \mathcal{L}(A) + \varepsilon/2$. There exists a compact (therefore closed) set $F \subseteq A$ with $\mathcal{L}(F) > \mathcal{L}(A) - \varepsilon/2$. Now $U \setminus F$ is open, hence measurable, and F is compact, hence measurable, so $\mathcal{L}(U) = \mathcal{L}(U \setminus F) + \mathcal{L}(F)$. Since the terms are all finite, we may subtract, and we get

$$\mathcal{L}(U \setminus F) = \mathcal{L}(U) - \mathcal{L}(F) < \mathcal{L}(A) + \varepsilon/2 - \mathcal{L}(A) + \varepsilon/2 = \varepsilon.$$

Now we take the case $\mathcal{L}(A) = \infty$. All of the sets $A \cap [-n, n]$ are measurable, so there exist open sets $U_n \supseteq A \cap [-n, n]$ and compact sets $F_n \subseteq A \cap [-n, n]$ with $\mathcal{L}(U_n \setminus F_n) < \varepsilon/2^{n+2}$. Define $U_n' = U_n \cap \left((-\infty, -n + 1 + \varepsilon/2^{n+2}) \cup (n - 1 - \varepsilon/2^{n+2}, \infty) \right)$ and $F_n' = F_n \cap \left([-n, -n + 1] \cup [n - 1, n] \right)$, so that U_n' is open, F_n' is compact, $U_n' \supseteq A \cap \left([-n, -n + 1] \cup [n - 1, n] \right) \supseteq F_n'$ and $\mathcal{L}(U_n' \setminus F_n') < 3\varepsilon/2^{n+2} < \varepsilon/2^n$. Now $U = \bigcup U_n'$ is open, and $F = \bigcup F_n'$ is closed (Exercise 2.2.27). We have $U \supseteq A \supseteq F$, and $U \setminus F \subseteq \bigcup_{n \in \mathbb{N}} (U_n' \setminus F_n')$, so that $\mathcal{L}(U \setminus F) \leq \sum \mathcal{L}(U_n' \setminus F_n') < \varepsilon$.

Conversely, suppose sets U and F exist. By Theorem 5.1.12 they are measurable. First assume $\overline{\mathcal{M}}(A) < \infty$. Then $\mathcal{L}(F) < \infty$, and $\mathcal{L}(U) \leq \mathcal{L}(U \setminus F) + \mathcal{L}(F) < \varepsilon + \mathcal{L}(F) < \infty$. Now $\overline{\mathcal{M}}(A) \leq \overline{\mathcal{M}}(U) = \mathcal{L}(U) < \mathcal{L}(F) + \varepsilon = \underline{\mathcal{L}}(F) + \varepsilon \leq \underline{\mathcal{L}}(A) + \varepsilon$. This is true for any $\varepsilon > 0$, so $\overline{\mathcal{M}}(A) = \underline{\mathcal{L}}(A)$, so A is measurable.

For the case $\overline{\mathcal{L}}(A) = \infty$, we have $U \cap (-n - \varepsilon, n + \varepsilon) \supseteq A \cap [-n, n] \supseteq F \cap [-n, n]$, and the previous case may be applied to these sets, using 3ε in place of ε.   □

Here are the basic algebraic properties of Lebesgue measurable sets.

Theorem 5.1.14. (1) *Both $\varnothing$ and $\mathbb{R}$ are Lebesgue measurable.*
(2) *If $A \subseteq \mathbb{R}$ is Lebesgue measurable, then so is its complement $\mathbb{R} \setminus A$.*
(3) *If A and B are measurable, then so are $A \cap B$, $A \cup B$, and $A \setminus B$.*
(4) *If A_n is measurable for $n \in \mathbb{N}$, then so are $\bigcup_{n \in \mathbb{N}} A_n$ and $\bigcap_{n \in \mathbb{N}} A_n$.*

Proof. For (1), note that $\overline{\mathcal{L}}(\varnothing) = 0$ and $\mathbb{R} \cap [-n, n]$ is measurable for all n.

For (2), note that if $F \subseteq A \subseteq U$, then $\mathbb{R} \setminus U \subseteq \mathbb{R} \setminus A \subseteq \mathbb{R} \setminus F$ and $(\mathbb{R} \setminus F) \setminus (\mathbb{R} \setminus U) = U \setminus F$.

For the intersection in (3), note that if $F_1 \subseteq A \subseteq U_1$ and $F_2 \subseteq B \subseteq U_2$, then $F_1 \cap F_2 \subseteq A \cap B \subseteq U_1 \cap U_2$ and $(U_1 \cap U_2) \setminus (F_1 \cap F_2) \subseteq (U_1 \setminus F_1) \cup (U_2 \setminus F_2)$. This is enough to show that $A \cap B$ is measurable. Now $A \cup B = \mathbb{R} \setminus ((\mathbb{R} \setminus A) \cap (\mathbb{R} \setminus B))$, so $A \cup B$ is measurable. And $A \setminus B = A \cap (\mathbb{R} \setminus B)$, so $A \setminus B$ is measurable.

Finally, for (4), note that by (3) we may find disjoint measurable sets B_n with the same union as A_n, so that Theorem 5.1.11 is applicable. The intersection follows by taking complements.   □

Note that (4) involves only *countable* unions and intersections.

Proposition 5.1.15. *The Lebesgue measure of the triadic Cantor dust is 0.*

Proof. The Cantor dust C is constructed on p. 2. The set $C_n \supseteq C$ consists of 2^n disjoint intervals of length 3^{-n}. Therefore $\mathcal{L}(C) \le 2^n \cdot 3^{-n}$. This has limit 0, so $\mathcal{L}(C) = 0$.   □

Carathéodory Measurability

Carathéodory provided an alternate definition of measurability. Its disadvantage is that the motivation is not as clear. Its advantage is (as we will see in Sect. 5.2) that it can be used in other situations.

A set $A \subseteq \mathbb{R}$ is **Carathéodory measurable** iff

$$\overline{\mathcal{L}}(E) = \overline{\mathcal{L}}(E \cap A) + \overline{\mathcal{L}}(E \setminus A)$$

for all sets $E \subseteq \mathbb{R}$.

Proposition 5.1.16. *A set $A \subseteq \mathbb{R}$ is Carathéodory measurable if and only if it is Lebesgue measurable.*

Proof. Suppose A is Lebesgue measurable. Let E be a test set. The inequality $\overline{\mathcal{L}}(E) \le \overline{\mathcal{L}}(E \cap A) + \overline{\mathcal{L}}(E \setminus A)$ is always true. Let $\varepsilon > 0$. There exist an open set U and a closed set F with $F \subseteq A \subseteq U$ and $\mathcal{L}(U \setminus F) < \varepsilon$. Let $V \supseteq E$ be an open set. Then

$$\begin{aligned}
\overline{\mathcal{L}}(E \setminus A) + \overline{\mathcal{L}}(E \cap A) &\le \mathcal{L}(V \setminus F) + \mathcal{L}(V \cap U) \\
&\le \mathcal{L}(V \setminus U) + \mathcal{L}(U \setminus F) + \mathcal{L}(V \cap U) \\
&< \mathcal{L}(V) + \varepsilon.
\end{aligned}$$

Now take the infimum over all such V, to get $\overline{\mathcal{L}}(E \cap A) + \overline{\mathcal{L}}(E \setminus A) < \overline{\mathcal{L}}(E) + \varepsilon$. Therefore $\overline{\mathcal{L}}(E \cap A) + \overline{\mathcal{L}}(E \setminus A) \le \overline{\mathcal{L}}(E)$. This proves that A is Carathéodory measurable.

Conversely, suppose A is Carathéodory measurable. Consider the case in which $\overline{\mathcal{L}}(A) < \infty$. Let $\varepsilon > 0$. Let $U \supseteq A$ satisfy $\mathcal{L}(U) < \overline{\mathcal{L}}(A) + \varepsilon$. Now we have

$$\overline{\mathcal{L}}(U) = \overline{\mathcal{L}}(U \cap A) + \overline{\mathcal{L}}(U \setminus A),$$

so that $\overline{\mathcal{L}}(U \setminus A) < \varepsilon$. So there is an open set $V \supseteq U \setminus A$ with $\mathcal{L}(V) < \varepsilon$. Then $U \setminus V$ is Lebesgue measurable, and $\mathcal{L}(U \setminus V) > \mathcal{L}(U) - \varepsilon$, so there is a closed set $F \subseteq U \setminus V \subseteq A$ with $\mathcal{L}(F) > \mathcal{L}(U) - \varepsilon$. Thus $F \subseteq A \subseteq U$ and $\mathcal{L}(U \setminus F) < \varepsilon$. Therefore A is Lebesgue measurable. □

Theorem 5.1.17. *Let $A \subseteq \mathbb{R}$ be Lebesgue measurable, and let a similarity $f \colon \mathbb{R} \to \mathbb{R}$ with ratio r be given. Then $f[A]$ is Lebesgue measurable and $\mathcal{L}(f[A]) = r\mathcal{L}(A)$.*

Proof. [Strictly speaking, "similarity" disallows $r = 0$, but even if $r = 0$ is allowed, this formula still works: If $r = 0$, then the range of f is a single point, so of course $f[A]$ is measurable and $\mathcal{L}(f[A]) = 0$.] Now suppose $r > 0$.

Consider an interval $I = [a, b)$. The image is an interval, either $[f(a), f(b))$ or $(f(b), f(a)]$, with length $|f(b) - f(a)| = r|b - a|$. Therefore $\overline{\mathcal{L}}(f[I]) = r|b - a|$. Now if $A \subseteq \bigcup_{j \in \mathbb{N}}[a_j, b_j)$, then $f[A] \subseteq \bigcup f[[a_j, b_j)]$, so $\overline{\mathcal{L}}(f[A]) \leq \sum \overline{\mathcal{L}}\big(f[[a_j, b_j)]\big) = r\sum(b_j - a_j)$. Therefore we have $\overline{\mathcal{L}}(f[A]) \leq r\overline{\mathcal{L}}(A)$. If we apply the same thing to the inverse map f^{-1}, which is a similarity with ratio $1/r$, we get $\overline{\mathcal{L}}(f[A]) \geq r\overline{\mathcal{L}}(A)$. Therefore $\overline{\mathcal{L}}(f[A]) = r\overline{\mathcal{L}}(A)$

Now f is a homeomorphism, so the image of an open set is open and the image of a closed set is closed. If $A \subseteq \mathbb{R}$ is measurable, then, for every $\varepsilon > 0$, there exist a closed set F and an open set U with $F \subseteq A \subseteq U$ and $\overline{\mathcal{L}}(U \setminus F) < \varepsilon$. So we have $f[F] \subseteq f[A] \subseteq f[U]$ and $\overline{\mathcal{L}}(f[U] \setminus f[F]) = \overline{\mathcal{L}}\big(f[U \setminus F]\big) < r\varepsilon$. So $f[A]$ is also measurable. □

Next is a preview of how measure theory is related to fractal dimension. In general, we do not yet know that the similarity dimension of a set is unique. However, we can now establish that in one situation we can determine the similarity dimension.

Exercise 5.1.18. Let $(r_1, r_2, \cdots, r_n)$ be a contracting ratio list; let s be its sim-value; let $(f_1, f_2, \cdots, f_n)$ be a corresponding iterated function system in $\mathbb{R}$; and let $A \subseteq \mathbb{R}$ be a nonempty measurable set. Suppose $\mathcal{L}(f_j[A] \cap f_k[A]) = 0$ for $j \neq k$, and

$$A = \bigcup_{j=1}^{n} f_j[A].$$

If $0 < \mathcal{L}(A) < \infty$, then $s = 1$.

Number Systems

Recall the situation from Sect. 1.6. Let b be a real number, and let D be a finite set of real numbers, including 0. We are interested in representing real numbers in the number system they define.

Write W for the set of "whole numbers"; that is, numbers of the form

$$\sum_{j=0}^{M} a_j b^j.$$

Write F for the set of "fractions"; that is numbers of the form

$$\sum_{j=-\infty}^{-1} a_j b^j.$$

We know (Proposition 3.2.21) that there is no number system that has a unique representation for every real number. So we will try to represent all real numbers, and arrange to have as few numbers as possible with more than one representation. One way to specify that the set with multiple representations is small is to require that it have Lebesgue measure 0.

If we analyze the size of the intersections $(F+w_1) \cap (F+w_2)$, $w_1, w_2 \in W$, $w_1 \neq w_2$, then we will know about all numbers with multiple representations:

Exercise 5.1.19. The set of all numbers with multiple representations is a countable union of sets, each of which is similar to one of the sets $(F+w_1) \cap (F+w_2)$, $w_1, w_2 \in W, w_1 \neq w_2$.

The set of all numbers that can be represented is $F + W$, a countable union of sets isometric to F. So if all real numbers can be represented, then $\mathcal{L}(F) > 0$. We know that F is a compact set, so also $\mathcal{L}(F) < \infty$. If the set of all real numbers with multiple representations has Lebesgue measure 0, then the sets $(F+w_1) \cap (F+w_2)$ have Lebesgue measure 0.

Suppose D has k elements. If F has positive Lebesgue measure, but the intersections $(F+w_1) \cap (F+w_2)$ have Lebesgue measure zero, then by Exercise 5.1.18, F has similarity dimension 1. But the similarity dimension is actually $\log k / \log |b|$. Therefore $|b| = k$.

5.2 Method I

We will need to discuss measures other than Lebesgue measure. The basics are contained in this section.

Measures and Outer Measures

A collection $\mathcal{F}$ of subsets of a set X is called an **algebra** on X iff:

(1) $\emptyset, X \in \mathcal{F}$;
(2) if $A \in \mathcal{F}$, then $X \setminus A \in \mathcal{F}$;
(3) if $A, B \in \mathcal{F}$, then $A \cup B \in \mathcal{F}$.

Note that $A \cap B = X \setminus ((X \setminus A) \cup (X \setminus B))$ and $A \setminus B = A \cap (X \setminus B)$, so an algebra is also closed under these two operations.

A collection $\mathcal{F}$ of subsets of a set X is called a σ-**algebra** on X iff:

(1) $\varnothing, X \in \mathcal{F}$;
(2) if $A \in \mathcal{F}$, then $X \setminus A \in \mathcal{F}$;
(3) if $A_1, A_2, \cdots \in \mathcal{F}$, then $\bigcup_{i \in \mathbb{N}} A_i \in \mathcal{F}$.

Of course (by Theorem 5.1.14), the collection of all Lebesgue measurable subsets of $\mathbb{R}$ is a σ-algebra on $\mathbb{R}$. Combining the clauses of the definition will produce a few more rules: For example, if $A_1, A_2, \cdots \in \mathcal{F}$, then $\bigcap_{i \in \mathbb{N}} A_i \in \mathcal{F}$; if $A, B \in \mathcal{F}$ then $A \cap B, A \cup B, A \setminus B \in \mathcal{F}$.

Theorem 5.2.1. *Let X be a set, and let $\mathcal{D}$ be any set of subsets of X. Then there is a set $\mathcal{F}$ of subsets of X such that*

(1) $\mathcal{F}$ *is a σ-algebra on X;*
(2) $\mathcal{F} \supseteq \mathcal{D}$;
(3) *if $\mathcal{G}$ is any σ-algebra on X with $\mathcal{G} \supseteq \mathcal{D}$, then $\mathcal{G} \supseteq \mathcal{F}$.*

Proof. First I claim that the intersection of any collection of σ-algebras on X is a σ-algebra. Let Γ be a collection of σ-algebras, and let $\mathcal{B} = \bigcap_{\mathcal{A} \in \Gamma} \mathcal{A}$ be the intersection. Then $\varnothing \in \mathcal{A}$ for all $\mathcal{A} \in \Gamma$, so $\varnothing \in \mathcal{B}$. Similarly $X \in \mathcal{B}$. If $A \in \mathcal{B}$, then $A \in \mathcal{A}$ for all $\mathcal{A} \in \Gamma$, so $X \setminus A \in \mathcal{A}$ for all $\mathcal{A} \in \Gamma$, and therefore $X \setminus A \in \mathcal{B}$. If $A_1, A_2, \cdots \in \mathcal{B}$, then each $A_n \in \mathcal{A}$ for all $\mathcal{A} \in \Gamma$, so $\bigcup_{n \in \mathbb{N}} A_n \in \mathcal{A}$ for all $\mathcal{A} \in \Gamma$ and therefore $\bigcup_{n \in \mathbb{N}} A_n \in \mathcal{B}$.

So suppose a set $\mathcal{D}$ of subsets of X is given. Let Γ be the collection of all σ-algebras $\mathcal{G}$ on X with $\mathcal{G} \supseteq \mathcal{D}$. (There is at least one such σ-algebra, namely the family of all subsets of X.) Then the intersection $\mathcal{F} = \bigcap_{\mathcal{G} \in \Gamma} \mathcal{G}$ is a σ-algebra on X. But clearly if $\mathcal{G}$ is any σ-algebra on X with $\mathcal{G} \supseteq \mathcal{D}$, then $\mathcal{G} \in \Gamma$, and therefore $\mathcal{G} \supseteq \mathcal{F}$. $\qquad\square$

We say that $\mathcal{F}$ is the **least σ-algebra** containing $\mathcal{D}$, or the σ-algebra **generated** by $\mathcal{D}$. Let S be a metric space. A subset of S is called a **Borel set** iff it belongs to the σ-algebra on S generated by the open sets.

Let X be a set, and let $\mathcal{F}$ be a σ-algebra of subsets of X. A **measure** on $\mathcal{F}$ is a set function* $\mathcal{M} \colon \mathcal{F} \to [0, \infty]$ such that:

(1) $\mathcal{M}(\varnothing) = 0$;
(2) If $A_n \in \mathcal{F}$ is a disjoint sequence of sets, then

$$\mathcal{M}\left(\bigcup_{n \in \mathbb{N}} A_n\right) = \sum_{n=1}^{\infty} \mathcal{M}(A_n).$$

We call (2) **countable additivity**.

Let X be a set. An **outer measure** on X is a function $\overline{\mathcal{M}}$ defined on all subsets of X, with values in the nonnegative extended real numbers $[0, \infty]$, satisfying:

---

* A **set function** is a function whose domain is a family of sets.

(1) $\overline{\mathcal{M}}(\varnothing) = 0$;
(2) if $A \subseteq B$, then $\overline{\mathcal{M}}(A) \leq \overline{\mathcal{M}}(B)$;
(3) $\overline{\mathcal{M}}\left(\bigcup_{n \in \mathbb{N}} A_n\right) \leq \sum_{n=1}^{\infty} \overline{\mathcal{M}}(A_n)$.

We call (3) *countable subadditivity.*

Defining Outer Measures

The Lebesgue outer measure was constructed in Sect. 5.1. The way in which the definition was formulated was not accidental. We will explore a general method for constructing outer measures, known as "method I." We begin with candidate values for the measures of some sets (like the lengths of the half-open intervals), and then attempt to produce an outer measure that is as large as possible, but no larger than the candidate values.

Let X be a set, and let $\mathcal{A}$ be a family of subsets of X that covers X. Let $\mathbf{c} \colon \mathcal{A} \to [0, \infty]$ be any function. The theorem on construction of outer measures is as follows:

Theorem 5.2.2 (Method I Theorem). *There is a unique outer measure $\overline{\mathcal{M}}$ on X such that*

(1) $\overline{\mathcal{M}}(A) \leq \mathbf{c}(A)$ *for all $A \in \mathcal{A}$;*
(2) *if $\overline{\mathcal{N}}$ is any outer measure on X with $\overline{\mathcal{N}}(A) \leq \mathbf{c}(A)$ for all $A \in \mathcal{A}$, then $\overline{\mathcal{N}}(B) \leq \overline{\mathcal{M}}(B)$ for all $B \subseteq X$.*

Proof. The uniqueness is easy: if two outer measures satisfy (1) and (2), then each is $\leq$ the other, so they are equal.

For any subset B of X, define

$$\overline{\mathcal{M}}(B) = \inf \sum_{A \in \mathcal{D}} \mathbf{c}(A), \tag{I}$$

where the infimum is over all countable covers $\mathcal{D}$ of B by sets of $\mathcal{A}$. (Recall that $\inf \varnothing = \infty$, so if there is no countable cover at all of B by sets of $\mathcal{A}$, then $\overline{\mathcal{M}}(B) = \infty$.)

I claim first that $\overline{\mathcal{M}}$ is an outer measure. First, $\overline{\mathcal{M}}(\varnothing) = 0$, since the empty set is covered by the empty cover, and the empty sum has value 0. If $B \subseteq C$, then any cover of C is also a cover of B, so $\overline{\mathcal{M}}(B) \leq \overline{\mathcal{M}}(C)$. Let $B_1, B_2, \cdots$ be given. I must prove

$$\overline{\mathcal{M}}\left(\bigcup_{n \in \mathbb{N}} B_n\right) \leq \sum_{n=1}^{\infty} \overline{\mathcal{M}}(B_n).$$

If $\overline{\mathcal{M}}(B_n) = \infty$ for some n, then the inequality is clear. So suppose $\overline{\mathcal{M}}(B_n) < \infty$ for all n. Let $\varepsilon > 0$. For each n, choose a countable cover $\mathcal{D}_n$ of B_n by sets of $\mathcal{A}$ with

$$\sum_{A \in \mathcal{D}_n} \mathbf{c}(A) \leq \overline{\mathcal{M}}(B_n) + 2^{-n}\varepsilon.$$

Now $\mathcal{D} = \bigcup_{n \in \mathbb{N}} \mathcal{D}_n$ is a countable cover of the union $\bigcup_{n \in \mathbb{N}} B_n$. Therefore

$$\overline{\mathcal{M}}\left(\bigcup_{n \in \mathbb{N}} B_n\right) \leq \sum_{A \in \mathcal{D}} \mathbf{c}(A)$$

$$\leq \sum_{n=1}^{\infty} \sum_{A \in \mathcal{D}_n} \mathbf{c}(A)$$

$$\leq \sum_{n=1}^{\infty} \overline{\mathcal{M}}(B_n) + \sum_{n=1}^{\infty} 2^{-n}\varepsilon$$

$$= \sum_{n=1}^{\infty} \overline{\mathcal{M}}(B_n) + \varepsilon.$$

Since ε was any positive number, we have

$$\overline{\mathcal{M}}\left(\bigcup_{n \in \mathbb{N}} B_n\right) \leq \sum_{n=1}^{\infty} \overline{\mathcal{M}}(B_n).$$

This completes the proof that $\overline{\mathcal{M}}$ is an outer measure.

Now we may check the two assertions of the theorem. For (1), note that for $A \in \mathcal{A}$, the singleton $\{A\}$ is a cover of A, so

$$\overline{\mathcal{M}}(A) \leq \sum_{B \in \{A\}} \mathbf{c}(B) = \mathbf{c}(A).$$

For (2), suppose that $\overline{\mathcal{N}}$ is any outer measure on X with $\overline{\mathcal{N}}(A) \leq \mathbf{c}(A)$ for all $A \in \mathcal{A}$. Then for any countable cover $\mathcal{D}$ of a set B by elements of $\mathcal{A}$ we have

$$\sum_{A \in \mathcal{D}} \mathbf{c}(A) \geq \sum_{A \in \mathcal{D}} \overline{\mathcal{N}}(A) \geq \overline{\mathcal{N}}\left(\bigcup_{A \in \mathcal{D}} A\right) \geq \overline{\mathcal{N}}(B).$$

Therefore $\overline{\mathcal{M}}(B) \geq \overline{\mathcal{N}}(B)$. □

When we say that an outer measure is to be constructed by **method I**, we are referring to this theorem. In practical terms, this means that the outer measure is defined by the formula (I).

Reduced Cover Classes

When a measure is defined by method I, it may be helpful to know that the covers $\mathcal{D}$ in (I) can be chosen from a smaller ("reduced") class of sets.

Proposition 5.2.3. *Let X be a set, and let $\mathbf{c}$ be a set function. For a collection of sets $\mathcal{A}$, let $\overline{\mathcal{M}}_{\mathcal{A}}$ be the method I outer measure defined using the class $\mathcal{A}$ of sets and the restriction of $\mathbf{c}$ to $\mathcal{A}$.*

(a) *If $B \subseteq A$, then $\overline{\mathcal{M}}_A \leq \overline{\mathcal{M}}_B$.*
(b) *Suppose that, for every $A \in \mathcal{A}$ and every $\varepsilon > 0$, there is $B \in \mathcal{B}$ with $B \supseteq A$ and $\mathbf{c}(B) \leq \mathbf{c}(A) + \varepsilon$. Then $\overline{\mathcal{M}}_B \leq \overline{\mathcal{M}}_A$.*
(c) *Let $C > 0$ be a constant, and suppose that, for every $A \in \mathcal{A}$, there is $B \in \mathcal{B}$ with $B \supseteq A$ and $\mathbf{c}(B) \leq C\,\mathbf{c}(A)$. Then $\overline{\mathcal{M}}_B \leq C\,\overline{\mathcal{M}}_A$.*

Proof. (a) The outer measure $\overline{\mathcal{M}}_B$ is the largest outer measure such that $\overline{\mathcal{M}}_B(E) \leq \mathbf{c}(E)$ for all $E \in \mathcal{B}$. But $\overline{\mathcal{M}}_A$ also has this property, so $\overline{\mathcal{M}}_A \leq \overline{\mathcal{M}}_B$.

(b) Let $\varepsilon > 0$ be given. Let $\mathcal{D} = \{A_1, A_2, \cdots\} \subseteq \mathcal{A}$ be a countable cover of a set E. For each A_j, choose $B_j \in \mathcal{B}$ with $B_j \supseteq A_j$ and $\mathbf{c}(B_j) \leq \mathbf{c}(A_j) + \varepsilon/2^j$. Then $\mathcal{D}' = \{B_1, B_2, \cdots\}$ is also a cover of E, and

$$\varepsilon + \sum_j \mathbf{c}(A_j) \geq \sum_j \mathbf{c}(B_j) \geq \overline{\mathcal{M}}_B(E).$$

Take the infimum over all countable $\mathcal{D} \subseteq \mathcal{A}$ that cover E to get

$$\varepsilon + \overline{\mathcal{M}}_A(E) \geq \overline{\mathcal{M}}_B(E).$$

This is true for all $\varepsilon > 0$, so $\overline{\mathcal{M}}_A(E) \geq \overline{\mathcal{M}}_B(E)$.

(c) Let $\mathcal{D} = \{A_1, A_2, \cdots\} \subseteq \mathcal{A}$ be a countable cover of a set E. For each A_j, choose $B_j \in \mathcal{B}$ with $B_j \supseteq A_j$ and $\mathbf{c}(B_j) \leq C\,\mathbf{c}(A_j)$. Then $\mathcal{D}' = \{B_1, B_2, \cdots\}$ is also a cover of E, and

$$C \sum_j \mathbf{c}(A_j) \geq \sum_j \mathbf{c}(B_j) \geq \overline{\mathcal{M}}_B(E).$$

Take the infimum over all countable $\mathcal{D} \subseteq \mathcal{A}$ that cover E to get

$$C\,\overline{\mathcal{M}}_A(E) \geq \overline{\mathcal{M}}_B(E). \qquad \square$$

When condition (b) holds, we will say that $\mathcal{B}$ is a **reduced cover class** for $\overline{\mathcal{M}}$. When condition (c) holds, we will say that $\mathcal{B}$ is a reduced cover class with factor C for $\overline{\mathcal{M}}$.

Here is an example. Lebesgue measure $\mathcal{L}$ on $\mathbb{R}$ is defined (p. 140) using the class of all intervals $[a, b)$. The **semi-dyadic intervals** are sets of the form $[(k-1)/2^n, (k+1)/2^n)$, where $n \in \mathbb{Z}$, $k \in \mathbb{Z}$. The class of all semi-dyadic intervals is a reduced cover class with factor 4 for Lebesgue measure. Indeed, if $a < b$, let n be the integer with $2^{-n-1} < b - a < 2^{-n}$ and k the integer with $k - 1 \leq a/2^n < k$, and compute $[k/2^n, (k+1)/2^n) \supseteq [b-a)$ and $(k+1)/2^n - (k-1)/2^n < 4(b-a)$.

The **dyadic net** is the class of intervals of the form $[k/2^n, (k+1)/2^n)$. It is not a reduced cover class by itself. Let $\mathcal{R}_n$ consist of the finite disjoint unions of dyadic intervals $[k/2^n, (k+1)/2^n)$ with denominator 2^n. The **dyadic ring** is $\mathcal{R} = \bigcup_n \mathcal{R}_n$.

Proposition 5.2.4. *Using the set function $\mathbf{c} \colon \mathcal{R} \to [0, \infty)$ defined by $\mathbf{c}(E) = \mathcal{L}(E)$, the dyadic ring $\mathcal{R}$ is a reduced cover class for Lebesgue measure.*

Proof. Let $a < b$ and let $\varepsilon > 0$. Let $n \in \mathbb{N}$ be so large that $2^{-n} < \varepsilon/2$. Let $j \in \mathbb{Z}$ be such that $j \le 2^n a < j+1$, and $m \ge j$ such that $m \le 2^n b < m+1$. Then

$$[a, b) \subseteq E := \bigcup_{k=j}^{m} \left[\frac{k}{2^n}, \frac{k+1}{2^n} \right),$$

and

$$\mathcal{L}(E) - (b-a) \le \mathcal{L}\left(\left[\frac{j}{2^n}, a \right) \right) + \mathcal{L}\left(\left[b, \frac{m+1}{2^n} \right) \right) \le \frac{2}{2^n} < \varepsilon. \qquad \square$$

Measurable Sets

Let $\overline{\mathcal{M}}$ be an outer measure on a set X. A set $A \subseteq X$ is $\overline{\mathcal{M}}$-*measurable* (in the sense of Carathéodory) iff $\overline{\mathcal{M}}(E) = \overline{\mathcal{M}}(E \cap A) + \overline{\mathcal{M}}(E \setminus A)$ for all sets $E \subseteq X$.

Theorem 5.2.5. *The collection $\mathcal{F}$ of $\overline{\mathcal{M}}$-measurable sets is a σ-algebra, and $\overline{\mathcal{M}}$ is countably additive on $\mathcal{F}$.*

Proof. First, $\varnothing \in \mathcal{F}$ since for any E, we have $\overline{\mathcal{M}}(E \cap \varnothing) + \overline{\mathcal{M}}(E \setminus \varnothing) = \overline{\mathcal{M}}(\varnothing) + \overline{\mathcal{M}}(E) = \overline{\mathcal{M}}(E)$. It is also easy to see that a set A belongs to $\mathcal{F}$ if and only if its complement $X \setminus A$ does.

Suppose $A_j \in \mathcal{F}$ for $j = 1, 2, \cdots$. Let E be any test set. Then

$$\overline{\mathcal{M}}(E) = \overline{\mathcal{M}}(E \cap A_1) + \overline{\mathcal{M}}(E \setminus A_1)$$
$$= \overline{\mathcal{M}}(E \cap A_1) + \overline{\mathcal{M}}((E \setminus A_1) \cap A_2) + \overline{\mathcal{M}}(E \setminus (A_1 \cup A_2))$$
$$= \cdots$$
$$= \sum_{j=1}^{k} \overline{\mathcal{M}}\left(\left(E \setminus \bigcup_{i=1}^{j-1} A_i \right) \cap A_j \right) + \overline{\mathcal{M}}\left(E \setminus \bigcup_{j=1}^{k} A_j \right).$$

Hence

$$\overline{\mathcal{M}}(E) \ge \sum_{j=1}^{k} \overline{\mathcal{M}}\left(\left(E \setminus \bigcup_{i=1}^{j-1} A_i \right) \cap A_j \right) + \overline{\mathcal{M}}\left(E \setminus \bigcup_{j \in \mathbb{N}} A_j \right),$$

so (let $k \to \infty$)

$$\overline{\mathcal{M}}(E) \ge \sum_{j=1}^{\infty} \overline{\mathcal{M}}\left(\left(E \setminus \bigcup_{i=1}^{j-1} A_i \right) \cap A_j \right) + \overline{\mathcal{M}}\left(E \setminus \bigcup_{j \in \mathbb{N}} A_j \right).$$

But

$$E \cap \bigcup_{j \in \mathbb{N}} A_j = \bigcup_{j \in \mathbb{N}} \left(\left(E \setminus \bigcup_{i=1}^{j-1} A_i \right) \cap A_j \right),$$

so

$$\overline{\mathcal{M}}(E) \le \overline{\mathcal{M}}\left(E \cap \bigcup_{j \in \mathbb{N}} A_j\right) + \overline{\mathcal{M}}\left(E \setminus \bigcup_{j \in \mathbb{N}} A_j\right)$$

$$\le \sum_{j=1}^{\infty} \overline{\mathcal{M}}\left(\left(E \setminus \bigcup_{i=1}^{j-1} A_i\right) \cap A_j\right) + \overline{\mathcal{M}}\left(E \setminus \bigcup_{j \in \mathbb{N}} A_j\right)$$

$$\le \overline{\mathcal{M}}(E).$$

Thus $\bigcup A_j \in \mathcal{F}$. This completes the proof that $\mathcal{F}$ is a σ-algebra.

Now if the sets $A_j \in \mathcal{F}$ are disjoint, we can let $E = \bigcup A_j$ in the previous computation, and we get

$$\overline{\mathcal{M}}\left(\bigcup_{j \in \mathbb{N}} A_j\right) = \sum_{j=1}^{\infty} \overline{\mathcal{M}}(A_j),$$

so $\overline{\mathcal{M}}$ is countably additive on $\mathcal{F}$.   □

We will write simply $\mathcal{M}$ for the restriction of $\overline{\mathcal{M}}$ to the σ-algebra $\mathcal{F}$ of measurable sets. It is a measure on $\mathcal{F}$. Thus we see that we have a generalization of Lebesgue measure as constructed in Sect. 5.1.

Corollary 5.2.6. *Every Borel set in $\mathbb{R}$ is Lebesgue measurable.*

Proof. Open sets are measurable by Theorem 5.1.12. The collection of measurable sets is a σ-algebra by Theorem 5.1.14.   □

5.3 Two-Dimensional Lebesgue Measure

We next define two-dimensional Lebesgue measure. This is a measure defined for subsets of the plane $\mathbb{R}^2$.

A **rectangle** in $\mathbb{R}^2$ is a set R of the form

$$R = [a, b) \times [c, d) = \{ (x, y) \in \mathbb{R}^2 : a \le x < b, c \le y < d \}$$

for some $a \le b$ and $c \le d$. The **area** of this rectangle is $\mathbf{c}(R) = (b-a)(d-c)$, as usual. In particular, if $a = b$ or $c = d$, we see that $\mathbf{c}(\varnothing) = 0$. **Two-dimensional Lebesgue outer measure** is the outer measure $\overline{\mathcal{L}}^2$ on $\mathbb{R}^2$ defined by method I from this function $\mathbf{c}$.

Theorem 5.3.1. *Two-dimensional Lebesgue outer measure $\overline{\mathcal{L}}^2$ is a metric outer measure.*

Proof. Suppose A and B are sets with positive separation. Since $\overline{\mathcal{L}}^2$ is an outer measure, we have $\overline{\mathcal{L}}^2(A \cup B) \leq \overline{\mathcal{L}}^2(A) + \overline{\mathcal{L}}^2(B)$. So I must prove the opposite inequality.

Let $\mathcal{D}$ be a cover of $A \cup B$ by rectangles. Now a rectangle $R = [a, b) \times [c, d)$ can be written as a union of the four rectangles

$$[a, (a + b)/2) \times [c, (c + d)/2)$$
$$[a, (a + b)/2) \times [(c + d)/2, d)$$
$$[(a + b)/2, b) \times [c, (c + d)/2)$$
$$[(a + b)/2, b) \times [(c + d)/2, d) \, ,$$

and the area of the large rectangle is the sum of the areas of the four small rectangles. So the sum

$$\sum_{R \in \mathcal{D}} \mathbf{c}(R)$$

is unchanged when we replace one of the rectangles by its four parts. Applying this repeatedly, we may assume that the diameters of the rectangles in $\mathcal{D}$ are all smaller than $\mathrm{dist}(A, B)$. Then no rectangle of $\mathcal{D}$ intersects both A and B. So $\mathcal{D}$ is a disjoint union of two families, $\mathcal{A}$ and $\mathcal{B}$, where $\mathcal{A}$ covers A and $\mathcal{B}$ covers B. But then

$$\sum_{R \in \mathcal{D}} \mathbf{c}(R) = \sum_{R \in \mathcal{A}} \mathbf{c}(R) + \sum_{R \in \mathcal{B}} \mathbf{c}(R) \geq \overline{\mathcal{L}}^2(A) + \overline{\mathcal{L}}^2(B).$$

So we conclude that $\overline{\mathcal{L}}^2(A \cup B) \geq \overline{\mathcal{L}}^2(A) + \overline{\mathcal{L}}^2(B)$.   □

The sets that are measurable in the sense of Carathéodory for $\overline{\mathcal{L}}^2$ are again called the **Lebesgue measurable sets**; the restriction of $\overline{\mathcal{L}}^2$ to this σ-algebra is called **two-dimensional Lebesgue measure**. We will write $\mathcal{L}^2$ for two-dimensional Lebesgue measure.

The fact that two-dimensional Lebesgue measure is not identically zero is left to you:

Exercise 5.3.2. If $a < b$ and $c < d$, then a rectangle of the form $R = [a, b) \times [c, d)$ is Lebesgue measurable and $\mathcal{L}^2(R) = (b - a)(d - c)$.

Now that we know that the Lebesgue measure of a square is what it should be, the usual scheme of approximating an area with a lot of little squares will show that the usual sets of Euclidean plane geometry have two-dimensional Lebesgue measure equal to their areas. In particular, a rectangle with sides not parallel to the coordinate axes has the right area. This should be enough to prove:

Exercise 5.3.3. Let $f: \mathbb{R}^2 \to \mathbb{R}^2$ be a similarity with ratio r. If $A \subseteq \mathbb{R}^2$ is Lebesgue measurable, then so is $f[A]$, and $\mathcal{L}^2(f[A]) = r^2 \mathcal{L}^2(A)$.

This will tell us something about the similarity dimension of a set $A \subseteq \mathbb{R}^2$, as in the one-dimensional case (Exercise 5.1.18).

Exercise 5.3.4. Let $(r_1, r_2, \cdots, r_n)$ be a contracting ratio list; let s be its sim-value; let $(f_1, f_2, \cdots f_n)$ be a corresponding iterated function system in $\mathbb{R}^2$; and let $A \subseteq \mathbb{R}^2$ be a nonempty Borel set. Suppose $\mathcal{L}^2(f_j[A] \cap f_k[A]) = 0$ for $j \neq k$, and $A = \bigcup_{j=1}^n f_j[A]$. If $0 < \mathcal{L}^2(A) < \infty$, then $s = 2$.

This result can be used for complex number systems in the same way as the corresponding result was used for real number systems in 5.1.18

Exercise 5.3.5. Let b be a complex number, and let D be a finite set of complex numbers, including 0. Suppose D has k elements. Suppose every complex number can be represented in the number system defined by base b and digit set D, and the set of complex numbers with multiple representations has two-dimensional Lebesgue measure 0. What does this mean about the relationship between b and k?

Higher Dimensions

Let d be a positive integer. In d-dimensional Euclidean space $\mathbb{R}^d$, we will consider **hyper-rectangles** of the form

$$R = [a_1, b_1) \times [a_2, b_2) \times \cdots \times [a_d, b_d),$$

where $a_j < b_j$ for all j. The "hyper-volume" of this hyper-rectangle R is

$$\mathbf{c}(R) = \prod_{j=1}^d (b_j - a_j).$$

We define d-**dimensional Lebesgue outer measure** to be the method I outer measure defined from this set function **c**. We define d-**dimensional Lebesgue measure** to be the restriction to the measurable subsets. As before, we use the notation $\overline{\mathcal{L}}^d$ and $\mathcal{L}^d$.

Exercise 5.3.6. The outer measure $\overline{\mathcal{L}}^d$ is a metric outer measure. If

$$R = [a_1, b_1) \times [a_2, b_2) \times \cdots \times [a_d, b_d),$$

where $a_j \leq b_j$ for all j, then

$$\mathcal{L}^d(R) = \prod_{j=1}^d (b_j - a_j).$$

Dyadic Cubes

A **semi-dyadic square** in $\mathbb{R}^2$ is a set of the form

$$\left[\frac{j-1}{2^n}, \frac{j+1}{2^n}\right) \times \left[\frac{k-1}{2^n}, \frac{k+1}{2^n}\right).$$

Exercise 5.3.7. Show that the class of semi-dyadic squares is a reduced cover class with factor 8 for $\mathcal{L}^2$.

A **dyadic square** in $\mathbb{R}^2$ is a set of the form

$$\left[\frac{j}{2^n}, \frac{j+1}{2^n}\right) \times \left[\frac{k}{2^n}, \frac{k+1}{2^n}\right).$$

For each n, let $\mathcal{R}_n$ be the set of finite disjoint unions of dyadic squares with denominator 2^n. Then $\mathcal{R} = \bigcup_n \mathcal{R}_n$ is called the **dyadic ring** in $\mathbb{R}^2$.

Exercise 5.3.8. Using the set function $\mathbf{c} \colon \mathcal{R} \to [0, \infty)$ defined by $\mathbf{c}(E) = \mathcal{L}^2(E)$, the dyadic ring $\mathcal{R}$ is a reduced cover class for 2-dimensional Lebesgue measure.

After you have completed the preceding two exercises, it should be easy to formulate the corresponding results for d-dimensional Lebesgue measure for any $d \in \mathbb{N}$. You would define semi-dyadic cubes, dyadic cubes, and the dyadic ring in $\mathbb{R}^d$.

5.4 Metric Outer Measure

Consider the following example of a method I outer measure on $\mathbb{R}$; the definition is very close to the definition used for Lebesgue measure. We begin with the collection $\mathcal{A} = \{[a, b) \colon a < b\}$ of half-open intervals and the set function $\mathbf{c}([a, b)) = \sqrt{b - a}$. Let $\overline{\mathcal{M}}$ be the corresponding method I outer measure. I claim that the interval $A = [0, 1]$ is not measurable.

Consider the measure of $[0, 1)$. Certainly the singleton $\{[0, 1)\}$ covers $[0, 1)$, so $\overline{\mathcal{M}}([0, 1)) \leq \mathbf{c}([0, 1)) = 1$. If $[0, 1) \subseteq \bigcup_{i \in \mathbb{N}} [a_i, b_i)$, then by what we know about Lebesgue measure, we must have $\sum_{i=1}^{\infty} (b_i - a_i) \geq 1$. So we have also

$$\left(\sum_{i=1}^{\infty} \sqrt{b_i - a_i}\right)^2 = \sum_{i=1}^{\infty} \left(\sqrt{b_i - a_i}\right)^2 + 2\sum_{i<j} \sqrt{b_i - a_i} \sqrt{b_j - a_j}$$

$$\geq \sum_{i=1}^{\infty} (b_i - a_i) \geq 1.$$

Therefore $\sum_{i=1}^{\infty} \sqrt{b_i - a_i} \geq 1$. This shows that $\overline{\mathcal{M}}([0, 1)) \geq 1$. So we have $\overline{\mathcal{M}}([0, 1)) = 1$.

Similarly $\overline{\mathcal{M}}\big([-1,0)\big) = 1$. The singleton $\{[-1,1)\}$ covers $[-1,1)$, so as before we have $\overline{\mathcal{M}}\big([-1,1)\big) \le \mathbf{c}\big([-1,1)\big) = \sqrt{2}$. So if $A = [0,1]$ and $E = [-1,1)$, we have

$$\overline{\mathcal{M}}(E \cap A) + \overline{\mathcal{M}}(E \setminus A) = 1 + 1 = 2 > \sqrt{2} \ge \overline{\mathcal{M}}(E).$$

This shows that $A = [0,1]$ is not measurable.

It is often desirable that the sets we work with be measurable sets. When we work with subsets of a metric space (as is common in this book), the sets are often open sets, closed sets, or sets constructed simply from open and closed sets. In particular, the sets are often Borel sets. There is a condition that will insure that all Borel sets are measurable sets.

Two sets A, B in a metric space have **positive separation** iff $\operatorname{dist}(A, B) > 0$; that is, there is $r > 0$ with $\varrho(x, y) \ge r$ for all $x \in A$ and $y \in B$. Let $\overline{\mathcal{M}}$ be an outer measure on a metric space S. We say that $\overline{\mathcal{M}}$ is a **metric outer measure** iff $\overline{\mathcal{M}}(A \cup B) = \overline{\mathcal{M}}(A) + \overline{\mathcal{M}}(B)$ for any pair A, B of sets with positive separation. (Theorem 5.1.6 shows that $\overline{\mathcal{L}}$ is a metric outer measure.) The measure $\mathcal{M}$ obtained by restricting a metric outer measure $\overline{\mathcal{M}}$ to its measurable sets will be called a **metric measure**.

The reason that metric outer measures are of interest is that open sets (and therefore all Borel sets) are measurable sets. Before I prove this, I formulate the lemma of Carathéodory.

Lemma 5.4.1. *Let $\overline{\mathcal{M}}$ be a metric outer measure on the metric space S. Let $A_1 \subseteq A_2 \subseteq \cdots$, and $A = \bigcup_{j \in \mathbb{N}} A_j$. Assume $\operatorname{dist}(A_j, A \setminus A_{j+1}) > 0$ for all j. Then $\overline{\mathcal{M}}(A) = \lim_{j \to \infty} \overline{\mathcal{M}}(A_j)$.*

Proof. For all j we have $\overline{\mathcal{M}}(A) \ge \overline{\mathcal{M}}(A_j)$, so $\overline{\mathcal{M}}(A) \ge \lim_{j \to \infty} \overline{\mathcal{M}}(A_j)$. (This inequality is true for any outer measure.) If $\lim_{j \to \infty} \overline{\mathcal{M}}(A_j) = \infty$, then the equation is true. So suppose $\lim_{j \to \infty} \overline{\mathcal{M}}(A_j) < \infty$.

Let $B_1 = A_1$ and $B_j = A_j \setminus A_{j-1}$ for $j \ge 2$. If $i \ge j + 2$, then $B_j \subseteq A_j$ and $B_i \subseteq A \setminus A_{i-1} \subseteq A \setminus A_{j+1}$, so B_i and B_j have positive separation. So

$$\overline{\mathcal{M}}\left(\bigcup_{k=1}^{m} B_{2k-1}\right) = \sum_{k=1}^{m} \overline{\mathcal{M}}(B_{2k-1})$$

$$\overline{\mathcal{M}}\left(\bigcup_{k=1}^{m} B_{2k}\right) = \sum_{k=1}^{m} \overline{\mathcal{M}}(B_{2k}).$$

Since $\lim_{j \to \infty} \overline{\mathcal{M}}(A_j) < \infty$, both of these converge (as $m \to \infty$). So

$$\overline{\mathcal{M}}(A) = \overline{\mathcal{M}}\left(\bigcup_{j \in \mathbb{N}} A_j\right) = \overline{\mathcal{M}}\left(A_j \cup \bigcup_{k \ge j+1} B_k\right)$$

$$\le \overline{\mathcal{M}}(A_j) + \sum_{k=j+1}^{\infty} \overline{\mathcal{M}}(B_k)$$

$$\leq \lim_{i \to \infty} \overline{\mathcal{M}}(A_i) + \sum_{k=j+1}^{\infty} \overline{\mathcal{M}}(B_k).$$

Now as $j \to \infty$, the tail of a convergent series goes to 0, so we get

$$\overline{\mathcal{M}}(A) \leq \lim_{i \to \infty} \overline{\mathcal{M}}(A_i). \qquad \square$$

Theorem 5.4.2. *Let $\overline{\mathcal{M}}$ be a metric outer measure on a metric space S. Then every Borel subset of S is $\overline{\mathcal{M}}$-measurable.*

Proof. Since the σ-algebra of Borel sets is the σ-algebra generated by the closed sets, and since the collection $\mathcal{F}$ of measurable sets is a σ-algebra, it is enough to show that every closed set F is measurable. Let A be any test set. I must show that $\overline{\mathcal{M}}(A) \geq \overline{\mathcal{M}}(A \cap F) + \overline{\mathcal{M}}(A \setminus F)$, since the opposite inequality is true for any outer measure.

Let $A_j = \{\, x \in A : \operatorname{dist}(x, F) \geq 1/j \,\}$. Then $\operatorname{dist}(A_j, F \cap A) \geq 1/j$, so

$$\overline{\mathcal{M}}(A \cap F) + \overline{\mathcal{M}}(A_j) = \overline{\mathcal{M}}\big((A \cap F) \cup A_j\big) \leq \overline{\mathcal{M}}(A). \qquad (1)$$

Now since F is closed, F contains all points of distance 0 from F, so $A \setminus F = \bigcup_{j \in \mathbb{N}} A_j$. We check the condition of the lemma: If $x \in \big(A \setminus (F \cup A_{j+1})\big)$, then there exists $z \in F$ with $\varrho(x, z) < 1/(j+1)$. If $y \in A_j$, then

$$\varrho(x, y) \geq \varrho(y, z) - \varrho(x, z) > \frac{1}{j} - \frac{1}{j+1}.$$

Thus

$$\operatorname{dist}\big(A \setminus (F \cup A_{j+1}), A_j\big) \geq \frac{1}{j} - \frac{1}{j+1} > 0.$$

Therefore, applying the lemma, we get $\overline{\mathcal{M}}(A \setminus F) \leq \lim_{j \to \infty} \overline{\mathcal{M}}(A_j)$. Taking the limit in (1), we get $\overline{\mathcal{M}}(A \cap F) + \overline{\mathcal{M}}(A \setminus F) \leq \overline{\mathcal{M}}(A)$, which completes the proof.

$\square$

Proposition 5.4.3. *Let $\mathcal{M}$ be a finite metric measure on a compact metric space S. Let $E \subseteq S$ be a Borel set. For any $\varepsilon > 0$, there exist a compact set K and an open set U with $U \supseteq E \supseteq K$ and $\mathcal{M}(U \setminus K) < \varepsilon$.*

Proof. Let $\mathcal{A}$ be the collection of all sets $E \subseteq S$ such that for all $\varepsilon > 0$, there exist a compact set K and an open set U with $U \supseteq E \supseteq K$ and $\mathcal{M}(U \setminus K) < \varepsilon$.

First I claim that all closed sets belong to $\mathcal{A}$. Let $F \subseteq S$ be closed and $\varepsilon > 0$. Now

$$U_n = \left\{\, x \in S : \operatorname{dist}(x, F) < \frac{1}{n} \,\right\}$$

defines open sets U_n with $U_1 \supseteq U_2 \supseteq \cdots$ and $\bigcap_{n \in \mathbb{N}} U_n = F$. So we have $\lim_{n \to \infty} \mathcal{M}(U_n) = \mathcal{M}(F)$. There is n so large that $\mathcal{M}(U_n) - \mathcal{M}(F) < \varepsilon$. Then: $U_n \supseteq F \supseteq F$, U_n is open, F is compact, and $\mathcal{M}(U_n \setminus F) < \varepsilon$.

Clearly $\varnothing \in \mathcal{A}$.

Next, $\mathcal{A}$ is closed under complements. Let $E \in \mathcal{A}$. Consider the complement $E' = S \setminus E$. Let $\varepsilon > 0$. There is an open U and a compact K with $U \supseteq E \supseteq K$ and $\mathcal{M}(U \setminus K) < \varepsilon$. But then $U' = S \setminus U$ is compact, $K' = S \setminus K$ is open, $K' \supseteq E' \supseteq U'$, and $\mathcal{M}(K' \setminus U') = \mathcal{M}(U \setminus K) < \varepsilon$.

Now I claim $\mathcal{A}$ is closed under countable unions. Let $E_n \in \mathcal{A}$ for $n \in \mathbb{N}$. Write $E = \bigcup_{n \in \mathbb{N}} E_n$, and let $\varepsilon > 0$. Then there exist open U_n and compact K_n with $U_n \supseteq E_n \supseteq K_n$ and $\mathcal{M}(U_n \setminus K_n) < \varepsilon/2^{n+1}$. Then $U = \bigcup_{n \in \mathbb{N}} U_n$ is open. Now

$$L_m = \bigcup_{n=1}^{m} K_n$$

is compact, increases with m, and $\bigcup_{m \in \mathbb{N}} L_m = \bigcup_{n \in \mathbb{N}} K_n$. There is m so large that

$$\mathcal{M}\left(\bigcup_{n \in \mathbb{N}} K_n \right) - \mathcal{M}(L_m) < \frac{\varepsilon}{2}.$$

So we have $U \supseteq E \supseteq L_n$ and

$$\mathcal{M}(U \setminus L_n) \leq \mathcal{M}\left(\bigcup_{n \in \mathbb{N}} (U_n \setminus K_n) \right) + \mathcal{M}\left(\bigcup_{n \in \mathbb{N}} K_n \right) - \mathcal{M}(L_m)$$

$$< \sum_{n=1}^{\infty} \frac{\varepsilon}{2^{n+1}} + \frac{\varepsilon}{2} = \varepsilon.$$

Therefore $\mathcal{A}$ includes at least the Borel sets.   □

Method II

We have seen that method I may fail to yield a measure where open sets are measurable. There is a related construction, called "method II" that will overcome this difficulty.

Let $\mathcal{A}$ be a family of subsets of a metric space S, and suppose, for every $x \in S$ and $\varepsilon > 0$, there exists $A \in \mathcal{A}$ with $x \in A$ and $\operatorname{diam} A \leq \varepsilon$. Suppose $\mathbf{c} \colon \mathcal{A} \to [0, \infty]$ is a given function. An outer measure will be constructed based on this data. For each $\varepsilon > 0$, let

$$\mathcal{A}_\varepsilon = \{ A \in \mathcal{A} : \operatorname{diam} A \leq \varepsilon \}.$$

Let $\overline{\mathcal{M}}_\varepsilon$ be the method I outer measure determined by $\mathbf{c}$ using the family $\mathcal{A}_\varepsilon$. Then by Proposition 5.2.3(a), for a given set E, when ε decreases, $\overline{\mathcal{M}}_\varepsilon(E)$ increases. Define

$$\overline{\mathcal{M}}(E) = \lim_{\varepsilon \to 0} \overline{\mathcal{M}}_\varepsilon(E) = \sup_{\varepsilon > 0} \overline{\mathcal{M}}_\varepsilon(E).$$

It is easily verified that $\overline{\mathcal{M}}$ is an outer measure. As usual, we will write $\mathcal{M}$ for the restriction to the measurable sets. This construction of an outer measure

$\overline{\mathcal{M}}$ from a set function $\mathbf{c}$ (and a measure $\mathcal{M}$ from $\overline{\mathcal{M}}$) is called **method II**. It is more complicated than method I, but (unlike method I) it insures that Borel sets are measurable:

Theorem 5.4.4. *The set function $\overline{\mathcal{M}}$ defined by method II is a metric outer measure.*

Proof. Let $A, B \subseteq S$ with $\operatorname{dist}(A, B) > 0$. Since $\overline{\mathcal{M}}$ is an outer measure, we have $\overline{\mathcal{M}}(A \cup B) \leq \overline{\mathcal{M}}(A) + \overline{\mathcal{M}}(B)$. So I must prove the opposite inequality.

Let $\varepsilon > 0$ so small that $\varepsilon < \operatorname{dist}(A, B)$. Let $\mathcal{D}$ be any countable cover of $A \cup B$ by sets of $\mathcal{A}_\varepsilon$. The sets $D \in \mathcal{D}$ have diameter less than $\operatorname{dist}(A, B)$, so such a set D intersects at most one of the sets A, B. Therefore, $\mathcal{D}$ may be divided into two disjoint collections, $\mathcal{D}_1$ and $\mathcal{D}_2$, where $\mathcal{D}_1$ covers A and $\mathcal{D}_2$ covers B. Then

$$\sum_{D \in \mathcal{D}} \mathbf{c}(D) = \sum_{D \in \mathcal{D}_1} \mathbf{c}(D) + \sum_{D \in \mathcal{D}_2} \mathbf{c}(D) \geq \overline{\mathcal{M}}_\varepsilon(A) + \overline{\mathcal{M}}_\varepsilon(B).$$

Now we may take the infimum over all covers, and conclude $\overline{\mathcal{M}}_\varepsilon(A \cup B) \geq \overline{\mathcal{M}}_\varepsilon(A) + \overline{\mathcal{M}}_\varepsilon(B)$. Then we may take the limit as $\varepsilon \to 0$ to conclude $\overline{\mathcal{M}}(A \cup B) \geq \overline{\mathcal{M}}(A) + \overline{\mathcal{M}}(B)$. □

Exercise 5.4.5. Let S be a metric space, and let $\mathbf{c}$ be a set function. For a collection of sets $\mathcal{A}$, let $\overline{\mathcal{M}}_\mathcal{A}$ be the method II outer measure defined using the class $\mathcal{A}$ of sets and the restriction of $\mathbf{c}$ to $\mathcal{A}$.

(a) If $\mathcal{B} \subseteq \mathcal{A}$, then $\overline{\mathcal{M}}_\mathcal{A} \leq \overline{\mathcal{M}}_\mathcal{B}$.
(b) Suppose that, for every $\eta > 0$ there is $\delta > 0$ such that for all $A \in \mathcal{A}$ with $\operatorname{diam} A \leq \delta$, and every $\varepsilon > 0$, there is $B \in \mathcal{B}$ with $\operatorname{diam} B \leq \eta$, $B \supseteq A$, and $\mathbf{c}(B) \leq \mathbf{c}(A) + \varepsilon$. Then $\overline{\mathcal{M}}_\mathcal{B} \leq \overline{\mathcal{M}}_\mathcal{A}$.
(c) Let $C > 0$ be a constant, and suppose that, for every $\eta > 0$ there is $\delta > 0$ such that for all $A \in \mathcal{A}$ with $\operatorname{diam} A \leq \delta$, there is $B \in \mathcal{B}$ with $\operatorname{diam} B \leq \eta$, $B \supseteq A$, and $\mathbf{c}(B) \leq C\,\mathbf{c}(A)$. Then $\overline{\mathcal{M}}_\mathcal{B} \leq C\,\overline{\mathcal{M}}_\mathcal{A}$.

When condition (b) holds, we will say that $\mathcal{B}$ is a **reduced cover class** for $\overline{\mathcal{M}}$. When condition (c) holds, we will say that $\mathcal{B}$ is a reduced cover class with factor C for $\overline{\mathcal{M}}$.

5.5 Measures for Strings

One of the useful ways we will employ the material of this chapter is by defining measures. Sometimes (for example Lebesgue measure or Hausdorff measure) we will define a measure on subsets of Euclidean space $\mathbb{R}^d$. But also measures will be defined on our string models and path models.

An Example

We will consider an easy example before we attack the more general case.

Begin with the two-letter alphabet $E = \{0, 1\}$. Consider, as usual, the metric space $E^{(\omega)}$ of infinite strings with metric $\varrho_{1/2}$. We will construct a measure on $E^{(\omega)}$. We begin with the family of "basic open sets":

$$\mathcal{A} = \Big\{ [\alpha] : \alpha \in E^{(*)} \Big\}$$

together with the set function $\mathbf{c} \colon \mathcal{A} \to [0, \infty)$ defined by

$$\mathbf{c}([\alpha]) = \frac{1}{2^{|\alpha|}}.$$

(Recall the notation $|\alpha|$ for the length of the string α.)

Proposition 5.5.1. *The method I outer measure $\overline{\mathcal{M}}_{1/2}$ constructed using this function $\mathbf{c}$ is a metric outer measure and satisfies $\overline{\mathcal{M}}_{1/2}([\alpha]) = \mathbf{c}([\alpha])$ for all $\alpha \in E^{(*)}$.*

Proof. Write $\mathcal{A} = \big\{ [\alpha] : \alpha \in E^{(*)} \big\}$, $\mathcal{A}_\varepsilon = \{ D \in \mathcal{A} : \operatorname{diam} D \leq \varepsilon \}$. Let $\overline{\mathcal{N}}_\varepsilon$ be the method I measure defined by the set function $\mathbf{c}$ restricted to $\mathcal{A}_\varepsilon$. If $D \in \mathcal{A}_\varepsilon$, then of course $\mathbf{c}(D) \geq \overline{\mathcal{M}}_{1/2}(D)$, so by the Method I theorem,

$$\overline{\mathcal{N}}_\varepsilon(A) \geq \overline{\mathcal{M}}_{1/2}(A)$$

for all A. Therefore the method II measure $\overline{\mathcal{N}}$ defined by

$$\overline{\mathcal{N}}(A) = \lim_{\varepsilon \to 0} \overline{\mathcal{N}}_\varepsilon(A)$$

satisfies $\overline{\mathcal{N}}(A) \geq \overline{\mathcal{M}}_{1/2}(A)$.

Note that, for any $\alpha \in E^{(*)}$, if k is the length $|\alpha|$, then we have

$$\mathbf{c}([\alpha]) = \frac{1}{2^k} = \frac{1}{2^{k+1}} + \frac{1}{2^{k+1}} = \mathbf{c}([\alpha 0]) + \mathbf{c}([\alpha 1]).$$

Applying this repeatedly, we see that, for any $\varepsilon > 0$, any set $D \in \mathcal{A}$ is a finite disjoint union $D_1 \cup D_2 \cup \cdots \cup D_n$ of sets in $\mathcal{A}_\varepsilon$ with $\mathbf{c}(D) = \sum \mathbf{c}(D_i)$. This means that $\overline{\mathcal{N}}_\varepsilon(D) \leq \mathbf{c}(D)$, so by the Method I theorem, $\overline{\mathcal{N}}_\varepsilon(A) \leq \overline{\mathcal{M}}_{1/2}(A)$ for all sets A. So $\overline{\mathcal{N}}(A) \leq \overline{\mathcal{M}}_{1/2}(A)$

Therefore $\overline{\mathcal{M}}_{1/2} = \overline{\mathcal{N}}$ is a method II outer measure, so it is a metric outer measure. □

Exercise 5.5.2. Let $h \colon E^{(\omega)} \to \mathbb{R}$ be the "base 2" addressing function defined on p. 14. If $A \subseteq E^{(\omega)}$ is a Borel set, then $\mathcal{M}_{1/2}(A) = \mathcal{L}(h[A])$.

Measures on String Spaces

Let E be a finite alphabet with at least two letters. Consider the space $E^{(\omega)}$ of infinite strings. This space is a metric space for many different metrics ϱ. But all of the metrics constructed according to the scheme in Proposition 2.6.5 produce the same open sets. A countable base for the open sets is $\{ [\alpha] : \alpha \in E^{(*)} \}$. Also $E^{(\omega)}$ is a compact ultrametric space. An important feature of all these metrics is $\lim_{k \to \infty} \operatorname{diam}[\sigma \restriction k] = 0$ for each $\sigma \in E^{(\omega)}$.

Exercise 5.5.3. It follows that $\lim_{k \to \infty} \left(\sup \left\{ \operatorname{diam}[\alpha] : \alpha \in E^{(k)} \right\} \right) = 0$.

Suppose a non-negative number w_α is given for each finite string α. Under what conditions is there a metric outer measure $\overline{\mathcal{M}}$ on $E^{(\omega)}$ with $\overline{\mathcal{M}}([\alpha]) = w_\alpha$ for all α? Since it is a metric outer measure, the open sets $[\alpha]$ are measurable, so $\overline{\mathcal{M}}$ is additive on them. Now the set $[\alpha]$ is the disjoint union of the sets $[\beta]$ as β ranges over the children of α (that is, $\beta = \alpha e$ for $e \in E$).

Theorem 5.5.4. *Suppose the non-negative numbers* w_α *satisfy*

$$w_\alpha = \sum_{e \in E} w_{\alpha e}$$

for $\alpha \in E^{(*)}$. *Then the method I outer measure defined by the set function* $\mathbf{c}([\alpha]) = w_\alpha$ *is a metric outer measure* $\overline{\mathcal{M}}$ *on* $E^{(\omega)}$ *with* $\overline{\mathcal{M}}([\alpha]) = w_\alpha$.

Proof. Write $\mathcal{A} = \left\{ [\alpha] : \alpha \in E^{(*)} \right\}$, $\mathcal{A}_\varepsilon = \{ D \in \mathcal{A} : \operatorname{diam} D \leq \varepsilon \}$. Let $\overline{\mathcal{N}}_\varepsilon$ be the method I measure defined by the set function $\mathbf{c}$ restricted to $\mathcal{A}_\varepsilon$. If $D \in \mathcal{A}_\varepsilon$, then of course $\mathbf{c}(D) \geq \overline{\mathcal{M}}(D)$, so by the Method I theorem,

$$\overline{\mathcal{N}}_\varepsilon(A) \geq \overline{\mathcal{M}}(A)$$

for all A. Therefore the method II measure $\overline{\mathcal{N}}$ defined by

$$\overline{\mathcal{N}}(A) = \lim_{\varepsilon \to 0} \overline{\mathcal{N}}_\varepsilon(A)$$

satisfies $\overline{\mathcal{N}}(A) \geq \overline{\mathcal{M}}(A)$.

Note that, for any $\alpha \in E^{(*)}$, we have

$$\mathbf{c}([\alpha]) = w_\alpha = \sum_{e \in E} w_{\alpha e} = \sum_{e \in E} \mathbf{c}([\alpha e]).$$

Applying this repeatedly, together with Exercise 5.5.3, we see that, for any $\varepsilon > 0$, any set $D \in \mathcal{A}$ is a finite disjoint union $D_1 \cup D_2 \cup \cdots \cup D_n$ of sets in $\mathcal{A}_\varepsilon$ with $\mathbf{c}(D) = \sum \mathbf{c}(D_i)$. This means that $\overline{\mathcal{N}}_\varepsilon(D) \leq \mathbf{c}(D)$, so by the Method I theorem, $\overline{\mathcal{N}}_\varepsilon(A) \leq \overline{\mathcal{M}}(A)$ for all sets A. So $\overline{\mathcal{N}}(A) \leq \overline{\mathcal{M}}(A)$.

Therefore $\overline{\mathcal{M}} = \overline{\mathcal{N}}$ is a method II outer measure. So we may conclude that it is a metric outer measure.   □

How should we formulate the corresponding theorem for the path spaces $E_v^{(\omega)}$ defined by a directed multigraph (V, E, i, t)? We will define measures for each of the spaces $E_v^{(\omega)}$. We only need to define one of them at a time.

Fix a vertex v. Suppose nonnegative numbers w_α are given, one for each $\alpha \in E_v^{(*)}$. They should (of course) satisfy

$$w_\alpha = \sum_{i(e)=t(\alpha)} w_{\alpha e}$$

for $\alpha \in E_v^{(*)}$. Note that this has consequences for the troublesome exceptional cases that came up when we were defining the metric. If α has no children, then (interpreting an empty sum as 0), we see that $w_\alpha = 0$. Similarly, if α has only one child β, then $w_\alpha = w_\beta$.

Exercise 5.5.5. Suppose the non-negative numbers w_α satisfy

$$w_\alpha = \sum_{i(e)=t(\alpha)} w_{\alpha e}$$

for $\alpha \in E_v^{(*)}$. Then the method I outer measure defined by the set function $\mathbf{c}([\alpha]) = w_\alpha$ is a metric outer measure $\overline{M}$ on $E_v^{(\omega)}$ with $\overline{M}([\alpha]) = w_\alpha$.

5.6 *Remarks

Henri Lebesgue's measure and integration theory dates from about 1900. It is one of the cornerstones of twentieth century mathematics. I have not discussed integration at all, in order to reduce the amount of material to its bare minimum. (My more advanced text [18] develops integration using the same ideas.) The more abstract measure theory was developed by many others, such as Constantin Carathéodory, during the early 1900's. Theorem 5.5.4 on the existence of measures on the string spaces is due essentially to A. N. Kolmogorov.

The σ-algebra $\mathcal{F}$ generated by a family $\mathcal{D}$ of sets (as in Theorem 5.2.1) is a complicated object to describe constructively. The proof given for 5.2.1 has the advantage of not requiring such a constructive description. Certainly $\mathcal{F}$ contains all countable unions

$$\bigcup_{i \in \mathbb{N}} D_i$$

with $D_i \in \mathcal{D}$; it contains complements of those unions; it contains countable intersections

$$\bigcap_{i \in \mathbb{N}} E_i$$

where each E_i is either a countable union or a complement of a countable union. But that may not be everything in $\mathcal{F}$. (See, for example, the proof of Theorem (10.23) in [34].)

An example of a set in the line not measurable for Lebesgue measure may be found in many texts. For example: [7, pp. 36–37], [11, Theorem 1.4.7], [34, (10.28)], or [58, Chap. 3, Sect. 4].

Exercise 5.3.5: $k = |b|^2$.

Exercise 5.5.2. Both measures are method I measures; use the Method I theorem twice, once to prove an inequality in each direction.

O, wiste a man how manye maladyes
Folwen of excesse and of glotonyes
He wolde been the moore mesurable
—G. Chaucer, *The Pardoner's Tale*

6

Fractal Dimension

Next we come to the fractal dimensions: in particular the Hausdorff dimension and the packing dimension. The surprising feature for these dimensions is that they need not be integers: they can be fractions. The Hausdorff dimension is the one singled out by Mandelbrot when he defined "fractal". The Hausdorff and packing dimensions are perhaps a bit more difficult to define than some of the other kinds of fractal dimension. But in recent years it has become clear that they are the most useful of the fractal dimensions.

6.1 Hausdorff Measure

Let S be a metric space. Consider a positive real number s, the candidate for the dimension. The *s-dimensional Hausdorff outer measure* is the method II outer measure defined from the set function $\mathbf{c}_s(A) = (\operatorname{diam} A)^s$. It is written $\overline{\mathcal{H}}^s$. The restriction to the measurable sets is called *s-dimensional Hausdorff measure*, and written $\mathcal{H}^s$. Since $\overline{\mathcal{H}}^s$ is constructed by method II, it is a metric outer measure. So all Borel sets are measurable (in particular, all open sets, closed sets, compact sets).

Recall that the Method I theorem gives a more explicit construction: A family $\mathcal{A}$ of subsets of S is called a *countable cover* of a set F iff

$$F \subseteq \bigcup_{A \in \mathcal{A}} A,$$

and $\mathcal{A}$ is a countable (possibly even finite) family of sets. Let ε be a positive number (presumably very small). The cover $\mathcal{A}$ is an *ε-cover* iff $\operatorname{diam} A \le \varepsilon$ for all $A \in \mathcal{A}$. Define

$$\overline{\mathcal{H}}^s_\varepsilon(F) = \inf \sum_{A \in \mathcal{A}} (\operatorname{diam} A)^s,$$

where the infimum is over all countable ε-covers $\mathcal{A}$ of the set F. (By convention, $\inf \varnothing = \infty$.) A computation shows that when ε gets smaller, $\overline{\mathcal{H}}^s_\varepsilon(F)$ gets larger. Finally:

$$\overline{\mathcal{H}}^s(F) = \lim_{\varepsilon \to 0} \overline{\mathcal{H}}^s_\varepsilon(F) = \sup_{\varepsilon > 0} \overline{\mathcal{H}}^s_\varepsilon(F)$$

is the s-dimensional Hausdorff outer measure of the set F. Figures 6.1.1 and 6.1.2 illustrate some of the ideas behind the definition.

There are variants in the definition of the Hausdorff measure that are sometimes useful. (i) Since the closure of a set has the same diameter as the set itself, we may use only closed sets in the covers $\mathcal{A}$. The class of closed sets is a (method II) reduced cover class for $\mathcal{H}^s$. (ii) If A is any set, it is contained in an open set with diameter as close as I like to the diameter of A. The class of open sets is a reduced cover class for $\mathcal{H}^s$. (iii) Any set of diameter r is contained in a closed ball of radius r (and diameter $\leq 2r$). The collection of open balls is a reduced cover class with factor 2^s for $\mathcal{H}^s$. (iv) In Euclidean space $\mathbb{R}^d$, the convex hull of any set has the same diameter as the set. The

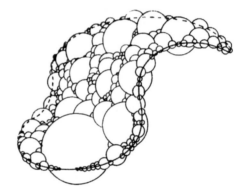

Fig. 6.1.1. The Hausdorff measure (area) of a piece of surface A is approximated by the cross-sections of little balls which cover it. (From [52])

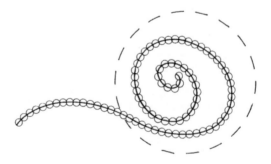

Fig. 6.1.2. One must cover by *small* sets to compute length accurately. Here the length of the spiral is well-estimated by the sum of the diameters of the tiny balls, but grossly under-estimated by the diameter of the huge ball. (From [52])

collection of convex sets is a reduced cover class for $\mathcal{H}^s$. (v) If a set K is compact, then every open cover of K has a finite subcover, so to compute the Hausdorff measure of a compact set K, we may use finite covers $\mathcal{A}$. (vi) If we replace a set in a cover $\mathcal{A}$ of a set F by a subset of itself, so that the result is still a cover of F, the sum

$$\sum_{A \in \mathcal{A}} (\operatorname{diam} A)^s$$

only becomes smaller. So if $F \subseteq T \subseteq S$, the value of $\overline{\mathcal{H}}_\varepsilon^s(F)$ when F is considered to be a subset of T is the same as when F is considered to be a subset of S. In particular, we may assume (if it is convenient) that the sets used in the covers $\mathcal{A}$ of the set F are subsets of F.

Exercise 6.1.3. If F is a finite set, then $\mathcal{H}^s(F) = 0$ for all $s > 0$.

Theorem 6.1.4. *In the metric space $\mathbb{R}$, the one-dimensional Hausdorff measure $\mathcal{H}^1$ coincides with the Lebesgue measure $\mathcal{L}$.*

Proof. If $A \subseteq \mathbb{R}$ has finite diameter r, then $\sup A - \inf A = r$, so A is contained in a closed interval I with length r, and $\overline{\mathcal{L}}(A) \leq \overline{\mathcal{L}}(I) = r$. But by the Method I theorem (5.2.2), $\overline{\mathcal{H}}_\varepsilon^1$ is the largest outer measure $\overline{\mathcal{M}}$ satisfying $\overline{\mathcal{M}}(A) \leq \operatorname{diam} A$ for all sets A with diameter less than ε. So $\overline{\mathcal{H}}_\varepsilon^1(F) \geq \overline{\mathcal{L}}(F)$ for all F. Therefore $\overline{\mathcal{H}}^1(F) \geq \overline{\mathcal{L}}(F)$.

Now if $[a, b)$ is a half-open interval and $\varepsilon > 0$, we may find points $a = x_0 < x_1 < \cdots < x_n = b$ with $x_j - x_{j-1} < \varepsilon$ for all j. Then $[a, b)$ is covered by the countable collection $\{ [x_{j-1}, x_j] : 1 \leq j \leq n \}$, and

$$\sum_{j=1}^n \operatorname{diam}[x_{j-1}, x_j] = \sum_{j=1}^n (x_j - x_{j-1}) = b - a.$$

Therefore $\overline{\mathcal{H}}_\varepsilon^1([a, b)) \leq b - a$. But by the Method I theorem, $\overline{\mathcal{L}}$ is the largest outer measure satisfying $\overline{\mathcal{L}}([a, b)) \leq b - a$ for all half-open intervals $[a, b)$. Therefore $\overline{\mathcal{L}}(F) \geq \overline{\mathcal{H}}^1(F)$. for all F.

The two outer measures $\overline{\mathcal{L}}$ and $\overline{\mathcal{H}}^1$ coincide. The measurable sets in each case are given by the criterion of Carathéodory, so the measures $\mathcal{L}$ and $\mathcal{H}^1$ also coincide. □

For a "zero-dimensional" Hausdorff measure, we can use the set function $\mathbf{c}_0$ defined by $\mathbf{c}_0(A) = 1$ for $A \neq \varnothing$ and $\mathbf{c}_0(\varnothing) = 0$.

Exercise 6.1.5. With this definition, $\mathcal{H}^0(A) = n$ if A has n elements, and $\mathcal{H}^0(A) = \infty$ if A is infinite.

Hausdorff Dimension

How does the Hausdorff measure $\mathcal{H}^s(F)$ behave as a function of s for a given set F? An easy calculation shows that when s increases, $\mathcal{H}^s(F)$ decreases. But much more is true.

Theorem 6.1.6. *Let F be a Borel set. Let $0 < s < t$. If $\mathcal{H}^s(F) < \infty$, then $\mathcal{H}^t(F) = 0$. If $\mathcal{H}^t(F) > 0$, then $\mathcal{H}^s(F) = \infty$.*

Proof. If diam $A \le \varepsilon$, then

$$\overline{\mathcal{H}}_\varepsilon^t(A) \le (\operatorname{diam} A)^t \le \varepsilon^{t-s}(\operatorname{diam} A)^s.$$

Therefore by the Method I theorem, $\overline{\mathcal{H}}_\varepsilon^t(F) \le \varepsilon^{t-s}\overline{\mathcal{H}}_\varepsilon^s(F)$ for all F. Now if $\mathcal{H}^s(F) < \infty$, then $\mathcal{H}^t(F) \le \lim_{\varepsilon \to 0} \varepsilon^{t-s}\,\overline{\mathcal{H}}_\varepsilon^s(F) = 0 \times \mathcal{H}^s(F) = 0$. The second assertion is the contrapositive. $\square$

This means that, for a given set F, there is a unique "critical value" $s_0 \in [0, \infty]$ such that:

$$\mathcal{H}^s(F) = \infty \qquad\qquad \text{for all } s < s_0;$$
$$\mathcal{H}^s(F) = 0 \qquad\qquad \text{for all } s > s_0.$$

This value s_0 is called the ***Hausdorff dimension*** of the set F. We will write $s_0 = \dim F$. Of course, it is possible that $\mathcal{H}^s(F) = 0$ for all $s > 0$; in that case $\dim F = 0$. In the same way, it is possible that $\mathcal{H}^s(F) = \infty$ for all s; in that case $\dim F = \infty$.

This idea of dimension is an abstraction of what we already know from elementary geometry. If A is a nice smooth rectifiable curve, then its length is a useful way to measure its size; but its "area" and "volume" are 0. The dimensions 2 and 3 are too large to help in measuring the size of A. If B is the surface of a sphere, then its area is positive and finite. We can say its "length" is infinite (for example, since it contains curves that are as long as we like which spiral around); its "volume" is 0, since it is contained in a solid shperical shell whose thickness is as small as we like. So for the set B, the dimension 1 is too small, the dimension 3 is too large, and the dimension 2 is just right. The s-dimensional Hausdorff measure give us a way of measuring the size of a set for dimensions s other than the integers $1, 2, 3, \cdots$.

Theorem 6.1.7. *Let A, B be Borel sets.*

(1) *If $A \subseteq B$, then $\dim A \le \dim B$.*
(2) *$\dim(A \cup B) = \max\{\dim A, \dim B\}$.*

Proof. (1) Suppose $A \subseteq B$. If $s > \dim B$, then $\mathcal{H}^s(A) \le \mathcal{H}^s(B) = 0$. Therefore $\dim A \le s$. This is true for all $s > \dim B$, so $\dim A \le \dim B$.

(2) Let $s > \max\{\dim A, \dim B\}$. Then $s > \dim A$, so $\mathcal{H}^s(A) = 0$. Similarly, $\mathcal{H}^s(B) = 0$. Then $\mathcal{H}^s(A \cup B) \le \mathcal{H}^s(A) + \mathcal{H}^s(B) = 0$. Therefore $\dim(A \cup B) \le s$. This is true for all $s > \max\{\dim A, \dim B\}$, so we have $\dim(A \cup B) \le \max\{\dim A, \dim B\}$. By (1), $\dim(A \cup B) \ge \max\{\dim A, \dim B\}$. $\square$

Exercise 6.1.8. Suppose $A_1, A_2, \cdots$ are Borel sets. Is it true that

$$\dim \bigcup_{k \in \mathbb{N}} A_k = \sup_k \dim A_k?$$

Theorem 6.1.9. *Let* $f: S \to T$ *be a similarity with ratio* $r > 0$, *let* s *be a positive real number, and let* $F \subseteq S$ *be a set. Then* $\overline{\mathcal{H}}^s(f[F]) = r^s\overline{\mathcal{H}}^s(F)$. *So* $\dim f[F] = \dim F$.

Proof. We may assume that $T = f[S]$. Then f has an inverse f^{-1}. A set $A \subseteq S$ satisfies diam $f[A] = r$ diam A. Therefore $(\text{diam } f[A])^s = r^s(\text{diam } A)^s$. By the Method I theorem (applied twice), $\overline{\mathcal{H}}^s_{r\varepsilon}(f[F]) = r^s\overline{\mathcal{H}}^s_{\varepsilon}(F)$. Therefore $\overline{\mathcal{H}}^s(f[F]) = r^s\overline{\mathcal{H}}^s(F)$ and $\dim f[F] = \dim F$. $\qquad\square$

Exercise 6.1.10. Suppose $f: S \to T$ is a function. Let $A \subseteq S$ be a Borel set. Prove or disprove: (1) If f is Lipschitz, then

$$\dim f[A] \leq \dim A.$$

(2) If f is inverse Lipschitz, then

$$\dim f[A] \geq \dim A.$$

Exercise 6.1.11. Suppose S is a metric space and $\dim S < \infty$. Does it follow that S is separable?

6.2 Packing Measure

In this section we define the packing measures and the packing dimension.

Mandelbrot says that his definition for "fractal" ($\text{Cov } S < \dim S$) is too broad, in that it admits "true geometric chaos". The sets that are of interest for applications (and in mathematics) are generally not the most general set, with few special properties. So it may be useful to restrict the term "fractal" so that the sets meeting the conditions have useful properties. One possible way to do this has been proposed by James Taylor. He proposed to apply the term "fractal" to (Borel) sets where the packing dimension is equal to the Hausdorff dimension.

Motivations

Before we formulate the definition of the packing measures, let us discuss some of the reasons for the definition, and why it has the form given.

Hausdorff measure is based on "covering" of a set. The set E to be measured is covered by small sets A_i. We attempt to make the covering "efficient" by minimizing

$$\sum_{i \in \mathbb{N}} \mathbf{c}(A_i),$$

subject to the constraint that the sets A_i cover E. When this sum is smaller, the cover of E by $\{A_i\}$ is considered to be more efficient.

Another possibility for "measuring" the set E is to do it by packing rather than covering. We want to put disjoint sets A_i inside E. We attempt to make this packing "efficient" by maximizing

$$\sum_{i \in \mathbb{N}} \mathbf{c}(A_i),$$

subject to the constraint that the sets A_i are disjoint subsets of E. When this sum is larger, the packing $\{A_i\}$ is considered to be more efficient.

For fractal measures, the set function $\mathbf{c}$ should be of the form $(\operatorname{diam} A)^s$, where $s > 0$ is the dimension we are interested in. But this leads to certain undesirable features if taken at face value. For example, in the plane, what if we pack a square as in Fig. 6.2.1(a)? By making the sets A_i narrow enough, we can make the sum

$$\sum_{i \in \mathbb{N}} (\operatorname{diam} A_i)^s,$$

as large as we like.

The way to avoid this is to pack only by sets of a special type. For example, in $\mathbb{R}$, packings with intervals cannot be beat. In Euclidean space, the choice is often to pack with cubes. In order to get a definition that applies in a general metric space, we will pack with balls.

Packing a set E with balls $A_i \subseteq E$ is fine when E is an open set, but other sets may contain no balls at all. So we drop the requirement that the balls be contained in E. But to make sure the packing measures the set E we require instead that the centers of the balls lie in E.

Let S be a metric space, $x \in S$ and $r > 0$. Recall the notation

$$B_r(x) = \{\, y \in S : \varrho(x,y) < r \,\}, \qquad \overline{B}_r(x) = \{\, y \in S : \varrho(x,y) \le r \,\}.$$

We will pack with closed balls. But open balls could be used just as well in our setting.

For two balls $\overline{B}_r(x), \overline{B}_s(y)$ in Euclidean space, we know that they are disjoint, $\overline{B}_r(x) \cap \overline{B}_s(y) = \varnothing$, if and only if $\varrho(x,y) > r + s$. In metric space other than Euclidean space, this equivalence may fail. But we do know that if $\varrho(x,y) > r + s$, then $\overline{B}_r(x) \cap \overline{B}_s(y) = \varnothing$. We will use $\varrho(x,y) > r + s$ for our definition of "packing".

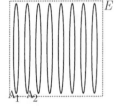

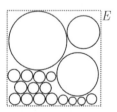

Fig. 6.2.1. (a) Packing with any sets   (b) packing with balls

In Euclidean space, two balls are equal, $\overline{B}_r(x) = \overline{B}_s(y)$, if and only if $x = y$ and $r = s$. In metric space other than Euclidean space, this equivalence may fail. For example, in an ultrametric space, every point of a ball is a center. This is a reason for our use of "constituents" rather than balls in the definition.

In Euclidean space, the diameter of the ball $\overline{B}_r(x)$ is $2r$. In metric space other than Euclidean space, this may not be true. For example, in an ultrametric space, diam $\overline{B}_r(x) \leq r$. We will use the set function $(2r)^s$ for our "radius-based" packing measure rather than the "diameter-based" option of $(\operatorname{diam} \overline{B}_r(x))^s$.

In some texts—including the first edition of this one—one or more of the above choices may be reversed. As noted, in Euclidean space this makes no difference.

Definition

Let S be a metric space. A **constituent** in S is a pair (x, r), where $x \in S$ and $r > 0$. We think of the constutuent (x, r) as standing for the closed ball $\overline{B}_r(x)$. We may even call x the "center" and r the "radius" of the constituent (x, r).

Let $E \subseteq S$. A **packing** of E is a countable collection Π of constituents, such that: (a) for all $(x, r) \in \Pi$, we have $x \in E$; (b) for all $(x, r), (y, s) \in \Pi$ with $(x, r) \neq (y, s)$, we have $\varrho(x, y) > r + s$.

For $\delta > 0$, we say that a packing Π is δ-**fine** iff for all $(x, r) \in \Pi$ we have $r \leq \delta$. Let $F \subseteq S$, and let $\delta, s > 0$. Define

$$\widetilde{\mathcal{P}}_\delta^s(F) = \sup \sum_{(x,r) \in \Pi} (2r)^s,$$

where the supremum is over all δ-fine packings Π of F. Note: because of the sup involved, we may restrict this to *finite* packings Π only.

When δ decreases to 0, the value $\widetilde{\mathcal{P}}_\delta^s(F)$ decreases, so we define

$$\widetilde{\mathcal{P}}_0^s(F) = \lim_{\delta \to 0} \widetilde{\mathcal{P}}_\delta^s(F) = \inf_{\delta > 0} \widetilde{\mathcal{P}}_\delta^s(F).$$

When we have done this, we get a family $\left(\widetilde{\mathcal{P}}_0^s\right)$ of set functions indexed by s. As before, there is a critical value:

Exercise 6.2.2. Let F be a set in a metric space. There is $s_0 \in [0, \infty]$ such that

$$\widetilde{\mathcal{P}}_0^s(F) = \infty \qquad \text{for all } s < s_0;$$
$$\widetilde{\mathcal{P}}_0^s(F) = 0 \qquad \text{for all } s > s_0.$$

This critical value s_0 will be called the **packing index** of the set F. However, the set functions $\widetilde{\mathcal{P}}_0^s$ are not really what we want. They are *not* outer measures. This is not unexpected, since the process used to construct them is not method II. Here is an illustration showing that $\widetilde{\mathcal{P}}_0^{1/2}$ fails to be an outer measure on $\mathbb{R}$:

Proposition 6.2.3. *Let K be the compact set $\{0, 1, 1/2, 1/3, 1/4, 1/5, \cdots\} \subseteq \mathbb{R}$. Then $\widetilde{\mathcal{P}}_0^{1/2}(K) > 0$.*

Proof. Let $k \in \mathbb{N}$ be odd, let $\varepsilon = 2^{-k}$, and let $n = 2^{(k-1)/2}$. Then

$$\frac{1}{n-1} - \frac{1}{n} > \frac{1}{n^2} = 2\varepsilon,$$

so the constituents with radius ε and centers $1, 1/2, 1/3, \cdots, 1/n$ form a packing. (That is, the balls with radius ε and centers $1, 1/2, 1/3, \cdots, 1/n$ are disjoint.) So

$$\widetilde{\mathcal{P}}_\varepsilon^{1/2}(K) \geq n\,(2\varepsilon)^{1/2} = 1,$$

and therefore $\widetilde{\mathcal{P}}_0^{1/2}(K) \geq 1$.   □

For many purposes it is unreasonable to claim that this countable set K has positive dimension. We know a good way (method I) to get an outer measure from a set function. So we apply method I to the set function $\widetilde{\mathcal{P}}_0^s$:

$$\overline{\mathcal{P}}^s(E) = \inf \sum_{C \in \mathcal{C}} \widetilde{\mathcal{P}}_0^s(C),$$

where the inf is over all countable covers $\mathcal{C}$ of the set E.

Theorem 6.2.4 (The closure theorem). *If C is a set and $\overline{C}$ is its closure, then $\widetilde{\mathcal{P}}_0^s(C) = \widetilde{\mathcal{P}}_0^s(\overline{C})$.*

Proof. Any packing of C is automatically a packing of $\overline{C}$. This shows that $\widetilde{\mathcal{P}}_\delta^s(C) \leq \widetilde{\mathcal{P}}_\delta^s(\overline{C})$ for all δ and thus $\widetilde{\mathcal{P}}_0^s(C) \leq \widetilde{\mathcal{P}}_0^s(\overline{C})$.

Conversely, let $\delta > 0$ and let Π be a finite δ-fine packing of $\overline{C}$. Write $\Pi = \{(x_1, r_1), \cdots, (x_n, r_n)\}$. For any $i \neq j$, we have $\varrho(x_i, x_j) - r_i - r_j > 0$, and there are only finitely many pairs i, j, so there is $\varepsilon > 0$ with $\varrho(x_i, x_j) - r_i - r_j > \varepsilon$ for all $i \neq j$. Now for each i, the point x_i belongs to the closure of C, so there is $y_i \in C$ with $\varrho(x_i, y_i) < \varepsilon/2$. But then $\Pi' = \{(y_1, r_1), \cdots, (y_n, r_n)\}$ is a packing of C, still δ-fine, and it has the same value $\sum(2r_i)^s$ as the packing Π. Therefore we get $\widetilde{\mathcal{P}}_\delta^s(C) \geq \widetilde{\mathcal{P}}_\delta^s(\overline{C})$ for all δ and thus $\widetilde{\mathcal{P}}_0^s(C) \geq \widetilde{\mathcal{P}}_0^s(\overline{C})$.   □

The class of closed sets is a reduced cover class for $\overline{\mathcal{P}}^s$:

Corollary 6.2.5. *Let $E \subseteq S$. Then*

$$\overline{\mathcal{P}}^s(E) = \inf \sum_{C \in \mathcal{C}} \widetilde{\mathcal{P}}_0^s(C),$$

where the inf is over all countable covers $\mathcal{C}$ of the set E by closed sets.

Lemma 6.2.6. *Let* $A, B \subseteq S$. *Then* $\widetilde{\mathcal{P}}_0^s(A \cup B) \leq \widetilde{\mathcal{P}}_0^s(A) + \widetilde{\mathcal{P}}_0^s(B)$. *If* $\mathrm{dist}(A, B)$
> 0, *then* $\widetilde{\mathcal{P}}_0^s(A \cup B) = \widetilde{\mathcal{P}}_0^s(A) + \widetilde{\mathcal{P}}_0^s(B)$.

Proof. Let $\delta > 0$ be given. Let Π be a δ-fine packing of $A \cup B$. Then Π is the disjoint union of

$$\Pi_1 = \{\, (x, r) \in \Pi : x \in A \,\} \quad \text{and} \quad \Pi_2 = \{\, (x, r) \in \Pi : x \notin A \,\}.$$

But Π_1 is a δ-fine packing of A and Π_2 is a δ-fine packing of B. So

$$\sum_{(x,r)\in\Pi} (2r)^s = \sum_{(x,r)\in\Pi_1} (2r)^s + \sum_{(x,r)\in\Pi_2} (2r)^s \leq \widetilde{\mathcal{P}}_\delta^s(A) + \widetilde{\mathcal{P}}_\delta^s(B).$$

Take the supremum over all δ-fine packings Π to get $\widetilde{\mathcal{P}}_\delta^s(A \cup B) \leq \widetilde{\mathcal{P}}_\delta^s(A) + $
$\widetilde{\mathcal{P}}_\delta^s(B)$. Let $\delta \to 0$ to get $\widetilde{\mathcal{P}}_0^s(A \cup B) \leq \widetilde{\mathcal{P}}_0^s(A) + \widetilde{\mathcal{P}}_0^s(B)$.
 Let $\mathrm{dist}(A, B) = \varepsilon > 0$. Then if $\delta < \varepsilon/2$, any δ-fine packing of $A \cup B$ is the disjoint union of a δ-fine packing of A and a δ-fine packing of B. And conversely, the union of a δ-fine packing of A and a δ-fine packing of B is a δ-fine packing of $A \cup B$. So $\widetilde{\mathcal{P}}_\delta^s(A \cup B) = \widetilde{\mathcal{P}}_\delta^s(A) + \widetilde{\mathcal{P}}_\delta^s(B)$. Let $\delta \to 0$ to get $\widetilde{\mathcal{P}}_0^s(A \cup B) = \widetilde{\mathcal{P}}_0^s(A) + \widetilde{\mathcal{P}}_0^s(B)$. $\square$

Theorem 6.2.7. *The set function* $\overline{\mathcal{P}}^s$ *is a metric outer measure.*

Proof. The only packing of the empty set is the empty packing, and an empty sum has the value 0. Therefore $\widetilde{\mathcal{P}}_\delta^s(\varnothing) = 0$ for all $\delta > 0$ and $\widetilde{\mathcal{P}}_0^s(\varnothing) = 0$. The empty set can be covered $\varnothing \subseteq \bigcup_{n \in \mathbb{N}} E_n$, where $E_n = \varnothing$ for all n, so $\overline{\mathcal{P}}^s(\varnothing) = 0$.
 If $A \subseteq B$, and $B \subseteq \bigcup_{n \in \mathbb{N}} E_n$, then also $A \subseteq \bigcup_{n \in \mathbb{N}} E_n$, so $\overline{\mathcal{P}}^s(A) \leq \overline{\mathcal{P}}^s(B)$.
 Suppose $A = \bigcup_{i \in \mathbb{N}} A_i$. We must show $\overline{\mathcal{P}}^s(A) \leq \sum_{i=1}^{\infty} \overline{\mathcal{P}}^s(A_i)$. If $\sum_i \overline{\mathcal{P}}^s(A_i)$ diverges, then there is nothing to do, so assume $\sum_{i=1}^{\infty} \overline{\mathcal{P}}^s(A_i) < \infty$. Let $\varepsilon > 0$ be given. For each i, there exist sets E_{ni}, $n \in \mathbb{N}$, so that $A_i \subseteq \bigcup_n E_{ni}$ and $\sum_n \widetilde{\mathcal{P}}_0^s(E_{ni}) < \overline{\mathcal{P}}^s(A_i) + \varepsilon/2^i$. Then $A \subseteq \bigcup_i \bigcup_n E_{ni}$ is a countable cover of A, so

$$\overline{\mathcal{P}}^s(A) \leq \sum_i \sum_n \widetilde{\mathcal{P}}_0^s(E_{ni}) < \sum_i \left(\overline{\mathcal{P}}^s(A_i) + \frac{\varepsilon}{2^i} \right) = \left(\sum_i \overline{\mathcal{P}}^s(A_i) \right) + \varepsilon.$$

This holds for any $\varepsilon > 0$, so $\overline{\mathcal{P}}^s(A) \leq \sum_i \overline{\mathcal{P}}^s(A_i)$.
 The metric property follows from Lemma 6.2.6. $\square$

 The restriction of $\overline{\mathcal{P}}^s$ to the measurable sets is a measure, called the *s-dimensional packing measure*, and written $\mathcal{P}^s$. As usual there is a critical value for s:

Exercise 6.2.8. Let F be a Borel set in a metric space. There is $s_0 \in [0, \infty]$ such that

$$\mathcal{P}^s(F) = \infty \qquad \text{for all } s < s_0;$$
$$\mathcal{P}^s(F) = 0 \qquad \text{for all } s > s_0.$$

This value s_0 is called the **packing dimension** of the set F. We will write $s_0 = \text{Dim } F$. It is a more reasonable quantity than the packing index defined above.

Elementary Properties

The packing dimension has many of the same properties as the Hausdorff dimension.

Theorem 6.2.9. *Let A, B be Borel sets.*

(1) *If $A \subseteq B$, then $\text{Dim } A \le \text{Dim } B$.*
(2) $\text{Dim}(A \cup B) = \max\{\text{Dim } A, \text{Dim } B\}$.

Proof. (1) Assume $A \subseteq B$. Let $s > \text{Dim } B$. then $\overline{\mathcal{P}}^s(B) = 0$. Therefore $\overline{\mathcal{P}}^s(A) = 0$. This shows $\text{Dim } A \le s$. This holds for all $s > \text{Dim } B$, so $\text{Dim } A \le \text{Dim } B$.

(2) By (1), $\text{Dim}(A \cup B) \ge \text{Dim } A$ and $\text{Dim}(A \cup B) \ge \text{Dim } B$. Therefore $\text{Dim}(A \cup B) \ge \max\{\text{Dim } A, \text{Dim } B\}$. If $s > \max\{\text{Dim } A, \text{Dim } B\}$, then $\overline{\mathcal{P}}^s(A) = 0$ and $\overline{\mathcal{P}}^s(B) = 0$. So by subadditivity, $\overline{\mathcal{P}}^s(A \cup B) = 0$. This shows $\text{Dim}(A \cup B) \le s$. It holds for all $s > \max\{\text{Dim } A, \text{Dim } B\}$, so $\text{Dim}(A \cup B) \le \max\{\text{Dim } A, \text{Dim } B\}$.   □

Exercise 6.2.10. Suppose $A_1, A_2, \cdots$ are Borel sets. Is it true that

$$\text{Dim} \bigcup_{k \in \mathbb{N}} A_k = \sup_k \text{Dim } A_k?$$

Theorem 6.2.11. *Let $f: S \to T$ be a similarity with ratio $r > 0$, let s be a positive real number, and let $E \subseteq S$ be a set. Then $\overline{\mathcal{P}}^s(f[E]) = r^s \overline{\mathcal{P}}^s(E)$. So $\text{Dim } f[E] = \text{Dim } E$.*

Proof. Let Π be a δ-fine packing of F. Then $\{ (f(x), rt) : (x, t) \in \Pi \}$ is an $r\delta$-fine packing of $f[F]$. So

$$\widetilde{\mathcal{P}}^s_{r\delta}(f[F]) \ge \sum_{(x,t) \in \Pi} (2rt)^s = r^s \sum_{(x,t) \in \Pi} (2t)^s.$$

This holds for all δ-fine packings of F, so

$$\widetilde{\mathcal{P}}^s_{r\delta}(f[F]) \ge r^s \widetilde{\mathcal{P}}^s_\delta(F).$$

Let $\delta \to 0$ to get

$$\widetilde{\mathcal{P}}_0^s\big(f[F]\big) \geq r^s \widetilde{\mathcal{P}}_0^s(F).$$

This holds for all subsets $F \subseteq S$.

Since $r > 0$, the map f is one-to-one, and maps F onto $f[F]$. Every $r\delta$-fine packing of $f[F]$ is of the form $\{ (f(x), rt) : (x, t) \in \Pi \}$ for some δ-fine packing of F. So the estimates hold in reverse, and we conclude

$$\widetilde{\mathcal{P}}_0^s\big(f[F]\big) = r^s \widetilde{\mathcal{P}}_0^s(F).$$

Now if $E \subseteq \bigcup_i A_i$ is a countable cover of a set E, then $f[E] \subseteq \bigcup_i f[A_i]$ is a cover of the image set $f[E]$. So

$$\sum_i \widetilde{\mathcal{P}}_0^s(A_i) = r^s \sum_i \widetilde{\mathcal{P}}_0^s(f[A_i]) \geq r^s \overline{\mathcal{P}}^s\big(f[E]\big).$$

This holds for all covers of E, so $\overline{\mathcal{P}}^s(E) \geq r^s \overline{\mathcal{P}}^s\big(f[E]\big)$.

If $f[E] \subseteq \bigcup B_i$ is a countable cover of $f[E]$, let $A_i = f^{-1}[B_i]$, so that $E \subseteq \bigcup A_i$ is a cover of E. Note $f[A_i] \subseteq B_i$. Now

$$r^s \sum_i \widetilde{\mathcal{P}}_0^s(B_i) \geq r^s \sum_i \widetilde{\mathcal{P}}_0^s(f[A_i]) = \sum_i \widetilde{\mathcal{P}}_0^s(A_i) \geq \overline{\mathcal{P}}^s(E).$$

This holds for all covers of $f[E]$, so $r^s \overline{\mathcal{P}}^s\big(f[E]\big) \geq \overline{\mathcal{P}}^s(E)$.

Therefore, we have $\overline{\mathcal{P}}^s(E) = r^s \overline{\mathcal{P}}^s\big(f[E]\big)$. And $\operatorname{Dim} f[E] = \operatorname{Dim} E$. □

Exercise 6.2.12. Suppose $f \colon S \to T$ is a function. Let $A \subseteq S$ be a Borel set. Prove or disprove: (1) If f is Lipschitz, then

$$\operatorname{Dim} f[A] \leq \operatorname{Dim} A.$$

(2) If f is inverse Lipschitz, then

$$\operatorname{Dim} f[A] \geq \operatorname{Dim} A.$$

Proposition 6.2.13. *In the metric space $\mathbb{R}$, the one-dimensional packing measure $\mathcal{P}^1$ coincides with Lebesgue measure $\mathcal{L}$.*

Proof. First consider a half-open interval, $[a, b)$. If Π is a finite packing of $[a, b)$, write $\Pi = \{(x_1, r_1), (x_2, r_2), \cdots, (x_n, r_n)\}$ with $x_1 < x_2 < \cdots < x_n$. Then all the balls $\overline{B}_{r_i}(x_i)$ are contained in the interval $[a - r_1, b + r_n]$ and are disjoint. By the additivity of Lebesgue measure (and the fact that intervals are Lebesgue measurable sets), we have

$$\sum_{i=1}^{n} (2r_i) \leq b - a + r_1 + r_n.$$

If Π is δ-fine, then

$$\sum_{i=1}^{n}(2r_i) \leq b - a + 2\delta.$$

Take the supremum on all δ-fine packings of $[a, b)$ to conclude

$$\widetilde{\mathcal{P}}_\delta^1([a, b)) \leq b - a + 2\delta.$$

Let $\delta \to 0$ to get

$$\widetilde{\mathcal{P}}_0^1([a, b)) \leq b - a.$$

On the other hand, given $\delta > 0$, choose n with $(b - a)/n < \delta$, then we can pack $[a, b)$ with n balls all of radius $r_i = (b - a)/(2n)$. So $\widetilde{\mathcal{P}}_\delta^1([a, b)) \geq b - a$. Take the limit to get $\widetilde{\mathcal{P}}_0^1([a, b)) \geq b - a$. Therefore $\widetilde{\mathcal{P}}_0^1([a, b)) = b - a$.

Now consider a finite disjoint union of half-open intervals $[a, b)$. If two of them are adjacent (the right endpoint of one is the left endpoint of the other), then they may be combined into a single interval. If two of them are not adjacent, then there is a gap of positive length between them. So by Lemma 6.2.6 we have $\widetilde{\mathcal{P}}_0^1(V) = \mathcal{L}(V)$ for all such finite disjoint unions. This holds in particular for the dyadic ring $\mathcal{R}$ defined on p. 150.

Fix a large $N > 0$ and consider $\mathcal{L}$ and $\overline{\mathcal{P}}^1$ for subsets of $[-N, N]$. We claim that $\mathcal{R}$ is a reduced cover class for $\overline{\mathcal{P}}^1$. Given any closed set $F \subseteq [-N, N]$ and any $\varepsilon > 0$, there is $V \in \mathcal{R}$ with $V \supseteq F$ and $\mathcal{L}(V \setminus F) < \varepsilon/2$; then there is $U \in \mathcal{R}$ with $U \supseteq V \setminus F$ and $\mathcal{L}(U \setminus (V \setminus F)) < \varepsilon/2$. Then

$$\widetilde{\mathcal{P}}_0^1(V) \leq \widetilde{\mathcal{P}}_0^1(F) + \widetilde{\mathcal{P}}_0^1(V \setminus F) \leq \widetilde{\mathcal{P}}_0^1(F) + \widetilde{\mathcal{P}}_0^1(U)$$
$$\leq \widetilde{\mathcal{P}}_0^1(F) + \mathcal{L}(U) \leq \widetilde{\mathcal{P}}_0^1(F) + \varepsilon.$$

Now the closed sets form a reduced cover class for $\overline{\mathcal{P}}^1$, so this shows that the dyadic ring $\mathcal{R}$ also forms a reduced cover class for $\overline{\mathcal{P}}^1$ in $[-N, N]$.

We have seen that $\widetilde{\mathcal{P}}_0^1$ and $\mathcal{L}$ agree on $\mathcal{R}$, so their method I measures also agree: $\overline{\mathcal{L}} = \overline{\mathcal{P}}^1$ for subsets of $[-N, N]$. For a general subset A of $\mathbb{R}$, take the increasing limit of $[-N, N] \cap A$. So $\overline{\mathcal{L}} = \overline{\mathcal{P}}^1$.   □

The packing dimension is related to the Hausdorff dimension.[*]

Proposition 6.2.14. *Let S be a metric space and $F \subseteq S$ a Borel set. Then $\mathcal{H}^s(F) \leq 2^s \mathcal{P}^s(F)$ and $\dim F \leq \operatorname{Dim} F$.*

Proof. I first show that $\overline{\mathcal{H}}_{4\varepsilon}^s(F) \leq 2^s \widetilde{\mathcal{P}}_\varepsilon^s(F)$. Now if $\widetilde{\mathcal{P}}_\varepsilon^s(F) = \infty$, then this is clear. So suppose $\widetilde{\mathcal{P}}_\varepsilon^s(F) < \infty$. If there were an *infinite* packing of F with all radii equal to ε, then $\widetilde{\mathcal{P}}_\varepsilon^s(F) = \infty$. So there is a maximal finite packing $\{(x_1, \varepsilon), (x_2, \varepsilon), \cdots, (x_n, \varepsilon)\}$ of F. Then $\widetilde{\mathcal{P}}_\varepsilon^s(F) \geq n(2\varepsilon)^s$. By the maximality,

---

[*] In the first edition, this was stated only in Euclidean space—one of the drawbacks of the diameter-based definition of the packing measure.

for any $x \in F$, there is some i between 1 and n with $\varrho(x, x_i) \leq 2\varepsilon$. So the collection $\left\{ \overline{B}_{2\varepsilon}(x_i) : 1 \leq i \leq n \right\}$ covers F, and

$$\overline{\mathcal{H}}^s_{4\varepsilon}(F) \leq \sum_{i=1}^n \left(\operatorname{diam} \overline{B}_{2\varepsilon}(x_i) \right)^s \leq n(4\varepsilon)^s = 2^s n(2\varepsilon)^s \leq 2^s \widetilde{\mathcal{P}}^s_\varepsilon(F).$$

Therefore $\overline{\mathcal{H}}^s_{4\varepsilon}(F) \leq 2^s \widetilde{\mathcal{P}}^s_\varepsilon(F)$.

Now take the limit as $\varepsilon \to 0$ and conclude $\mathcal{H}^s(F) \leq 2^s \widetilde{\mathcal{P}}^s_0(F)$. So by the Method I theorem, $\mathcal{H}^s(F) \leq 2^s \mathcal{P}^s(F)$.

Now if $s < \dim F$, then $\mathcal{H}^s(F) = \infty$, so $\mathcal{P}^s(F) = \infty$, and therefore $s \leq \operatorname{Dim} F$. We therefore conclude that $\dim F \leq \operatorname{Dim} F$. $\qquad\square$

A set $F \subseteq \mathbb{R}^d$ is a **fractal** (in the sense of Taylor) iff $\dim F = \operatorname{Dim} F$.

6.3 Examples

According to Mandelbrot's definition, a fractal is a set A with $\operatorname{Cov} A < \dim A$. According to Taylor's definition, a fractal is a set A with $\dim A = \operatorname{Dim} A$. In order for these definitions to be useful, we will have to be able to compute the dimensions involved. In some cases this is not easy to do.

In this section, we will do a few examples directly from the definitions. We will carry out the calculations in great detail. In Sect. 6.4 we will discuss self-similar sets in general.

Binary Tree

Here is our first official example of a fractal. We computed $\operatorname{ind}\{0,1\}^{(\omega)} = 0$ in Theorem 3.4.4. For $\{0,1\}^{(\omega)}$ we will use the metric $\varrho_{1/2}$ defined on p. 44 and the measure $\mathcal{M}_{1/2}$ defined on p. 160. Recall the notation $[\alpha]$ for cylinders from p. 13.

Proposition 6.3.1. *Let $E = \{0,1\}$ be a two-letter alphabet, let $E^{(\omega)}$ be the space of all infinite strings using E, and let $\varrho_{1/2}$ be the metric for $E^{(\omega)}$. Then $\mathcal{H}^1 = \mathcal{M}_{1/2}$ and $\dim E^{(\omega)} = 1$.*

Proof. To prove that $\mathcal{H}^1 = \mathcal{M}_{1/2}$, we will use two applications of the Method I theorem.

If a set $A \subseteq E^{(\omega)}$ has positive diameter, then (Proposition 2.6.7) there is a string $\alpha \in E^{(*)}$ with $A \subseteq [\alpha]$ and $\operatorname{diam} A = \operatorname{diam}[\alpha]$. So $\mathcal{M}_{1/2}(A) \leq \mathcal{M}_{1/2}([\alpha]) = \operatorname{diam}[\alpha] = \operatorname{diam} A$. But $\overline{\mathcal{H}}^1_\varepsilon$ is the largest outer measure with $\overline{\mathcal{H}}^1_\varepsilon(A) \leq \operatorname{diam} A$ for all sets A of diameter $\leq \varepsilon$. So $\overline{\mathcal{M}}_{1/2} \leq \overline{\mathcal{H}}^1_\varepsilon$. This is true for all $\varepsilon > 0$, so $\overline{\mathcal{M}}_{1/2} \leq \overline{\mathcal{H}}^1$.

On the other hand, let $\alpha \in E^{(*)}$ be a finite string, and $\varepsilon > 0$. There is n so large that $2^{-n} < \varepsilon$ and $n \geq |\alpha|$, the length of α. Then the basic open set $[\alpha]$ is the disjoint union of all sets $[\beta]$, where $\beta \geq \alpha$ and $|\beta| = n$. There are $2^{n-|\alpha|}$ of these sets. Then

$$\overline{\mathcal{H}}^1_\varepsilon([\alpha]) \leq \sum_{\substack{\beta \geq \alpha \\ |\beta| = n}} \mathrm{diam}[\beta] = \sum_{\substack{\beta \geq \alpha \\ |\beta| = n}} 2^{-n} = 2^{-|\alpha|}.$$

But $\overline{\mathcal{M}}_{1/2}$ is the largest outer measure with $\overline{\mathcal{M}}_{1/2}([\alpha]) \leq 2^{-|\alpha|}$ for all $\alpha \in E^{(*)}$. So $\overline{\mathcal{H}}^1_\varepsilon \leq \overline{\mathcal{M}}_{1/2}$, and thus $\overline{\mathcal{H}}^1 \leq \overline{\mathcal{M}}_{1/2}$.

Therefore $\overline{\mathcal{H}}^1 = \overline{\mathcal{M}}_{1/2}$. The measurable sets in both cases are given by the criterion of Carathéodory, so $\mathcal{H}^1 = \mathcal{M}_{1/2}$.

Now since $0 < \mathcal{H}^1(E^{(\omega)}) < \infty$, we conclude that $\dim E^{(\omega)} = 1$. $\qquad\square$

So we know that $\{0,1\}^{(\omega)}$ is a fractal in the sense of Mandelbrot. It is also a fractal in the sense of Taylor:

Proposition 6.3.2. Let $E = \{0,1\}$ be a two-letter alphabet, let $E^{(\omega)}$ be the space of all infinite strings using E, and let $\varrho_{1/2}$ be the metric for $E^{(\omega)}$. Then $\mathcal{P}^1 = 4\mathcal{M}_{1/2}$ and $\mathrm{Dim}\, E^{(\omega)} = 1$.

Proof. The outer measure $\overline{\mathcal{M}} = \overline{\mathcal{M}}_{1/2}$ is the largest outer measure such that $\overline{\mathcal{M}}([\alpha]) \leq 2^{-|\alpha|}$ for all $\alpha \in E^{(*)}$. In fact, $\overline{\mathcal{M}}([\alpha]) = 2^{-|\alpha|}$.

We will describe the balls in $E^{(\omega)}$. Let $\sigma \in E^{(\omega)}$ and let r satisfy $0 < r < 1$. There is a unique $n \in \mathbb{N}$ with $2^{-n} \leq r < 2^{-n+1}$. The prefix $\alpha = \sigma{\restriction}n$ of length n defines a cylinder $[\alpha]$. I claim that $\overline{B}_r(\sigma) = [\alpha]$. To see this, note that any string $\tau \in [\alpha]$ agrees with σ at least for the first n letters, so $\varrho(\sigma, \tau) \leq 2^{-n} \leq r$. And any string $\tau \notin [\alpha]$ disagrees with σ somewhere in the first n letters, so the longest common prefix is shorter than n, and thus $\varrho(\sigma, \tau) \geq 2^{-n+1} > r$. The measure of the ball is $\mathcal{M}(\overline{B}_r(\sigma)) = \mathcal{M}([\alpha]) = 2^{-n}$, so $\mathcal{M}(\overline{B}_r(\sigma)) \leq r$, $\mathcal{M}(\overline{B}_r(\sigma)) > r/2$, and $r < 2\mathcal{M}(\overline{B}_r(\sigma))$.

(a) First we prove $\widetilde{\mathcal{P}}^1_0(E^{(\omega)}) \leq 4$. Let $\delta > 0$, and let Π be a δ-packing of $E^{(\omega)}$. Then the corresponding closed balls $\{\, \overline{B}_r(\sigma) : (\sigma, r) \in \Pi \,\}$ are disjoint. So

$$\sum_{(\sigma,r)\in\Pi} (2r) \leq 4 \sum_{(\sigma,r)\in\Pi} \mathcal{M}(\overline{B}_r(\sigma)) \leq 4.$$

This is true for all δ-packings, so $\widetilde{\mathcal{P}}^1_\delta(E^{(\omega)}) \leq 4$. This holds for all $\delta > 0$, so $\widetilde{\mathcal{P}}^1_0(E^{(\omega)}) \leq 4$.

Because $E^{(\omega)}$ is a one-element cover of itself, we have also $\mathcal{P}^1(E^{(\omega)}) \leq 4$.

(b) Now we prove* $\widetilde{\mathcal{P}}^1_0(E^{(\omega)}) \geq 4$. Fix $N \in \mathbb{N}$, $N \geq 2$. Write $\eta = 1 - 2^{-N}$, so $0 < \eta < 1$. Note $(1 + 2^{-N})\eta = 1 - 2^{-2N} < 1$. Let $\delta > 0$ be given. We will construct a δ-fine packing Π. Choose $M \in \mathbb{N}$ so that $2^{-M} < \delta/2$. Define

---

* A technical note: packing by balls all the same size is *not* the most efficient packing in this case!

$$N_0 = M, N_1 = M + N, \cdots, N_k = M + kN, \cdots.$$

For $k = 0, 1, \cdots$, let Π_k be the set of all constituents (σ, r) where $r = 2^{-N_k+1}\eta$ and the string σ, counting from the beginning, has letter 0 in locations $N_0, N_1, \cdots, N_{k-1}$, letter 1 in location N_k and all 0s beyond location N_k. Pictorially,

for Π_0

$$\underbrace{xx\cdots x1}_{M}\,000\cdots,$$

for Π_1

$$\underbrace{xx\cdots x0}_{M}\,\underbrace{xx\cdots x1}_{N}\,000\cdots,$$

and in general for Π_k

$$\underbrace{xx\cdots x0}_{M}\,\underbrace{xx\cdots x0}_{N}\,\underbrace{xx\cdots x0}_{N}\,\cdots\,\underbrace{xx\cdots x0}_{N}\,\underbrace{xx\cdots x1}_{N}\,000\cdots,$$

where there are k blocks of length N. In the the locations x, arbitrary letters are allowed. The number of elements in Π_k is determined by the number of locations where the letter may be freely chosen, so Π_k has $2^{M-1+k(N-1)} = 2^{N_k-k-1}$ elements.

Let $\Pi = \bigcup_k \Pi_k$. We claim that Π is a packing. Let $(\sigma, r), (\sigma', r') \in \Pi$. We must show $\varrho(\sigma, \sigma') > r + r'$. First suppose that $(\sigma, r), (\sigma', r')$ are in the same Π_k. So $r = r' = 2^{-N_k+1}\eta$. Strings σ, σ' differ somewhere in the first $N_k - 1$ places, so their longest common prefix has length at most $N_k - 2$, and $\varrho(\sigma, \sigma') \geq 2^{-N_k+2}$. On the other hand, $r + r' = 2^{-N_k+1}\eta + 2^{-N_k+1}\eta = 2^{-N_k+2}\eta < 2^{-N_k+2}$, as required. Now suppose $(\sigma, r) \in \Pi_k$, $(\sigma', r') \in \Pi_{k'}$, $k' > k$. Strings σ, σ' differ in location N_k, so their longest common prefix has length at most $N_k - 1$, so $\varrho(\sigma, \sigma') \geq 2^{-N_k+1}$. And $r + r' = 2^{-N_k+1}\eta + 2^{-N_{k'}+1}\eta \leq 2^{-N_k+1}(1 + 2^{-N})\eta < 2^{-N_k+1}$ as required.

Note that the packing Π is δ-fine, since for any k, the radius $2^{-N_k+1}\eta \leq 2^{-M+1}\eta < 2^{-M+1} \leq \delta$. Now compute

$$\widetilde{\mathcal{P}}_\delta^1(E^{(\omega)}) \geq \sum_{(\sigma, r) \in \Pi} (2r) = \sum_{k=0}^{\infty} \sum_{(\sigma, r) \in \Pi_k} (2r)$$

$$= \sum_{k=0}^{\infty} 2^{N_k-k-1} 2\eta 2^{-N_k+1} = 2\eta \sum_{k=0}^{\infty} 2^{-k} = 4\eta.$$

This holds for all $\delta > 0$, so $\widetilde{\mathcal{P}}_0^1(E^{(\omega)}) \geq 4\eta$. Now let $N \to \infty$ so that $\eta \to 1$, to get $\widetilde{\mathcal{P}}_0^1(E^{(\omega)}) \geq 4$.

From (a) and (b), we have $\widetilde{\mathcal{P}}_0^1(E^{(\omega)}) = 4 = 4\mathcal{M}(E^{(\omega)})$.

(c) Next: if $[\alpha]$ is a cylinder, then $\widetilde{\mathcal{P}}_0^1([\alpha]) = 4 \times 2^{-|\alpha|}$. The right shift $\sigma \mapsto \alpha\sigma$ is a similarity with ratio $2^{-|\alpha|}$, so this follows from the case already proved. Then we have also $\mathcal{P}^1([\alpha]) \leq 4\mathcal{M}([\alpha])$. The method I outer measure $\overline{\mathcal{M}}$ is the largest outer measure such that $\overline{\mathcal{M}}([\alpha]) \leq 2^{-|\alpha|}$ for all cylinders. So we conclude $\overline{\mathcal{P}}^1 \leq 4\overline{\mathcal{M}}$.

The clopen sets in $E^{(\omega)}$ are exactly the finite disjoint unions of cylinders $[\alpha]$. Two disjoint cylinders have positive distance. Therefore, by Lemma 6.2.6, for any clopen set V, we have $\widetilde{\mathcal{P}}_0^1(V) = 4\mathcal{M}(V)$.

(d) Now we claim that the class of clopen sets is a reduced cover class for $\overline{\mathcal{P}}^1$. Let F be a closed set and $\varepsilon > 0$. Now $\overline{\mathcal{M}}$ is a metric outer measure, so by Proposition 5.4.3, there is an open set $V \supseteq F$ with $\mathcal{M}(U \setminus F) < \varepsilon/4$; while V is a union of cylinders, so by compactness we may replace it by a finite union. Then applying Lemma 6.2.6 we have

$$\widetilde{\mathcal{P}}_0^1(V) \leq \widetilde{\mathcal{P}}_0^1(F) + \widetilde{\mathcal{P}}_0^1(V \setminus F) \leq \widetilde{\mathcal{P}}_0^1(F) + 4\mathcal{M}(V \setminus F) \leq \widetilde{\mathcal{P}}_0^1(F) + \varepsilon.$$

So the collection of clopen sets is a reduced cover class. This means, in the application of method I to define $\overline{\mathcal{P}}^1$, we may use only covers by clopen sets. But we have $\widetilde{\mathcal{P}}_0^1(V) = 4\mathcal{M}(V)$ for all clopen sets V, and $\overline{\mathcal{P}}^1$ is the largest outer measure such that $\overline{\mathcal{P}}^1(V) \leq \widetilde{\mathcal{P}}_0^1(V)$ for all clopen V. So $4\overline{\mathcal{M}} \leq \overline{\mathcal{P}}^1$.

Thus we get $\overline{\mathcal{P}}^1 = 4\overline{\mathcal{M}}$. Then in particular, $0 < \mathcal{P}^1(E^{(\omega)}) < \infty$, so $\operatorname{Dim} E^{(\omega)} = 1$. $\qquad\square$

The Line

Next is our first official example of a non-fractal. We proved $\operatorname{Cov}\mathbb{R} = 1$ in Theorem 3.2.15.

Proposition 6.3.3. *The Hausdorff dimension of the line $\mathbb{R}$ is 1.*

Proof. By Theorem 6.1.4, we have $\mathcal{H}^1([0,1]) = \mathcal{L}([0,1]) = 1$. Therefore $\dim[0,1] = 1$. Now $[0,1] \subseteq \mathbb{R}$, so $\dim \mathbb{R} \geq \dim[0,1] = 1$. If $s > 1$, then $\mathcal{H}^s([0,1]) = 0$. The intervals $[n, n+1]$ are isometric to $[0,1]$, so it follows that $\mathcal{H}^s([n, n+1]) = 0$. Therefore

$$\mathcal{H}^s(\mathbb{R}) \leq \sum_{n=-\infty}^{\infty} \mathcal{H}^s([n, n+1]) = 0.$$

This means that $\dim \mathbb{R} \leq s$. But this is true for any $s > 1$, so $\dim \mathbb{R} \leq 1$. Therefore we have seen that $\dim \mathbb{R} = 1$. $\qquad\square$

Exercise 6.3.4. The packing dimension of the line $\mathbb{R}$ is 1.

Lebesgue Measure vs. Hausdorff Measure

Since the Lebesgue measure was useful in computing $\dim \mathbb{R}$, it is easy to guess that $\mathcal{L}^2$ is useful in computing $\dim \mathbb{R}^2$.

Proposition 6.3.5. *The Hausdorff dimension of two-dimensional Euclidean space* $\mathbb{R}^2$ *is 2.*

Proof. Consider the (half-open) unit square $Q = [0, 1) \times [0, 1)$. It is covered by n^2 small squares with side $1/n$, so if $\varepsilon \geq \sqrt{2}/n$, we have $\overline{\mathcal{H}}_\varepsilon^2(Q) \leq n^2(\sqrt{2}/n)^2 = 2$. Therefore $\mathcal{H}^2(Q) \leq 2$, so $\dim Q \leq 2$.

On the other hand, if $\mathcal{A}$ is any cover of Q by closed sets, then (since any set A of diameter r is contained in a closed square Q_A with side $\leq r$),

$$\sum_{A \in \mathcal{A}} (\operatorname{diam} A)^2 \geq \sum_{A \in \mathcal{A}} \mathcal{L}^2(Q_A)$$

$$\geq \mathcal{L}^2 \left(\bigcup_{A \in \mathcal{A}} Q_A \right)$$

$$\geq \mathcal{L}^2(Q) = 1.$$

Therefore $\mathcal{H}^2(Q) \geq 1$, so $\dim Q \geq 2$.

For $\mathbb{R}^2$, since $Q \subseteq \mathbb{R}^2$, we have $\dim \mathbb{R}^2 \geq \dim Q = 2$. If $s > 2$, then $\mathcal{H}^s(Q) = 0$; but $\mathbb{R}^2$ can be covered by a countable collection $\{ Q_n : n \in \mathbb{N} \}$ of squares of side 1, so $\mathcal{H}^s(\mathbb{R}^2) \leq \sum_n \mathcal{H}^s(Q_n) = 0$. This shows that $\dim \mathbb{R}^2 \leq s$. Therefore $\dim \mathbb{R}^2 \leq 2$. □

Note that the proof showed $0 < \mathcal{H}^2(Q) < \infty$, where Q is the unit square.

What is the relation between the two measures $\mathcal{L}^2$ and $\mathcal{H}^2$ on $\mathbb{R}^2$? In fact, one of them is just a constant multiple of the other.

Theorem 6.3.6. *There is a positive constant c such that*

$$\mathcal{H}^2(B) = c\mathcal{L}^2(B)$$

for all Borel sets $B \subseteq \mathbb{R}^2$.

Proof. Let $Q = [0, 1) \times [0, 1)$ be the unit square. Let $c = \mathcal{H}^2(Q)$. (We have seen that $1 \leq c \leq 2$.) First, if $B = rQ = [0, r) \times [0, r)$, then $\mathcal{H}^2(B) = r^2 \mathcal{H}^2(Q) = r^2 c = c\mathcal{L}^2(B)$. Next, the same is true for a translate of such a square.

Both measures are metric measures, and these squares are Borel sets. So we have $\mathcal{H}^2(V) = c\mathcal{L}^2(V)$ for any finite disjoint union of squares. In particular, this holds for V belonging to the dyadic ring $\mathcal{R}$ (see p. 155).

Now consider $\mathcal{H}^2$ and $\mathcal{L}^2$ restricted to subsets of a large square $K = [-N, N] \times [-N, N]$. The dyadic ring $\mathcal{R}$ is a reduced measure class for $\mathcal{L}^2$ on K. And $\mathcal{H}^2(V) = c\mathcal{L}^2(V)$ it follows that $\mathcal{R}$ is also a reduced measure class for $\mathcal{H}^2$ on K. Since $\mathcal{H}^2$ and $\mathcal{L}^2$ agree on $\mathcal{R}$, their method I extensions also agree. Thus $\mathcal{L}^2$ and $\mathcal{H}^2$ agree on K.

By countable additivity, $\mathcal{H}^2$ and $\mathcal{L}^2$ agree on the whole plane $\mathbb{R}^2$. □

Let d be a positive integer. The same method may be used to prove that there exists a positive constant c_d such that $\mathcal{H}^d(B) = c_d \mathcal{L}^d(B)$ for all Borel sets $B \subseteq \mathbb{R}^d$.

Exercise 6.3.7. If $B \subseteq \mathbb{R}^d$, then $\dim B \leq d$. If B contains an open ball, then $\dim B = d$.

Arc Length

Let $f: [0, 1] \to S$ be a continuous curve in S. The **arc length** of the curve is

$$\sup \sum_{i=1}^{n} \varrho\big(f(x_{i-1}), f(x_i)\big),$$

where the supremum is over all finite subdivisions

$$0 = x_0 < x_1 < \cdots < x_n = 1$$

of the interval $[0, 1]$. If the arc length is finite, then we say that the curve is **rectifiable**.

Theorem 6.3.8. *Let* $f: [0, 1] \to S$ *be a continuous curve, let l be its arc length, and write* $C = f\big[[0, 1]\big]$.

(a) $l \geq \mathcal{H}^1(C);$
(b) *If f is one-to-one, then* $l = \mathcal{H}^1(C)$.

Proof. (a) Let $\varepsilon > 0$. Now f is uniformly continuous (Theorem 2.3.21), so there is $\delta > 0$ such that $\varrho\big(f(x), f(y)\big) < \varepsilon$ whenever $|x - y| < \delta$. Choose a subdivision

$$0 = x_0 < x_1 < \cdots < x_n = 1$$

of $[0, 1]$ with $|x_i - x_{i-1}| < \delta$ for all i. Then the sets

$$A_i = f\Big[[x_{i-1}, x_i]\Big]$$

cover C. (But diam A_i may not be $\varrho\big(f(x_{i-1}), f(x_i)\big)$.) By the compactness of $[x_{i-1}, x_i]$, there exist y_i, z_i with $x_{i-1} \leq y_i < z_i \leq x_i$ such that diam $A_i = \varrho\big(f(y_i), f(z_i)\big)$. Now we may use the subdivision

$$0 \leq y_1 \leq z_1 \leq y_2 \leq z_2 \leq \cdots \leq y_n \leq z_n \leq 1$$

to estimate the length. So

$$l \geq \sum_{i=1}^{n} \varrho\big(f(y_i), f(z_i)\big) = \sum_{i=1}^{n} \operatorname{diam} A_i \geq \overline{\mathcal{H}}_{\varepsilon}^1(C).$$

Now let $\varepsilon \to 0$ to obtain $l \geq \mathcal{H}^1(C)$.

(b) First, I claim that if $0 \leq a < b \leq 1$, then $\mathcal{H}^1\big(f\big[[a, b]\big]\big) \geq \varrho\big(f(a), f(b)\big)$. To see this, consider the function $h: f\big[[a, b]\big] \to \mathbb{R}$ defined by $h(u) = \varrho\big(f(a), u\big)$. Now h is continuous, and h has values $h(a) = 0$ and $h(b) = \varrho\big(f(a), f(b)\big)$, so by the intermediate value theorem, applied to the continuous function $h \circ f: [a, b] \to \mathbb{R}$, we know that h also has all values between. Now h satisfies the Lipschitz condition $|h(u) - h(v)| \leq \varrho(u, v)$, and we have

$$\mathcal{H}^1\Big(f[[a,b]]\Big) \geq \mathcal{H}^1\Big(h[f[[a,b]]]\Big)$$
$$\geq \mathcal{H}^1\Big([0, \varrho(f(a), f(b))]\Big)$$
$$= \varrho(f(a), f(b)).$$

This proves the claim.

Now we apply this inequality. If we have a subdivision

$$0 = x_0 < x_1 < \cdots < x_n = 1$$

of $[0, 1]$, then the set $f[[x_{i-1}, x_i)] = f[[x_{i-1}, x_i]] \setminus \{f(x_i)\}$ is the difference of two compact sets, hence measurable. The sets $f[[x_{i-1}, x_i)]$ are disjoint, since f is one-to-one. So

$$\sum_{i=1}^n \varrho(f(x_{i-1}), f(x_i)) \leq \sum_{i=1}^n \mathcal{H}^1\Big(f[[x_{i-1}, x_i)]\Big)$$
$$= \mathcal{H}^1\left(\bigcup_{i=1}^n f[[x_{i-1}, x_i)]\right)$$
$$= \mathcal{H}^1\Big(f[[0, 1)]\Big) \leq \mathcal{H}^1(C).$$

This is true for all subdivisions, so $l \leq \mathcal{H}^1(C)$. □

Exercise 6.3.9. What is the relation between the surface area (of a surface in $\mathbb{R}^3$) and its two-dimensional Hausdorff measure?

Fractal Dimension vs. Topological Dimension

We will see in the next section that it is possible for the Hausdorff dimension $\dim F$ to have a non-integer value. But it is not completely unrelated to the topological dimension.

Theorem 6.3.10. Let S be a metric space. Then $\operatorname{Cov} S \leq \dim S$.

A complete proof of this result can be found, for example, in [18, Sect. 3.1]. (The proof uses Lebesgue integration, which we have avoided in this book.) Here, we will prove it for compact spaces:

Theorem 6.3.11. Let S be a compact metric space. Then $\operatorname{Cov} S \leq \dim S$.

Proof. Let $n = \operatorname{Cov} S$. This means that $\operatorname{Cov} S \leq n - 1$ is false. So there exist open sets $U_1, U_2, \cdots, U_{n+1}$ such that $\bigcup_{i=1}^{n+1} U_i = S$, but for any closed sets $F_i \subseteq U_i$ with $\bigcup_{i=1}^{n+1} F_i = S$, we must have $\bigcap_{i=1}^{n+1} F_i \neq \varnothing$.

Define functions on S as follows:

$$d_i(x) = \text{dist}(x, S \setminus U_i), \qquad 1 \leq i \leq n+1$$
$$d(x) = d_1(x) + d_2(x) + \cdots + d_{n+1}(x).$$

The functions are continuous—in fact, Lipschitz:

$$|d_i(x) - d_i(y)| \leq \varrho(x, y),$$
$$|d(x) - d(y)| \leq (n+1)\varrho(x, y).$$

Since the sets U_i cover S, we have $d(x) > 0$ for all x. So since S is compact, there exist positive constants a, b such that $a \leq d(x) \leq b$ for all $x \in S$. Now define $h \colon S \to \mathbb{R}^{n+1}$ by

$$h(x) = \left(\frac{d_1(x)}{d(x)}, \frac{d_2(x)}{d(x)}, \cdots, \frac{d_{n+1}(x)}{d(x)} \right).$$

The function h is Lipschitz:

$$\left| \frac{d_i(x)}{d(x)} - \frac{d_i(y)}{d(y)} \right| = \frac{|d(x)d_i(y) - d(y)d_i(x)|}{d(x)d(y)}$$
$$\leq \frac{d(x)\,|d_i(y) - d_i(x)| + d_i(x)\,|d(x) - d(y)|}{d(x)d(y)}$$
$$\leq \frac{b(n+2)}{a^2}\,\varrho(x, y),$$

and therefore

$$|h(x) - h(y)| \leq \frac{b(n+1)(n+2)}{a^2}\,\varrho(x, y).$$

Now I claim that $h[S]$ includes the simplex

$$T = \left\{ (t_1, t_2, \cdots, t_{n+1}) \in \mathbb{R}^{n+1} : t_i > 0, \sum_{i=1}^{n+1} t_i = 1 \right\}.$$

Given $(t_1, t_2, \cdots, t_{n+1}) \in T$, consider the sets

$$F_i = \left\{ x : \frac{d_i(x)}{d(x)} \geq t_i \right\}.$$

Then F_i is closed, $F_i \subseteq U_i$, and $\bigcup_{i=1}^{n+1} F_i = S$ since $\sum_i d_i(x)/d(x) = 1$. So we know by hypothesis that $\bigcap_{i=1}^{n+1} F_i \neq \varnothing$. That is, there exists a point $x \in S$ with $d_i(x)/d(x) \geq t_i$ for all i. But since $\sum_i d_i(x)/d(x) = 1$ we have $d_i(x)/d(x) = t_i$ for all i. That is, $h(x) = (t_1, t_2, \cdots t_{n+1})$. So $h[S] \supseteq T$.

Now T is isometric to an open set in $\mathbb{R}^n$. By Theorem 6.1.7 and Exercise 6.1.10, we have $\dim S \geq \dim T = n$. $\qquad\square$

Exercise 6.3.12. Let S be a metric space. If $\operatorname{Cov} S \geq 1$, then $\dim S \geq 1$.

6.4 Self-Similarity

Self-similarity is one of the easiest ways to produce examples of fractals. This section deals with the question of when the similarity dimension can be used to compute the Hausdorff dimension. When the two coincide we have a desirable situation: the similarity dimension is easy to compute, and the Hausdorff dimension is more generally applicable and has many useful properties.

Let $(r_1, r_2, \cdots, r_n)$ be a contracting ratio list. Let $(f_1, f_2, \cdots, f_n)$ be an iterated function system of similarities realizing the ratio list in a complete metric space S. Let s be the sim-value for the iterated function system. Let K is the invariant set for the iterated function system. Of course, K is a measurable set, since it is compact. Does it follow that $\dim K = s$? In general, the answer is no. There is always an inequality $\dim K \leq s$. But simple examples show that if there is "too much" overlap among the pieces $f_i[K]$, then $\dim K < s$ is possible.

String Models

Hausdorff and packing measures are often easy to compute for the string models we use. Or if not easy to compute exactly, easy to estimate. Often estimates are good enough, since to compute the fractal dimensions dim and Dim it is enough to know merely whether $\mathcal{H}^s$ or $\mathcal{P}^s$ is positive or finite.

When computing $\mathcal{H}^s$ or $\mathcal{P}^s$ in our string spaces, we often already have a candidate measure $\mathcal{M}$. This helps in the computation.

Lemma 6.4.1. *Let E be a finite alphabet, let $E^{(\omega)}$ be the space of all infinite strings constructed from E. Let $s > 0$ and let $\overline{\mathcal{M}}$ be a finite metric outer measure on $E^{(\omega)}$. (i) If $\widetilde{\mathcal{P}}_0^s([\alpha]) = \mathcal{M}([\alpha])$ for all $\alpha \in E^{(*)}$, then $\overline{\mathcal{P}}^s = \overline{\mathcal{M}}$. (ii) If $\widetilde{\mathcal{P}}_0^s([\alpha]) \leq \mathcal{M}([\alpha])$ for all $\alpha \in E^{(*)}$, then $\overline{\mathcal{P}}^s \leq \overline{\mathcal{M}}$.*

Proof. The balls in $E^{(\omega)}$ are the cylinders $[\alpha]$. The clopen sets in $E^{(\omega)}$ are the finite disjoint unions of cylinders. Write $\mathcal{R}$ for the class of clopen sets in $E^{(\omega)}$. Note that $\mathcal{R}$ is an algebra of sets. By Lemma 6.2.6, we have $\widetilde{\mathcal{P}}_0^s(V) = \mathcal{M}(V)$ for all $V \in \mathcal{R}$ in case (i) and $\widetilde{\mathcal{P}}_0^s(V) \leq \mathcal{M}(V)$ in case (ii).

Let $F \subseteq E^{(\omega)}$ be closed, and let $\varepsilon > 0$. Then there is an open set $U \supseteq F$ with $\mathcal{M}(U \setminus F) < \varepsilon$. The open set U is a union of cylinders, so by compactness of F there is a finite union V of cylinders with $F \subseteq V \subseteq U$. (Alternatively, think of this as the fact that $E^{(\omega)}$ is zero-dimensional.) We conclude (as in the proof of Proposition 6.3.2) that $\widetilde{\mathcal{P}}_0^s(V) \leq \widetilde{\mathcal{P}}_0^s(F) + \varepsilon$. So $\mathcal{R}$ is a reduced cover class for $\mathcal{P}^s$. For any set $A \subseteq E^{(\omega)}$ we have

$$\overline{\mathcal{P}}^s(A) = \sup\left\{\sum_n \widetilde{\mathcal{P}}_0^s(F_n) : A \subseteq \bigcup_n F_n, \ F_n \text{ closed}\right\}$$

$$= \sup\left\{\sum_n \widetilde{\mathcal{P}}_0^s(V_n) : A \subseteq \bigcup_n V_n, \ V_n \in \mathcal{R}\right\}$$

$$= \sup\left\{\sum_n \mathcal{M}(V_n) : A \subseteq \bigcup_n V_n, \ V_n \in \mathcal{R}\right\} = \overline{\mathcal{M}}(A)$$

in case (i) and inequality $\leq$ in case (ii). $\square$

Exercise 6.4.2. Let E be a finite alphabet, let $E^{(\omega)}$ be the space of all infinite strings constructed from E. Let $s > 0$ and let $\overline{\mathcal{M}}$ be a finite metric outer measure on $E^{(\omega)}$. If $\overline{\mathcal{H}}^s([\alpha]) = \mathcal{M}([\alpha])$ for all $\alpha \in E^{(*)}$, then $\overline{\mathcal{H}}^s = \overline{\mathcal{M}}$. If $\overline{\mathcal{H}}^s([\alpha]) \leq \mathcal{M}([\alpha])$ for all $\alpha \in E^{(*)}$, then $\overline{\mathcal{H}}^s \leq \overline{\mathcal{M}}$.

The Natural Measure

Begin with a contracting ratio list $(r_1, r_2, \cdots, r_n)$, with $n > 1$. Then the sim-value s associated with it is the unique positive number s satisfying

$$\sum_{i=1}^n r_i^s = 1.$$

Let E be an n-letter alphabet, and let $E^{(\omega)}$ be the string model. The metric ϱ on E is defined so that the right shifts realize the given ratio list. We define $r(\alpha)$ recursively, starting with the empty string Λ, by:

$$r(\Lambda) = 1,$$
$$r(\alpha e) = r(\alpha)\, r_e,$$

then define ϱ so that $\operatorname{diam}[\alpha] = r(\alpha)$.

We will also need a measure defined to fit the ratio list. The basis is the equation defining the sim-value s:

$$\sum_{i=1}^n r_i^s = 1.$$

It follows from this that

$$\sum_{i=1}^n \left(r(\alpha)r_i\right)^s = r(\alpha)^s.$$

That is, the expression $r(\alpha)^s$ satisfies the additivity condition for a metric outer measure (Theorem 5.5.4). The measure $\mathcal{M}$ in question is defined on the string space $E^{(\omega)}$, and satisfies $\mathcal{M}([\alpha]) = r(\alpha)^s$ for all α. Of course, it is no coincidence that s was chosen so that $\mathcal{M}([\alpha]) = (\operatorname{diam}[\alpha])^s$.

Theorem 6.4.3. *Let $E^{(\omega)}$ have metric ϱ and measure $\mathcal{M}$ defined from ratio list (r_e) with sim-value s. Then (a) $\mathcal{H}^s = \mathcal{M}$, (b) there is a constant $c > 0$ so that $\mathcal{P}^s = c\mathcal{M}$, and thus (c) $\dim E^{(\omega)} = \operatorname{Dim} E^{(\omega)} = s$.*

Proof. Write $r_{\max} = \max_e r_e$ and $r_{\min} = \min_e r_e$.

(a) If a set $A \subseteq E^{(\omega)}$ has positive diameter, then (Proposition 2.6.7) there is a string $\alpha \in E^{(*)}$ with $A \subseteq [\alpha]$ and $\operatorname{diam} A = \operatorname{diam}[\alpha]$. So $\mathcal{M}(A) \leq \mathcal{M}([\alpha]) = (\operatorname{diam}[\alpha])^s = (\operatorname{diam} A)^s$. But $\overline{\mathcal{H}}_\varepsilon^s$ is the largest outer measure with $\overline{\mathcal{H}}_\varepsilon^s(A) \leq (\operatorname{diam} A)^s$ for all sets A of diameter $\leq \varepsilon$. So $\overline{\mathcal{M}} \leq \overline{\mathcal{H}}_\varepsilon^s$. This is true for all $\varepsilon > 0$, so $\overline{\mathcal{M}} \leq \overline{\mathcal{H}}^s$.

On the other hand, let $\alpha \in E^{(*)}$ be a finite string, and $\varepsilon > 0$. There is n so large that $r_{\max}^n < \varepsilon$, $n \geq |\alpha|$, and so $r(\beta) < \varepsilon$ for all $\beta \in E^{(n)}$. The cylinder $[\alpha]$ is the disjoint union of all sets $[\beta]$, where $\beta \geq \alpha$ and $|\beta| = n$. Then

$$\overline{\mathcal{H}}_\varepsilon^s([\alpha]) \leq \sum_{\substack{\beta \geq \alpha \\ |\beta| = n}} \left(\operatorname{diam}[\beta]\right)^s = \sum_{\substack{\beta \geq \alpha \\ |\beta| = n}} \mathcal{M}([\beta]) = \mathcal{M}([\alpha]).$$

Let $\varepsilon \to 0$ to get $\overline{\mathcal{H}}^s([\alpha]) \leq \mathcal{M}([\alpha])$. Therefore $\overline{\mathcal{H}}^s \leq \overline{\mathcal{M}}$.

(b) Now consider the packing measure. We will show that $c = \widetilde{\mathcal{P}}_0^s(E^{(\omega)})$ satisfies the condition. Note that

$$\widetilde{\mathcal{P}}_0^s(E^{(\omega)}) \geq \mathcal{P}^s(E^{(\omega)}) > 0$$

by part (a) and Proposition 6.2.14.

We will describe the balls in $E^{(\omega)}$. Let $\sigma \in E^{(\omega)}$ and $0 < t < 1$. Consider the ball $\overline{B}_t(\sigma)$. The ratios $r(\sigma{\restriction}n)$ go to 0 as $n \to \infty$, and $r(\Lambda) = 1$. So there is a unique n with $r(\sigma{\restriction}n) \leq t < r(\sigma{\restriction}(n-1))$. Then as in the proof of Proposition 6.3.2 we have $\overline{B}_t(\sigma) = [\alpha]$ where $\alpha = \sigma{\restriction}n$. And $\mathcal{M}(\overline{B}_t(\sigma)) = \mathcal{M}([\alpha]) = r(\alpha)^s$. Estimate from above: $\mathcal{M}(\overline{B}_t(\sigma)) = r(\alpha)^s \leq t^s$. Estimate from below: $\mathcal{M}(\overline{B}_t(\sigma)) = r(\sigma{\restriction}n)^s \geq \left(r_{\min} r(\sigma{\restriction}(n-1))\right)^s > (r_{\min}/2)^s (2t)^s$.

Now let Π be a packing of $E^{(\omega)}$. The corresponding balls $\overline{B}_t(\sigma), (\sigma, t) \in \Pi$, are disjoint. So

$$\sum_{(\sigma,t)\in\Pi} (2t)^s < \frac{2^s}{r_{\min}^s} \sum_{(\sigma,t)\in\Pi} \mathcal{M}(\overline{B}_t(\sigma)) \leq \frac{2^s}{r_{\min}^s}.$$

This holds for all δ-fine packings, so $\widetilde{\mathcal{P}}_\delta^s(E^{(\omega)}) \leq 2^s/r_{\min}^s$. This holds for all $\delta > 0$, so $\widetilde{\mathcal{P}}_0^s(E^{(\omega)}) \leq 2^s/r_{\min}^s < \infty$.

Thus $c = \widetilde{\mathcal{P}}_0^s(E^{(\omega)})$ is positive and finite. Now if $\alpha \in E^{(*)}$, then the right shift $\sigma \mapsto \alpha\sigma$ is a similarity with ratio $r(\alpha)$, and it maps $E^{(\omega)}$ onto $[\alpha]$. Therefore $\widetilde{\mathcal{P}}_0^s([\alpha]) = r(\alpha)^s c = c\mathcal{M}([\alpha])$. And thus, by Lemma 6.4.1(i), we have $\mathcal{P}^s = c\mathcal{M}$. $\qquad\square$

Exercise 6.4.4. Let s be any positive real number. There is a metric space S with $\dim S = s$.

Exercise 6.4.5. Prove versions of Lemma 6.4.1, Exercise 6.4.2, and Theorem 6.4.3 for the path spaces $E_v^{(\omega)}$ defined by a directed multigraph (V, E, i, t).

Exercise 6.4.6. Compute the value of the constant c in Theorem 6.4.3(b).

Cantor Dust

Let us consider the fractal dimension of the triadic Cantor dust (defined on p. 2). The ratio list for this set is $(1/3, 1/3)$. The string model is the set $E^{(\omega)}$ of infinite strings from the alphabet $E = \{0, 1\}$, together with the metric $\varrho_{1/3}$. The two similarities on the model space are the right shifts, say θ_0 and θ_1, defined as follows:

$$\theta_0(\sigma) = 0\sigma$$
$$\theta_1(\sigma) = 1\sigma.$$

Thus (θ_0, θ_1) is a realization of the ratio list $(1/3, 1/3)$, with invariant set $E^{(\omega)}$.

Proposition 6.4.7. *The Hausdorff dimension and packing dimension for $E^{(\omega)}$ with metric $\varrho_{1/3}$ are both $\log 2 / \log 3$.*

Proof. For the sim-value, solve $2\, (1/3)^s = 1$ for s to get $s = \log 2 / \log 3$. By Theorem 6.4.3 we have dim $=$ Dim $= s$.   □

Corollary 6.4.8. *The Cantor dust has Hausdorff dimension and packing dimension $\log 2 / \log 3$.*

Proof. A lipeomorphism preserves the Hausdorff dimension (Exercise 6.1.10) and the packing dimension (Exercise 6.2.12). The addressing function h from $E^{(\omega)}$ onto the triadic Cantor dust C is a lipeomorphism (Proposition 2.6.3).   □

Sierpiński gasket

Next we discuss a slightly more difficult example, the Sierpiński gasket (see p. 8).

Let S be the Sierpiński gasket. It is the invariant set for an iterated function system with ratio list $(1/2, 1/2, 1/2)$. Let s $[= \log 3 / \log 2]$ be the sim-value of the ratio list. Let $E = \{L, U, R\}$ be the appropriate three-letter alphabet. Next we describe the natural metric and measure defined on $E^{(\omega)}$ from the ratio list.

Let ϱ be the metric on $E^{(\omega)}$ for the ratio list $(1/2, 1/2, 1/2)$. That is, ϱ is defined so that $\operatorname{diam}[\alpha] = 2^{-|\alpha|}$ for all $\alpha \in E^{(*)}$. Then the right shifts realize the ratio list:

$$\varrho(L\sigma, L\tau) = \frac{1}{2}\, \varrho(\sigma, \tau),$$

$$\varrho(U\sigma, U\tau) = \frac{1}{2}\, \varrho(\sigma, \tau),$$

$$\varrho(R\sigma, R\tau) = \frac{1}{2}\, \varrho(\sigma, \tau).$$

The measure $\mathcal{M}$ is specified by $\mathcal{M}([\alpha]) = 3^{-|\alpha|}$. Each node in $E^{(*)}$ has exactly 3 children, so these numbers satisfy the required additivity (Theorem 5.5.4). The fact to notice is this:

$$\mathcal{M}([\alpha]) = \big(\operatorname{diam}[\alpha]\big)^s$$

for all $\alpha \in E^{(*)}$, where $s = \log 3 / \log 2$. By Theorem 6.4.3, the Hausdorff dimension of the string space $E^{(\omega)}$ is $s = \log 3 / \log 2$.

The dimension calculation for the string model will be used to help with the dimension calculation of the Sierpiński gasket S itself.

Let $h \colon E^{(\omega)} \to \mathbb{R}^2$ be the addressing function that sends $E^{(\omega)}$ onto the gasket S. If the iterated function system in $\mathbb{R}^2$ is $(f_\mathsf{L}, f_\mathsf{U}, f_\mathsf{R})$, then

$$h(\mathsf{L}\sigma) = f_\mathsf{L}\big(h(\sigma)\big),$$
$$h(\mathsf{U}\sigma) = f_\mathsf{U}\big(h(\sigma)\big),$$
$$h(\mathsf{R}\sigma) = f_\mathsf{R}\big(h(\sigma)\big).$$

Proposition 6.4.9. *The Sierpiński gasket has Hausdorff dimension and packing dimension at most* $\log 3 / \log 2$.

Proof. The addressing function h is Lipschitz (Exercise 4.2.1). By Exercise 6.1.10, we have $\dim S \leq \operatorname{Dim} S \leq \operatorname{Dim} E^{(\omega)} = \log 3 / \log 2$. □

For the general iterated function system, the upper bound is proved in the same way.

Theorem 6.4.10. *Let* $(r_e)_{e \in E}$ *be a contracting ratio list. Let* s *be its sim-value, and let* $(f_e)_{e \in E}$ *be a realization in a complete metric space* S. *Let* K *be the invariant set. Then* $\dim K \leq \operatorname{Dim} K \leq s$.

Proof. The string model $E^{(\omega)}$ with the natural metric ϱ has $\operatorname{Dim} E^{(\omega)} = s$ (Theorem 6.4.3). The addressing function $h \colon E^{(\omega)} \to K$ is Lipschitz. Therefore $\operatorname{Dim} K \leq s$. □

Lower Bound

The addressing function for the Sierpiński gasket is not inverse Lipschitz. In fact, it is not even one-to-one. (This is the answer to Exercise 4.2.2.) So we will need a bit more effort to prove the lower bound for the fractal dimension of S. Pay attention to the ingredients of the proof, since they will be used again for the general case. To simplify the notation, we will write $\mathsf{L}(x)$ in place of $f_\mathsf{L}(x)$, and similarly for the other two letters, and write $\alpha(x)$ for a finite string α.

Proposition 6.4.11. *The Sierpiński gasket* S *has Hausdorff dimension equal to the similarity dimension* $\log 3 / \log 2$.

Proof. Let V be the interior of the first triangle S_0 approximating the Sierpiński gasket S. Then $\mathcal{L}^2(V) = \sqrt{3}/4$, and if $|\alpha| = |\beta|$, $\alpha \neq \beta$, then $\alpha[V] \cap \beta[V] = \varnothing$. Also, $h[[\alpha]] = \overline{\alpha[V]} \cap S$. The set S_k approximating S is the union

$$\bigcup_{\alpha \in E^{(k)}} \overline{\alpha[V]}.$$

Given a set $A \subseteq S$, let k be the positive integer satisfying

$$2^{-k} < \operatorname{diam} A \leq 2^{-k+1}.$$

Let

$$T = \left\{ \alpha \in E^{(k)} : \overline{\alpha[V]} \cap A \neq \varnothing \right\}.$$

I claim that T has at most 100 elements. Let m be the number of elements in T. A set $\alpha[V]$ is the image of V under a similarity with ratio 2^{-k}, so it has area

$$\mathcal{L}^2(\alpha[V]) = 4^{-k}\frac{\sqrt{3}}{4}.$$

The sets $\alpha[V]$ with $\alpha \in T$ are all disjoint. If x is a point of A, then all of the elements of all of the sets $\alpha[V]$ with $\alpha \in T$ are within distance $\operatorname{diam} A + 2^{-k} \leq 3 \cdot 2^{-k}$ of x. So m disjoint sets of area $4^{-k}\sqrt{3}/4$ are contained in the ball with center x and radius $3 \cdot 2^{-k}$. Therefore

$$m4^{-k}\frac{\sqrt{3}}{4} \leq \pi(3 \cdot 2^{-k})^2.$$

Solving for m, we get $m \leq 36\pi/\sqrt{3}$, which is smaller than 100.

Next I claim $\mathcal{M}(h^{-1}[A]) \leq 100 \, (\operatorname{diam} A)^s$ for all Borel sets $A \subseteq S$. Given A, let k and T be as above. Then $A \subseteq \bigcup_{\alpha \in T} \overline{\alpha[V]}$, so $h^{-1}[A] \subseteq \bigcup_{\alpha \in T}[\alpha]$. Therefore

$$\mathcal{M}(h^{-1}[A]) \leq \sum_{\alpha \in T} \mathcal{M}([\alpha])$$
$$\leq 100 \times 3^{-k} = 100 \, (2^{-k})^s$$
$$\leq 100 \, (\operatorname{diam} A)^s.$$

By the Method I theorem, $\mathcal{M}(h^{-1}[A]) \leq 100, \mathcal{H}^s(A)$ for all Borel sets A. So $1 \leq 100, \mathcal{H}^s(S)$, and therefore $\dim S \geq s$. □

Exercise 6.4.12. Improve the estimate 100.

6.5 The Open Set Condition

Let $(r_1, \cdots, r_n)$ be a contracting ratio list with dimension s. Let $(f_1, \cdots, f_n)$ be a corresponding iterated function system of similarities in $\mathbb{R}^d$. Suppose K is the invariant set for the iterated function system. Write s for the sim-value.

In general it is not true that $\dim K = s$. For example, consider the iterated function system(f_L, f_U, f_R) for the Sierpiński gasket, realizing the ratio list $(1/2, 1/2, 1/2)$. Now the iterated function system(f_L, f_L, f_U, f_R) has the same invariant set, of course, but it realizes the longer ratio list $(1/2, 1/2, 1/2, 1/2)$. The Hausdorff dimension of the invariant set K is $\log 3 / \log 2$, but the sim-value of the iterated function system is 2.

Of course, the problem is that the first two images $f_L[K]$ and $f_L[K]$ overlap too much. Now we might require that the images do not overlap at all, as in the Cantor dust. But that would rule out many of the most interesting examples, such as the Sierpiński gasket itself, where the overlap sets like $f_L[K] \cap f_U[K]$ are nonempty.

We do have inequality between the Hausdorff dimension, packing dimension, and similarity dimension. If s is the similarity dimension, then the string model has packing dimension s and the addressing function is Lipschitz, so $\dim K \leq \mathrm{Dim}\, K \leq s$.

Lower Bound

Now we turn to the "lower bound" proof. That is, we want to show $\dim K \geq$ something or $\mathrm{Dim}\, K \geq$ something. Generally we do this by showing $\mathcal{H}^s(K) > 0$ or $\mathcal{P}^s(K) > 0$.

The iterated function system $(f_1, f_2, \cdots, f_n)$ satisfies **Moran's open set condition** iff there exists a nonempty open set U for which we have $f_i[U] \cap f_j[U] = \varnothing$ for $i \neq j$ and $U \supseteq f_i[U]$ for all i. Such an open set U will be called a **Moran open set** for the iterated function system.

For example, consider the Cantor dust. The similarities are

$$f_0(x) = \frac{x}{3},$$

$$f_1(x) = \frac{x+2}{3}.$$

The open set $U = (0, 1)$ is a Moran open set: the two images are $(0, 1/3)$ and $(2/3, 1)$, which are disjoint and contained in U.

Or, consider the Sierpiński gasket (Fig. 1.2.1). The interior* U of the large triangle S_0 is a Moran open set. The three images are three small triangles, contained in U, and disjoint.

For a third example, consider the Koch curve (Fig. 1.5.1). The interior of the triangle L_0 is a Moran open set.

The fourth example to consider is Heighway's dragon. This time the open set condition is not quite as trivial. The interior U of Heighway's dragon itself will serve. The fact that the two images are contained in U is a consequence of the fact that Heighway's dragon itself is the invariant set of the iterated function system. The fact that the two images are disjoint is a consequence of the

---

* The interior of a set consists of all the interior points of the set.

fact that the approximating polygon never crosses itself (Proposition 1.5.7). The verification requires some work, but it is left to the reader:

Exercise 6.5.1. The interior of Heighway's dragon is a Moran open set for the iterated function system realizing Heighway's dragon.

In one case, the open set condition is easily verified:

Exercise 6.5.2. If the invariant set K for an iterated function system $\{f_i\}$ satisfies $f_i[K] \cap f_j[K] = \varnothing$ for $i \neq j$, then Moran's open set condition is satisfied.

The proof of the lower bound will proceed following the same technique as Proposition 6.4.11. The area is replaced by the d-dimensional volume, namely $\mathcal{L}^d$. You may find it instructive to compare this argument with the proof of the special case in Proposition 6.4.11.

Let E be an alphabet with n letters. Write the ratio list as $(r_e)_{e \in E}$ and the iterated function system as $(f_e)_{e \in E}$. To simplify the notation, we will write $e(x)$ in place of $f_e(x)$, and similarly $\alpha(x)$ for a finite string α. With this notation, the model map $h \colon E^{(\omega)} \to \mathbb{R}^d$ satisfies $h(e\alpha) = e(h(\alpha))$ for $\alpha \in E^{(*)}$ and $e \in E$.

The open set condition implies that $\alpha[U] \cap \beta[U] = \varnothing$ for two strings $\alpha, \beta \in E^{(*)}$ unless one is an initial segment of the other. If α is a string with length $k \geq 1$, we will write α^- for the parent of α; that is: $\alpha^- = \alpha{\restriction}(k-1)$.

Lemma 6.5.3. *Let $(r_e)_{e \in E}$ be a contracting ratio list. Let s be its sim-value, and let $(f_e)_{e \in E}$ be a realization in $\mathbb{R}^d$. Let K be the invariant set. Let U be a Moran open set for (f_e). Then there is a constant $c > 0$ such that: if $A \subseteq K$, then the set*

$$T = \left\{ \alpha \in E^{(*)} : \overline{\alpha[U]} \cap A \neq \varnothing, \ \operatorname{diam} \alpha[U] < \operatorname{diam} A \leq \operatorname{diam} \alpha^-[U] \right\}$$

has at most c elements.

Proof. As α ranges over T, the sets $\alpha[U]$ are disjoint, since no such α is an initial segment of another. The map f_α is a similarity with ratio equal to $\operatorname{diam}[\alpha]$, so if w is the diameter of U, then $w \operatorname{diam}[\alpha]$ is the diameter of $\alpha[U]$. Write $r_{\min} = \min r_e$. Then

$$\operatorname{diam} \alpha[U] = w \operatorname{diam}[\alpha] \geq w r_{\min} \operatorname{diam}[\alpha^-]$$
$$= r_{\min} \operatorname{diam} \alpha^-[U] \geq r_{\min} \operatorname{diam} A.$$

If $p = \mathcal{L}^d(U)$ is the volume of U, then the volume of $\alpha[U]$ is

$$\mathcal{L}^d(\alpha[U]) = p \left(\frac{\operatorname{diam} \alpha[U]}{\operatorname{diam} U} \right)^d \geq \frac{p r_{\min}^d}{w^d} (\operatorname{diam} A)^d$$

If x is a point of A, then every point of every set $\alpha[U]$ for $\alpha \in T$ is within distance $\operatorname{diam} A + \operatorname{diam} \alpha[U] \leq 2 \operatorname{diam} A$ of x. If m is the number of elements of T, then we have m disjoint sets $\alpha[U]$, all with volume at least $(pr_{\min}^d/w^d)(\operatorname{diam} A)^d$, contained within a ball of radius $2 \operatorname{diam} A$. So if $t = \mathcal{L}^d(B_1(0))$ is the volume of the unit ball, we have

$$\frac{mpr_{\min}^d}{w^d} (\operatorname{diam} A)^d \leq t(2 \operatorname{diam} A)^d.$$

Solving for m yields

$$m \leq \frac{tw^d 2^d}{pr_{\min}^d}.$$

Summary: We may use the constant $c = tw^d 2^d/pr_{\min}^d$, where $r = \min r_e$, t is the volume of the unit ball, p is the volume of the Moran open set U, and w is the diameter of U.   □

Theorem 6.5.4. *Let $(r_e)_{e \in E}$ be a contracting ratio list. Let s be its sim-value, and let $(f_e)_{e \in E}$ be a realization in $\mathbb{R}^d$. Let K be the invariant set. If Moran's open set condition is satisfied, then $\dim K = s$.*

Proof. Let c be a constant as in the lemma. I claim there is a positive constant b so that for any Borel set $A \subseteq K$, we have

$$\mathcal{M}(h^{-1}[A]) \leq b (\operatorname{diam} A)^s.$$

Let U be a Moran open set, and let $w = \operatorname{diam} U$. Given A, let T be as in the lemma. So $A \subseteq \bigcup_{\alpha \in T} \overline{\alpha[U]}$, and $h^{-1}[A] \subseteq \bigcup_{\alpha \in T}[\alpha]$. If $\alpha \in T$, then $\mathcal{M}([\alpha]) = (\operatorname{diam}[\alpha])^s = ((1/w) \operatorname{diam} \alpha[U])^s \leq (1/w^s)(\operatorname{diam} A)^s$. Therefore

$$\mathcal{M}(h^{-1}[A]) \leq \sum_{\alpha \in T} \mathcal{M}([\alpha])$$
$$\leq c(1/w^s)(\operatorname{diam} A)^s.$$

So $b = c/w^s$ will work.

Therefore, by the Method I theorem, $1 = \mathcal{M}(h^{-1}[K]) \leq b\mathcal{H}^s(K)$, so we have $\dim K \geq s$.   □

The proof given above clearly used the properties of Lebesgue measure in $\mathbb{R}^d$. What happens in other metric spaces? Readers who know about some exotic metric spaces may like to attempt this:

Exercise 6.5.5. Let S be a complete metric space (other than $\mathbb{R}^d$). Let $(f_1, f_2, \cdots, f_n)$ be a realization in S of a contracting ratio list $(r_1, r_2, \cdots, r_n)$ with dimension s. Let K be the invariant set. Suppose Moran's open set condition is satisfied. Does it follow that $\dim K = s$?

Exercise 6.5.6. Let $(r_1, r_2, \cdots, r_n)$ be a contracting ratio list with dimension s. Let $(f_1, f_2, \cdots, f_n)$ be an iterated function system consisting not of similarities, but of maps $f_i \colon \mathbb{R}^d \to \mathbb{R}^d$ satisfying

$$\varrho\big(f_i(x), f_i(y)\big) \ge r_i\, \varrho(x,y).$$

Suppose the open set condition holds, and suppose there is an invariant set K. Does it follow that $\dim K \ge s$?

Heighway Dragon Boundary

Heighway's Dragon (p. 20) is a set P in the plane with nonempty interior (a space-filling curve) which tiles the plane (p. 74). We have $\operatorname{Cov} P = \dim P = \operatorname{Dim} P = 2$, so it is not a fractal in the sense of Mandelbrot. We now have the tools to analyze the boundary of P. It turns out that ∂P is a fractal. So the tile P is an example of what we have called a fractile.

Recall the discussion on p. 74 showing that P tiles the plane. Let us continue with this line of reasoning. In Plate 16 we see a black segment from A to B that produces a sequence P_n of polygons that converge to P, the black tile in Plate 17.

First note that the point A in P_0 also belongs to all P_n and is a boundary point of P, since point A lies not only in the black tile P, but also in the brown, gray, and red tiles. Similarly point B is a boundary point of P. We will write $\partial P = U \cup V$ as follows. Set U is the portion of the boundary to the left of curve AB—that is, the points that belong not only to the black tile, but also to at least one of the brown, blue, or yellow tiles. Set V is the portion of the boundary to the right of curve AB—that is, the points that belong not only to the black tile, but also to at least one of the red, green, or cyan tiles. (In fact, the cyan tile never meets the black tile, as we can see by looking at P_1 in Plate 16.) No other tiles can touch the black tile, because the plane has topological dimension 2, so they would have to cross one of the curves shown to reach the black tile.

Consider set U. According to P_1 in Plate 16, the "midpoint" C of P is a boundary point, since it lies in both the black and blue tiles. The portion of U between A and C is a copy of U shrunk by factor $1/\sqrt{2}$. The portion of U between B and C is a copy of V shrunk by factor $1/\sqrt{2}$.

Now consider set V. Look at P_2 in Plate 16. The "three-quarter" point D of P is a boundary point, because it lies in both the black and red tiles. The portion of V between B and D is a copy of U shrunk by factor $1/2$. The portion of V between D and A is the boundary between black and red; looking at it from the red point of view, we see it is a copy of U shrunk by factor $1/2$.

Set V is made up of two copies of U, so $\dim U = \dim V$ and $\operatorname{Dim} U = \operatorname{Dim} V$. Set U is made up of one copy of U (shrunk by factor $1/\sqrt{2}$) and one copy of V (shrunk by factor $1/\sqrt{2}$). That copy of V is made up of two copies of U each shrunk by a further factor of $1/2$. So the complete decomposition has U made up of three copies of itself, with ratio list $(2^{-1/2}, 2^{-3/2}, 2^{-3/2})$.

This ratio list has sim-value $s \approx 1.52$ given by $s = 2 \log \lambda / \log 2$, where $\lambda \approx 1.6956$ is a solution of $\lambda^3 - \lambda^2 - 2 = 0$. In order to claim that this value s is also the Hausdorff and packing dimensions of U, we need an open set condition. Define an open set G as follows: in Plate 16 start with the four segments: black, yellow, blue, brown. The curve U lies in the union of the four tiles they produce: see the black, yellow, blue, brown in Plate 17. The open set G will be the interior of this union. Set G is shown in blue in Plate 18, with U shown in yellow. The images of G under the three maps that make up the iterated function system for U are shown in the second picture. The large red set is an image of G shrunk by factor $1/\sqrt{2}$. The cyan and green sets are images of G shrunk by factor $2^{-3/2}$. These (open) images are disjoint, since the tiles generated from different segments in Plate 16 have disjoint interiors. The three images descend from the edges bordering the like-colored squares shown in P_3 of Plate 18. This completes our description of the open set condition.

Eisenstein Boundary

Let us consider next the set F of "fractions" for the Eisenstein number system (Fig. 1.6.11). The base is $b = -2$, and the digit set consists of 0, 1, $A = \omega$, and $B = \omega^2$. The set F may be done in the same way as the twindragon (p. 33). But let us proceed in a more direct way. The first set L_0 is just the point 0. The next set L_1 consists of the four points $(.0)_{-2}$, $(.1)_{-2}$, $(.A)_{-2}$, and $(.B)_{-2}$. The set L_2 contains 16 points, all that can be represented with two digits in this system. The illustrations in Fig. 6.5.7 show these approximations. The set F obtained this time has (of course) similarity dimension 2.

To see that F is a "fractile" we need to analyze its boundary. The Eisenstein boundary is made up of six congruent parts, as shown in Fig. 6.5.8. The individual parts are self-similar. Each consists of three copies of itself, shrunk by factor $1/2$. See Fig. 6.5.9(a).

Drawings in Logo can be done. When it is done as a curve, we need to take into account that some parts are drawn backward and/or reflected.

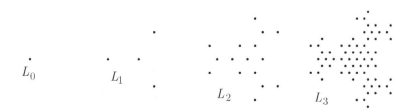

Fig. 6.5.7. Eisenstein fractions

Fig. 6.5.8. The Eisenstein boundary consists of six congruent parts

```
to EBforward "depth "size "parity
  if :depth = 0 [forward :size stop] [
    left (60 * :parity)
    EBbackward (:depth - 1) (:size / 2) (-:parity)
    right (120 * :parity)
    EBforward (:depth - 1) (:size / 2) :parity
    left (60 * :parity)
    EBforward (:depth - 1) (:size / 2) (-:parity)]
end
to EBbackward "depth "size "parity
  if :depth = 0 [forward :size stop] [
    EBbackward (:depth - 1) (:size / 2) (-:parity)
    right (60 * :parity)
    EBbackward (:depth - 1) (:size / 2) :parity
    left (120 * :parity)
    EBforward (:depth - 1) (:size / 2) (-:parity)
    right (60 * :parity)]
end
```

The iterated function system and the open set condition for this curve are both illustrated by Fig. 6.5.9(b)(c). So we may conclude that the Eisenstein boundary has fractal dimension s (both Hausdorff and packing) satisfying $3\,(1/2)^s = 1$, so $s = \log 3/\log 2$. This is > 1, so this boundary is a fractal. That is, F is a fractile.

Examples for the Reader

Here is another dragon curve (Fig. 6.5.10, Plate 11), known as the ***terdragon***.

Fig. 6.5.9. (a) Decomposition   (b)(c) Iterated function system and OSC

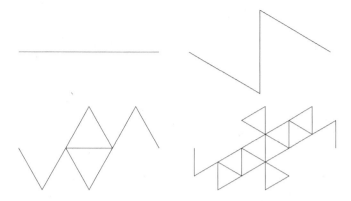

Fig. 6.5.10. Terdragon

```
make "shrink 1 / sqrt 3
to ter :depth :size
  if :depth = 0 [forward :size stop] [
    right 30
    ter :depth - 1 :size * :shrink
    left 120
    ter :depth - 1 :size * :shrink
    right 120
    ter :depth - 1 :size * :shrink
    left 30]
end
```

This is a space-filling curve. (Six copies of the terdragon exactly fit around a point; Plate 12.) It is not a fractal. But what about its boundary? Is it a "fractile"?

Exercise 6.5.11. Compute the Hausdorff and packing dimensions for the boundary of the terdragon.

A variant of the dragon that constructs the McWorter pentigree (p. 24) is shown in Fig. 6.5.12. Five copies of the limit set fit together to form a certain "dendrite". This will be called the ***pentadendrite***. (Fig. 6.5.13).

Exercise 6.5.14. Compute the topological, Hausdorff, and packing dimensions of the pentadendrite.

The set of "fractions" for the number system with base $-1 + i$ and digit set $\{0, 1\}$ is the twindragon (Fig. 1.6.8).

Exercise 6.5.15. Compute the Hausdorff and packing dimensions of the boundary of the twindragon.

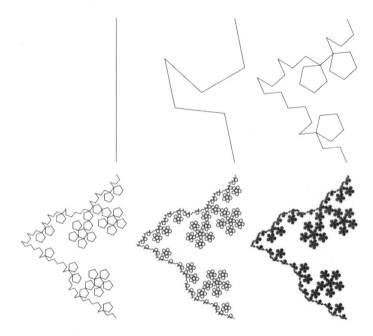

Fig. 6.5.12. Pentadendrite construction

Fig. 6.5.13. Complete pentadendrite

Exercise 6.5.16. Compute the Hausdorff and packing dimensions for the limit of the sets constructed by the following program (p. 116). Warning: it is not self-similar.

```
to Schmidt :depth :size
   if :depth = 0 [stop] [
      repeat 3 [
         forward :size
         Schmidt :depth - 1 :size / 2
         right 120] ]
end
```

Another example of the same kind is the "I" fractal, p. 135.

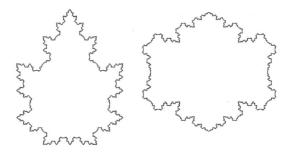

Fig. 6.5.17. Leaf outline; leaf lake

Barnsley Leaf Outline

Recall Barnsley's leaf B (p. 26). Because of the overlap among the parts of the iterated function system, the open set condition fails, so the methods of this section cannot compute its dimension (as far as I know). Discussion of overlap is in Sect. 7.1. But there are some related dimensions that can be done now. When all of the surrounded areas of B are filled in, we get a solid leaf with fractal boundary. Call that boundary the "leaf outline" J. Or consider one of the small regions surrounded by the leaf: call that a "leaf lake" K.

We will do a "deconstruction" of the leaf in Sect. 7.1 and conclude that there is a set H so that B consists of 8 copies of H (and its reflection H') arranged as shown in Fig. 6.5.18: four copies of the tile H on the right, and four copies of the reflected tile H' on the left.

Accordingly, the outline J is made up of 8 copies of the "upper left" edge of H (when oriented as shown). Tile H is deconstructed as in Fig. 6.5.19(a); the portion in the lower right is irrelevant for us now. Thus the segment of the leaf outline obeys an iterated function system also represented by 6.5.19(a). The picture also provides a Moran open set. The ratio list is $(2^{-1}, 2^{-3/2}, 2^{-3/2})$. The fractal dimension is $-2 \log r / \log 2 \approx 1.21076$, where $r \approx 0.657298$ is a solution of the cubic $r^2 + 2r^3 = 1$.

6.6 Graph Self-Similarity

Next we consider evaluation of the Hausdorff dimension connected with graph self-similar sets.

Fig. 6.5.18. Deconstruction of Barnsley's leaf

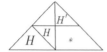   

Fig. 6.5.19. (a) Decomposition of H (b) Self-similarity of the segment

The Two-Part Dust

We begin with a simple example, the **two-part dust**. It has been "rigged" so that the calculation of the dimension is easier than the general case. The Mauldin–Williams graph is as illustrated in Fig. 6.6.1. Here is a description of the realization in $\mathbb{R}^2$ that will be considered. The map a has ratio $1/2$, fixed point $(0,0)$, and rotation 30 degrees counterclockwise. The map b has ratio $1/4$, fixed point $(1,0)$, and rotation 60 degrees clockwise. The map c has ratio $1/2$, fixed point $(0,0)$, and rotation 90 degrees counterclockwise. The map d has ratio $3/4$, fixed point $(1,0)$, and rotation 120 degrees clockwise.

As we know, there is a unique pair of nonempty compact sets $U, V \subseteq \mathbb{R}^2$ satisfying

$$U = \mathsf{a}[U] \cup \mathsf{b}[V]$$
$$V = \mathsf{c}[U] \cup \mathsf{d}[V].$$

This pair of sets is the **two-part dust**. A sequence of approximations is pictured in Fig. 6.6.2. They converge in the Hausdorff metric. We may start with any two nonempty compact sets U_0 and V_0 in $\mathbb{R}^2$. In this case, both have been chosen as the line segment from the point $(0,0)$ to the point $(1,0)$. Then further approximations are defined recursively:

$$U_{n+1} = \mathsf{a}[U_n] \cup \mathsf{b}[V_n],$$
$$V_{n+1} = \mathsf{c}[U_n] \cup \mathsf{d}[V_n].$$

The sequence (U_n) converges in the Hausdorff metric to a nonempty compact set U, and the sequence (V_n) converges in the Hausdorff metric to a nonempty compact set V. This pair of sets is the required invariant list.

Here is the Logo program for the pictures.

```
; two-part dust
to U :depth :size
```

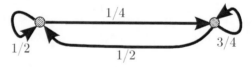

Fig. 6.6.1. Graph for the two-part dust

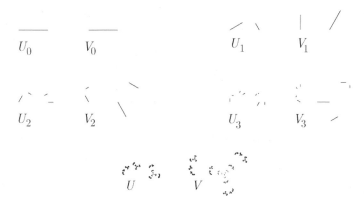

Fig. 6.6.2. Two-part dust

```
if :depth = 0 [forward :size stop] [
   left 30
   U :depth - 1 :size / 2
   penup
       back :size / 2
       right 30
       forward :size
       right 60
       back :size / 4
   pendown
   V :depth - 1 :size / 4 left 60]
end
to V :depth :size
  if :depth=0 [forward :size stop] [
   left 90
   U :depth - 1 :size / 2
   penup
       back :size / 2
       right 90
       forward :size
       right 120
       back :size * 0.75
   pendown
   V :depth - 1 :size * 0.75
   left 120]
end
```

We are interested in computing the Hausdorff (and packing) dimensions of the sets U and V. Since each of the sets is similar to a subset of the other, their dimensions must be the same. As usual, we begin by computing the dimension of the path models corresponding to the Mauldin–Williams graph.

We will need to use the **Perron numbers** of the graph. In this case, the Perron numbers (one for each node) are $q_U = 1/3$ and $q_V = 2/3$. The important facts about these numbers are: they are positive and they satisfy equations

$$q_U = r(a)q_U + r(b)q_V,$$
$$q_V = r(c)q_U + r(d)q_V. \tag{1}$$

By Proposition 4.3.16, the graph has* sim-value 1. I want to show that U and V have Hausdorff dimension 1. We will use the Perron numbers to assign diameters to the nodes of the path forest. Two ultrametrics ϱ, one on each of the two path spaces $E_U^{(\omega)}$, $E_V^{(\omega)}$, will be defined so that the diameters of the basic open sets $[\alpha]$ are as follows: Begin with $\mathrm{diam}[\Lambda_U] = q_U$ and $\mathrm{diam}[\Lambda_V] = q_V$. If α is a path and e is an edge with $t(e) = i(\alpha)$, then the diameter for the string $e\alpha$ is $\mathrm{diam}([e\alpha]) = r(e)\,\mathrm{diam}([\alpha])$.

Next, we will use the same numbers to define measures.

There is a metric measure $\mathcal{M}$ on each of the path spaces $E_v^{(\omega)}$ such that $\mathcal{M}([\alpha]) = \mathrm{diam}([\alpha])$ for all finite paths α. The additivity condition (Exercise 5.5.5) is true by equations (1). Of course we can repeat the steps from Lemma 6.4.1 and Exercise 6.4.2 to conclude that the path spaces $E_U^{(\omega)}$, $E_V^{(\omega)}$ both have Hausdorff and packing dimension 1. (Therefore, we say that U and V have [graph] similarity dimension 1.)

But we are more interested in the Hausdorff and packing dimensions of U and V. We will need an "open set condition." A little experimentation with a graphics program reveals that this may be satisfied by the two sets pictured in Fig. 6.6.3. The two sets are a rectangle and an irregular hexagon. (The images of these sets under the maps are appropriately disjoint and contained in the appropriate sets.) The dimension is now easy to check.

Exercise 6.6.4. Let U and V be the two parts of the two part dust. Show that the addressing functions

$$h_U \colon E_U^{(\omega)} \to \mathbb{R}^2$$
$$h_V \colon E_V^{(\omega)} \to \mathbb{R}^2$$

are lipeomorphisms.

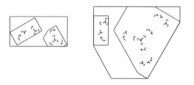

Fig. 6.6.3. Open set condition

---

* Solution to Exercise 4.3.14.

As consequences of this we have $\dim U = \operatorname{Dim} U = \dim E_U^{(\omega)} = 1$ and $\dim V = \operatorname{Dim} V = \dim E_V^{(\omega)} = 1$.

Perron Numbers

To compute the Hausdorff dimension for the other examples with graph self-similarity, we need only find the proper sort of "Perron numbers" in those cases. (It will not be quite as simple as the rigged example above if the dimension is not 1.)

Consider a Mauldin-Williams graph (V, E, i, t, r). We will consider only the case when the graph is strongly connected (p. 80). This will mean that when the invariant set list is found, each of the sets will be similar to a subset of each of the others. So they will all have the same fractal dimension.

We are interested in assigning metrics to the spaces $E_v^{(\omega)}$ of strings. (There is one space for each tree in the path forest.) The realization consists, as usual, of the right shifts. For an edge $e \in E_{uv}$, the function θ_e defined by

$$\theta_e(\sigma) = e\sigma$$

maps $E_v^{(\omega)}$ to $E_u^{(\omega)}$. The metrics should be chosen in such a way that θ_e is a similarity with ratio $r(e)$. We are also interested in defining measures (one for each space $E_v^{(\omega)}$) that will make the computation of the Hausdorff dimension easy.

In order to do this, we need the proper Perron numbers. If s is a positive real number, then s-dimensional **Perron numbers** for the graph are positive numbers q_v, one for each vertex $v \in V$, such that

$$q_u^s = \sum_{\substack{v \in V \\ e \in E_{uv}}} r(e)^s\, q_v^s$$

for all $u \in V$.

There is exactly one positive number s such that s-dimensional Perron numbers exist. This unique number s will be called the **sim-value** of the Mauldin-Williams graph. The existence and uniqueness of the sim-value were proved for the case of a 2 node graph in Sect 4.3. For the general graph, the proof requires some linear algebra. See Theorem 6.9.5.

We can proceed even without the proof of this result: if we can find Perron numbers, then we will be able to do the computations. When the set V of nodes is small, finding Perron numbers can often be done by trial and error.

Fractal Dimension

Now that all of the ingredients have been specified, we may proceed to analyze the Hausdorff and packing dimensions in this case. Suppose that (V, E, i, t, r)

is a strongly connected, contracting Mauldin-Williams graph. Let $s > 0$ be such that s-dimensional Perron numbers q_v exist. We will suppose that the graph is strictly contracting so that $r(e) < 1$ for all e. We will compute the dimension for the path model. There is one path space $E_v^{(\omega)}$ for each node $v \in V$.

First we need metrics for the path spaces. We want the right shifts to realize similarities with the ratios assigned by the Mauldin-Williams graph. For each finite path α, let $r(\alpha)$ be the product of the numbers $r(e)$, for all the edges e in α. For $\alpha \in E_{uv}^{(*)}$, we want the diameter of $[\alpha]$ to be $r(\alpha)q_v$.

Ultrametrics ϱ exist with these diameters. (One for each space $E_v^{(*)}$.) They satisfy

$$\varrho(e\sigma, e\tau) = r(e)\, \varrho(\sigma, \tau)$$

for $\sigma, \tau \in E_v^{(*)}$ and $e \in E_{uv}$.

Next we want to define measures on the path spaces. (The measures will all be called $\mathcal{M}$.) Because of the equations satisfied by the Perron numbers, we see that the values $(\mathrm{diam}[\alpha])^s$ satisfy the additivity condition (Theorem 5.5.4), namely

$$(\mathrm{diam}[\alpha])^s = \sum_{i(e)=t(\alpha)} (\mathrm{diam}[\alpha e])^s.$$

So there exists a metric measure on each of the spaces $E_v^{(\omega)}$ satisfying $\mathcal{M}([\alpha]) = (\mathrm{diam}[\alpha])^s$ for all $\alpha \in E_v^{(*)}$.

We can easily find an upper bound for the packing dimension. This is done in much the same way as has been done in previous cases. By Lemma 6.4.1, there is a positve constant c such that $\mathcal{P}^s(E_v^{(\omega)}) = c\mathcal{M}(E_v^{(\omega)}) = cq_v^s$. And by Exercise 6.4.2, $\mathcal{H}^s(E_v^{(\omega)}) = \mathcal{M}(E_v^{(\omega)}) = q_v^s$. But $0 < q_v^s < \infty$, so $\dim E_v^{(\omega)} = \mathrm{Dim}\, E_v^{(\omega)} = s$.

Once we know the Hausdorff and packing dimensions of the path spaces, we can try to apply it to the sets in $\mathbb{R}^d$ that we are really interested in. Let $(f_e)_{e \in E}$ be an iterated function system realizing the Mauldin-Williams graph (V, E, i, t, r) in $\mathbb{R}^d$. Let $(K_v)_{v \in V}$ be the unique invariant list of nonempty compact sets. As usual, we may construct the addressing functions

$$h_v \colon E_v^{(\omega)} \to \mathbb{R}^d,$$

one for each $v \in V$, such that

$$h_u(e\sigma) = f_e(h_v(\sigma)),$$

for $\sigma \in E_v^{(*)}$ and $e \in E_{uv}$. Then $h_v[E_v^{(\omega)}] = K_v$ for $v \in V$. These are Lipschitz maps, so the upper bound for the fractal dimensions is easy: $\dim K_v \leq \mathrm{Dim}\, K_v \leq \mathrm{Dim}\, E_v^{(\omega)} = s$.

6.7 Graph Open Set Condition

For the lower bound, we need to limit the overlap. We will use an open set condition.

Definition 6.7.1. If (f_e) is a realization of (V, E, i, t, r) in $\mathbb{R}^d$, then we say it satisfies the **open set condition** iff there exist nonempty open sets U_v, one for each $v \in V$, with

$$U_u \supseteq f_e[U_v]$$

for all $u, v \in V$ and $e \in E_{uv}$; and

$$f_e[U_v] \cap f_{e'}[U_{v'}] = \varnothing$$

for all $u, v, v' \in V$, $e \in E_{uv}$, $e' \in E_{uv'}$ with $e \neq e'$.

Now the argument proceeds as before. Fix a node $v \in V$. I want to show that $\dim K_v \geq s$. As before, if α is a finite (nonempty) string, write α^- for its parent. Also, we will use the notation $e(x)$ for $f_e(x)$, and similarly for strings: $\alpha(x)$.

First, I claim that there is a constant $c > 0$ such that: if $A \subseteq K_v$, then the set

$$T = \left\{ \alpha \in E_v^{(*)} : \overline{\alpha[U_{t(\alpha)}]} \cap A \neq \varnothing, \right.$$

$$\left. \operatorname{diam} \alpha[U_{t(\alpha)}] < \operatorname{diam} A \leq \operatorname{diam} \alpha^-[U_{t(\alpha^-)}] \right\}$$

has at most c elements.

Writing

$$w_{\max} = \max_{u \in V} \operatorname{diam} U_u, \qquad w_{\min} = \min_{u \in V} \operatorname{diam} U_u, \qquad r_{\min} = \min_{e \in E} r_e,$$

we have for $\alpha \in T$:

$$\operatorname{diam} \alpha[U_{t(\alpha)}] = r(\alpha) \operatorname{diam} U_{t(\alpha)} \geq w_{\min} r_{\min} r(\alpha^-)$$

$$\geq \frac{w_{\min} r_{\min}}{w_{\max}} \operatorname{diam} \alpha^-[U_{t(\alpha^-)}] \geq \frac{w_{\min} r_{\min}}{w_{\max}} \operatorname{diam} A.$$

Now if $p = \min_{u \in V} \mathcal{L}^d(U_u)$, we have the volume calculation

$$\mathcal{L}^d(\alpha[U_{t(\alpha)}]) \geq p \left(\frac{\operatorname{diam} \alpha[U_{t(\alpha)}]}{\operatorname{diam} U_{t(\alpha)}} \right)^d \geq p \left(\frac{w_{\min} r_{\min}}{w^{*2}} \right)^d (\operatorname{diam} A)^d.$$

Now if $x \in A$, then every point of every set $\alpha[U_{t(\alpha)}]$ for $\alpha \in T$ is within distance $\operatorname{diam} A + \operatorname{diam} \alpha[U_{t(\alpha)}] < 2 \operatorname{diam} A$ of x. The sets $\alpha[U_{t(\alpha)}]$ are disjoint, so if T has m elements, then there are m disjoint sets, with volume at least $p(w_{\min} r_{\min}/w^{*2})^d (\operatorname{diam} A)^d$ inside a ball with radius $2 \operatorname{diam} A$. If $t = \mathcal{L}^d(B_1(0))$, we have

$$mp \left(\frac{w_{\min} r_{\min}}{w^{*2}} \right)^d (\operatorname{diam} A)^d \leq t(2 \operatorname{diam} A)^d.$$

Solving for m, we get

$$m \leq \frac{t}{p} \left(\frac{2w^{*2}}{w_{\min} r_{\min}} \right)^d.$$

Next, I claim that there is a constant $b > 0$ so that for any Borel set $A \subseteq K_v$, we have

$$\mathcal{M}\big(h_v^{-1}[A]\big) \leq b(\operatorname{diam} A)^s.$$

Given A, let T be as before. Write $q = \max_{u \in V} q_u$. Then $h_v^{-1}[A] \subseteq \bigcup_{\alpha \in T}[\alpha]$. If $\alpha \in T$, then

$$\mathcal{M}([\alpha]) \leq r(\alpha)^s q^s \leq q^s \left(\frac{\operatorname{diam} \alpha[U_{t(\alpha)}]}{w_{\min}} \right)^s \leq \frac{q^s (\operatorname{diam} A)^s}{w_{\min}^s}.$$

Therefore

$$\mathcal{M}\big(h_v^{-1}[A]\big) \leq \sum_{\alpha \in T} \mathcal{M}([\alpha]) \leq \frac{cq^s}{w_{\min}^s}(\operatorname{diam} A)^s.$$

Then we conclude from the Method I theorem that $\mathcal{M}(h_v^{-1}[A]) \leq b\mathcal{H}^s(A)$ for all Borel sets A. In particular,

$$\mathcal{H}^s(K_v) \geq \frac{\mathcal{M}(h_v^{-1}[K_v])}{b} = \frac{\mathcal{M}\big(E_v^{(\omega)}\big)}{b} = \frac{q_v^s}{b} > 0.$$

Therefore $\dim K_v \geq s$. And of course $\operatorname{Dim} K_v \geq \dim K_v$.

We have established the result:

Theorem 6.7.2. *Let (V, E, i, t, r) be a strongly connected contracting Mauldin-Williams graph describing the graph self-similarity of a list $(K_v)_{v \in V}$ of nonempty compact sets in $\mathbb{R}^d$. Let $s > 0$ be such that s-dimensional Perron numbers exist. Then $\dim K_v \leq \operatorname{Dim} K_v \leq s$ for all v. If, in addition, the realization satisfies the open set condition, then $\dim K_v = \operatorname{Dim} K_v = s$.*

Exercise 6.7.3. Let (V, E, i, t, r) be a strongly connected contracting Mauldin-Williams graph. Let $(f_e)_{e \in E}$ be a family of maps on $\mathbb{R}^d$ satisfying

$$\varrho(f_e(x), f_e(y)) \leq r(e)\varrho(x, y).$$

Formulate the proper analog of Theorem 6.7.2.

Exercise 6.7.4. Let (V, E, i, t, r) be a Mauldin-Williams graph. Let $(f_e)_{e \in E}$ be a family of maps on $\mathbb{R}^d$ satisfying $\varrho(f_e(x), f_e(y)) \geq r(e)\varrho(x, y)$. Formulate the proper analog of Theorem 6.7.2.

Exercise 6.7.5. Discuss Hausdorff dimension for graph self-similar sets with Mauldin-Williams graph not strongly connected.

Li Lake

The graph self-similar set called Li's Lace is seen on p. 84. Its description is on p. 126 shows it is is made up of isosceles right trianglar blocks of two kinds, called P and Q. An open space completely surrounded by the fractal is called a **lake**. The boundary of one lake is shown in Fig. 6.7.6(b). This boundary can also be described in the language of graph self-similarity. Fig. 6.7.7 identifies two boundary parts A and B in tile P. The lake boundary is made up of eight parts A.

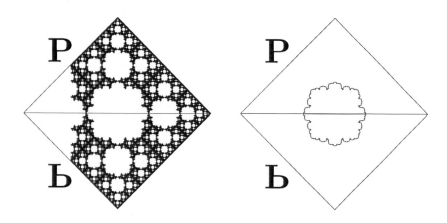

Fig. 6.7.6. (a) Two tiles P   (b) Lake boundary

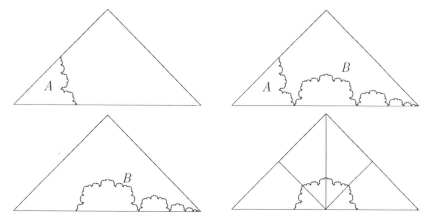

Fig. 6.7.7. (a) Parts A and B (b) Four parts A = half of boundary

Exercise 6.7.8. The parts A and B may be described using the descriptions of P and Q from Fig. 4.3.3. Set A is made up of two copies of A shrunk by factor $1/4$ and one copy of B shrunk by factor $1/4$. Set B is made up of four copies of A shrunk by factor $1/2$ and one copy of B shrunk by factor $1/2$. Compute the corresponding sim-value for sets A and B. Verify the OSC to conclude this is the Hausdorff and packing dimension for the lake boundary.

Pentigree Outline

Recall the construction on p. 25 of the second form of McWorter's pentigree. The **pentigree outline** is what we get if we fill in all the "lakes", and take the boundary of the result (Fig. 6.7.9, see also p. XIII).

Here is the program used to draw Fig. 6.7.9.

```
; pentigree outline
make "shrink (3 + sqrt 5)/2
to pent :depth :size
    repeat 5 [A :depth :size]
end
to A :depth :size
  if :depth = 0 [forward :size right 72 stop] [
    B :depth - 1 :size / :shrink
    A :depth - 1 :size / :shrink
    A :depth - 1 :size / :shrink
    BR :depth - 1 :size / :shrink]
end
to B :depth :size
  if :depth = 0 [forward :size left 36 stop] [
    C :depth - 1 :size / :shrink
    A :depth - 1 :size / :shrink
    BR :depth - 1 :size / :shrink]
end
to BR :depth :size
```

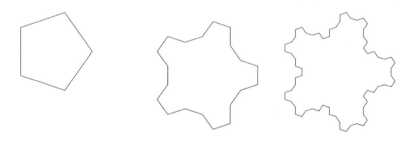

Fig. 6.7.9. Pentigree outline

```
    if :depth = 0 [forward :size left 36 stop] [
       B :depth - 1 :size / :shrink
       A :depth - 1 :size / :shrink
       C :depth - 1 :size / :shrink]
  end
  to C :depth :size
    if :depth = 0 [forward :size left 72 stop] [
       B :depth - 1 :size / :shrink
       BR :depth - 1 :size / :shrink]
  end
```

Exercise 6.7.10. Does this really converge to the outline of McWorter's pentigree?

Exercise 6.7.11. Determine the Mauldin-Williams graph describing the self-similarity of curve A.

Exercise 6.7.12. Compute the Hausdorff dimension of the pentigree outline.

Number Systems

Let b be a complex number, $|b| > 1$, and let D be a finite set of complex numbers, including 0. We are interested in the numbers that can be represented in the form

$$\sum_{j=1}^{\infty} a_j b^{-j}.$$

In some cases, the set of representations may be restricted to allow only certain combinations of digits. Consider $b = 3$ and $D = \{0, 1, 2\}$. Let F be the set of all numbers x of the form

$$x = \sum_{j=1}^{\infty} a_j b^{-j},$$

where each a_j is in the set D, and such that $a_j + a_{j+1} \leq 2$ for all j. This set is graph self-similar.

Let $F(d_1)$ be the set of numbers where the representation has $a_1 = d_1$. Let $F(d_1, d_2)$ be those numbers where the representation has $a_1 = d_1$ and $a_2 = d_2$. Then $F = F(0) \cup F(1) \cup F(2)$. The set $F(0)$ is similar to F, with ratio $1/3$. The set $F(1) = F(1, 0) \cup F(1, 1)$, since $F(1, 2) = \emptyset$. But $F(1, 0)$ is similar to F with ratio $1/9$ and $F(1, 1)$ is similar to $F(1)$ with ratio $1/3$. Finally, $F(2) = F(2, 0)$ is similar to F with ratio $1/9$. The graph is shown in Fig. 6.7.13.

Exercise 6.7.14. Compute the Hausdorff dimension of the set F.

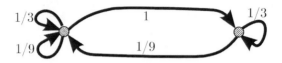

Fig. 6.7.13. Graph

Exercise 6.7.15. Let $b = (1 + \sqrt{5})/2$ and $D = \{0, 1\}$. Let F be the set of numbers of the form

$$\sum_{j=1}^{\infty} a_j b^{-j},$$

where each $a_j \in D$, and two consecutive digits 1 are not allowed. Describe the set F.

Topological Dimension

The addressing function, or model map, which has been developed here for the purpose of computing the fractal dimension of a [graph] self-similar set, can also sometimes be used for the topological dimension as well. The addressing functions $h_v \colon E_v^{(\omega)} \to K$ are continuous and surjective. The spaces $E_v^{(\omega)}$ are compact. When the overlap is small, the characterization of topological dimension of Theorem 3.4.19 is often applicable. Let us do some examples.

The addressing function for the Cantor dust is one-to-one. Therefore the Cantor dust is zero-dimensional.

The addressing function for the Sierpiński gasket maps at most 2 strings to each point. Therefore the Sierpiński gasket has small inductive dimension ≤ 1. It contains line segments, so it has dimension exactly 1.

Exercise 6.7.16. Show that McWorter's pentigree has small inductive dimension 1.

If you have not solved Exercise 1.6.5 yet, you can now do so painlessly.

Theorem 3.4.19 yields only an inequality. The addressing function for the Menger sponge (using the construction suggested in Fig. 4.1.9) is 4-to-1 at some points, so we obtain the uninteresting result that the topological dimension is ≤ 3. In fact, the topological dimension is 1. Is there another way to produce the Menger sponge as a self-similar set such that the model map is only at most 2-to-one?

6.8 *Other Fractal Dimensions

According to Mandelbrot, a ***fractal*** is a set S with $\dim S > \operatorname{Cov} S$. We will consider (for the moment) only nonempty compact sets in Euclidean space.

---

* Optional section.

Mandelbrot expressed dissatisfaction with the definition for two reasons: (1) "borderline fractals" are excluded; and (2) "true geometric chaos" is included.

Borderline Fractal

What might be meant by a "borderline fractal"? This will be a set K with the usual features of fractals that we have seen often, but where $\dim K = \operatorname{Cov} K$ anyway. To illustrate this, we will consider a curve, with a dragon-like construction. We begin with a sequence of positive numbers w_k, with $w_0 = 1$, $w_k > w_{k+1} > w_k/2$, and $\lim_{k \to \infty} w_k = 0$. The first set is a line segment P_0 with length $w_0 = 1$. If the polygon P_k has been constructed, consisting of many line segments of length w_k, then to construct P_{k+1}, we replace each of those line segments by two segments of length w_{k+1}. (It is possible to do this, and still have a polygon, since $w_{k+1} > w_k/2$.) If $w_k \to 0$ fast enough, we can avoid having the curve cross itself (even in the limit) by alternating between sides of the curve, as shown in the illustration. Then the limit will be homeomorphic to an interval, and therefore have small inductive dimension 1.

If we choose w_k that satisfy $w_k^s = 1/2^k$, for some $s > 1$, then this is a self-similar dragon curve that we have seen before. In the binary tree, if α is a finite string of length k, then when we use the metric and measure appropriate for the tree, we have $\operatorname{diam}[\alpha] = w_k$ and $\mathcal{M}([\alpha]) = 1/2^k = (\operatorname{diam}[\alpha])^s$. The usual calculation shows that the Hausdorff dimension for the curve will be s.

But suppose we have w_k that satisfy

$$\frac{w_k}{\log(1/w_k)} = \frac{1}{2^k}.$$

This means that w_k goes to zero more rapidly than $(1/2^k)^{1/s}$ for any $s > 1$, but more slowly than $1/2^k$ itself.

Exercise 6.8.2. When

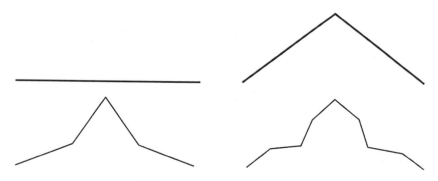

Fig. 6.8.1. Borderline dragon

$$\frac{w_k}{\log(1/w_k)} = \frac{1}{2^k},$$

the topological, Hausdorff, and packing dimensions are all 1.

Why would we call this a "borderline fractal"?

Box Dimensions

We will next discuss some other fractal dimensions known as box dimensions. We will begin with $\mathbb{R}^2$ for simplicity. But first consider a variant of the Hausdorff dimension.

If $r > 0$, then the **square net** of side r consists of all squares of the form

$$A = \big[(m-1)r, mr\big) \times \big[(n-1)r, nr\big),$$

where $m, n \in \mathbb{Z}$. Write $\mathcal{S}_r$ for this set of squares. So the plane $\mathbb{R}^2$ is the disjoint union of this countable collection of squares. Write $\mathcal{S} = \bigcup_{r>0} \mathcal{S}_r$.

For $s > 0$, consider the method II outer measure $\overline{\mathcal{M}}^s$ defined using the set function $\mathbf{c} \colon \mathcal{S} \to [0, \infty)$ defined by: $\mathbf{c}(A) = r^s$ if $A \in \mathcal{S}_r$. Now any set $B \subseteq \mathbb{R}^2$ of diameter r is contained in the union of at most 4 sets of $\mathcal{S}_r$. On the other hand, a square with side r has diameter $\sqrt{2}\,r$. This means that

$$2^{-s/2}\,\overline{\mathcal{H}}^s(F) \leq \overline{\mathcal{M}}^s(F) \leq 4\,\overline{\mathcal{H}}^s(F).$$

Therefore $s_0 = \dim F$ is the critical value for which

$$\overline{\mathcal{M}}^s(F) = \infty \qquad \text{for all } s < s_0;$$
$$\overline{\mathcal{M}}^s(F) = 0 \qquad \text{for all } s > s_0.$$

Some calculations involving the Hausdorff dimension are simplified by using this alternative to the definition (for example [23, Chapt. 5]).

Now we discuss another variant. Fix a number $r > 0$, and cover only by sets of $\mathcal{S}_r$; then let $r \to 0$. It should be emphasized that this is not method II. Now if $\mathcal{A} \subseteq \mathcal{S}_r$ covers a set F, then $\sum_{A \in \mathcal{A}} r^s$ is just $N r^s$, where N is the number of elements of $\mathcal{A}$. So the definition may be phrased as follows. Let $N_r(F)$ be the number of sets of the square net $\mathcal{S}_r$ that intersect F. Define

$$\widetilde{\mathcal{K}}_r^s(F) = N_r(F)\,r^s,$$
$$\widetilde{\mathcal{K}}^s(F) = \liminf_{r \to 0} \widetilde{\mathcal{K}}_r^s(F).$$

As usual there is a critical value for s.

Exercise 6.8.3. Let

$$s_0 = \liminf_{r \to 0} \frac{\log N_r(F)}{\log(1/r)}.$$

Then

$$\widetilde{\mathcal{K}}^s(F) = \infty \qquad \text{for all } s < s_0;$$
$$\widetilde{\mathcal{K}}^s(F) = 0 \qquad \text{for all } s > s_0.$$

The critical value s_0 will be called the **lower box dimension** or **lower box-counting dimension** or **lower entropy index** of F. We will write $\underline{\dim}_B F$. The set functions $\widetilde{\mathcal{K}}^s$ have the same shortcoming as $\widetilde{\mathcal{P}}^s$:

Exercise 6.8.4. There is a countable compact set K with positive lower box dimension.

Because of this undesirable property, we can apply Method I to $\widetilde{\mathcal{K}}^s$ to get a metric outer measure. Or we can modify the dimension directly. The **modified lower box dimension** is

$$\underline{\dim}_{MB} F = \inf \sup_i \underline{\dim}_B F_i$$

where the infimum is over all countable covers $F \subseteq \bigcup_{i \in \mathbb{N}} F_i$ of F.

A variant is obtained by replacing $\liminf$ with $\limsup$. The set functions are then

$$\limsup_{r \to 0} \widetilde{\mathcal{K}}^s_r(F).$$

The critical value for s, called the the **upper box dimension**,* is given by

$$\overline{\dim}_B F = \limsup_{r \to 0} \frac{\log N_r(F)}{\log(1/r)}.$$

Again the set function is not an outer measure, and $\overline{\dim}_B$ is not countably stable, so define the **modified upper box dimension** by

$$\overline{\dim}_{MB} F = \inf \sup_i \overline{\dim}_B F_i$$

where the infimum is over all countable covers $F \subseteq \bigcup_{i \in \mathbb{N}} F_i$ of F.

Exercise 6.8.5. Let $F \subseteq \mathbb{R}^2$. Then

$$\dim F \leq \underline{\dim}_B F \leq \overline{\dim}_B F \leq \text{packing index of } F,$$
$$\dim F \leq \underline{\dim}_{MB} F \leq \overline{\dim}_{MB} F \leq \text{Dim } F.$$

So if F is a fractal in the sense of Taylor, these four fractal dimensions all coincide. (In fact, we will see below that $\overline{\dim}_{MB} F = \text{Dim } F$.)

Exercise 6.8.6. Define the set functions $\widetilde{\mathcal{K}}^s$ and $\overline{\mathcal{K}}^s$ in the space $\mathbb{R}^d$. Prove analogs of Exercise 6.8.5.

---

* Barnsley [3] uses the term "fractal dimension" for this value.

Boxes do not make sense in a general metric space, but a box dimension can be defined there anyway. Let S be a metric space, let $F \subseteq S$, and let $r > 0$. Define $\dot{N}_r(F)$ to be the maximum number of disjoint closed balls with radius r and center in F. (Actually, instead of "disjoint" we can take the sense used for the packing measure: the centers x_i satisfy $\varrho(x_i, x_j) > 2r$ for all $i \neq j$.)

Proposition 6.8.7. *Let $F \subseteq \mathbb{R}^2$. Then*

$$\liminf_{r \to 0} \frac{\log N_r(F)}{\log(1/r)} = \liminf_{r \to 0} \frac{\log \dot{N}_r(F)}{\log(1/r)}$$

$$\limsup_{r \to 0} \frac{\log N_r(F)}{\log(1/r)} = \limsup_{r \to 0} \frac{\log \dot{N}_r(F)}{\log(1/r)}$$

Proof. Let $\delta > 0$. There are $\dot{N}_\delta(F)$ disjoint balls with radius δ and center in F. No two of those centers have distance $\leq 2\delta$, so no two of those centers are in the same square of $\mathcal{S}_{\delta\sqrt{2}}$. So $\dot{N}_\delta(F) \leq N_{\delta\sqrt{2}}(F)$.

Let $r > 0$. In each square of $\mathcal{S}_r$ that meets F, choose one point of F, and use it as the center of a closed ball of radius $r/4$. Any such ball intersects at most 4 squares of $\mathcal{S}_r$, and $N_r(F) \leq 4\dot{N}_{r/4}(E)$.

Therefore, for any $r > 0$ we have

$$\frac{\log N_r(F)}{\log(1/r)} \leq \frac{\log(4\dot{N}_{r/4}(F))}{\log((1/4)(4/r))} = \frac{\log \dot{N}_{r/4}(F) + \log 4}{\log(4/r) - \log 4}$$

$$\frac{\log N_r(F)}{\log(1/r)} \geq \frac{\log \dot{N}_{r/\sqrt{2}}(F)}{\log((1/\sqrt{2})(\sqrt{2}/r))} = \frac{\log \dot{N}_{r/\sqrt{2}}(F)}{\log(\sqrt{2}/r) - \log \sqrt{2}}.$$

But as $r \to 0$, we have $\log(1/r) \to \infty$, so we get the lim sup and lim inf results claimed. $\square$

In a metric space S, define the upper and lower box dimension using $\dot{N}$ in place of N. Then define the upper and lower modified box dimensions as before. It doesn't have "boxes" in the defintion any more. Sometimes $\overline{\dim}_B$ or $\underline{\dim}_B$ may be called **Bouligand dimension** or **Minkowski dimension** instead.

Proposition 6.8.8. *Let S be a metric space and $F \subseteq S$. Then $\overline{\dim}_{MB} F = \text{Dim } F$.*

Proof. First we claim: $\text{Dim } F \leq \overline{\dim}_B F$. Let $t < s < \text{Dim } F$. Then $\overline{\mathcal{P}}^s(F) = \infty$, so $\tilde{\mathcal{P}}_0^s(F) = \infty$ and $\tilde{\mathcal{P}}_\delta^s(F) = \infty$ for all $\delta > 0$. Now let δ be given with $0 < \delta < 1$. Then there is a δ-fine packing Π of F with $\sum_{(x,r) \in \Pi} r^s > 1$. For $k \in \mathbb{N}$, let n_k be the number of $(x, r) \in \Pi$ with $2^{-k-1} < r \leq 2^{-k}$. Then there is some k with $n_k > 2^{kt}(1 - 2^{t-2})$, since if not we would have

$$1 < \sum_{(x,r) \in \Pi} r^s \leq \sum_{k=0}^{\infty} n_k 2^{-ks} \leq (1 - 2^{t-s}) \sum_{k=0}^{\infty} 2^{k(t-s)} = 1.$$

For that k, we get n_k disjoint balls centered in F with radius 2^{-k-1}, so

$$\dot{N}_{2^{-k-1}}(F) \geq n_k > 2^{kt}(1 - 2^{t-2}),$$

$$\frac{\log \dot{N}_{2^{-k-1}}(F)}{\log(2^{k+1})} \geq \frac{kt \log 2 + \log(1 - 2^{t-s})}{(k+1) \log 2}.$$

Now the right-hand side goes to t as $k \to \infty$, and for any δ there is k as given with $2^{-k} < \delta$, so we conclude that $\overline{\dim}_{\mathrm{B}} F \geq t$. But t was any value $< \operatorname{Dim} F$, so in fact $\overline{\dim}_{\mathrm{B}} F \geq \operatorname{Dim} F$.

Next we claim $\operatorname{Dim} F \leq \overline{\dim}_{\mathrm{MB}} F$. Let $F \subseteq \bigcup_i F_i$ be a countable cover of F. Then

$$\operatorname{Dim} F \leq \sup_i \operatorname{Dim} F_i \leq \sup_i \overline{\dim}_{\mathrm{B}} F_i.$$

Take the infimum over all such covers to get $\operatorname{Dim} F \leq \overline{\dim}_{\mathrm{MB}} F$.

And finally we claim $\overline{\dim}_{\mathrm{MB}} F \leq \operatorname{Dim} F$. Let $s > \operatorname{Dim} F$. Then $\overline{\mathcal{P}}^s(F) = 0$, so there is a countable cover $F \subseteq \bigcup F_i$ with $\sum_i \widetilde{\mathcal{P}}_0^s(F_i) < \infty$ for all i. Fix an i. There is $\delta > 0$ with $\widetilde{\mathcal{P}}_\delta^s(F_i) < \infty$. Then since $\dot{N}_\delta(F_i)\delta^s \leq \widetilde{\mathcal{P}}_\delta^s(F_i)$, it remains bounded as $\delta \to 0$, so $\overline{\dim}_{\mathrm{B}} F_i \leq s$. This is true for all i, so $\sup_i \overline{\dim}_{\mathrm{B}} F_i \leq s$. Take the infimum on all covers to get $\overline{\dim}_{\mathrm{MB}} F \leq s$. This is true for all $s > \operatorname{Dim} F$, so $\overline{\dim}_{\mathrm{MB}} F \leq \operatorname{Dim} F$. $\qquad\square$

Lipschitz Condition of Order p

Let S, T be metric spaces, and let $p > 0$. We say that a function $f \colon S \to T$ satisfies a **Lipschitz condition** of order p if there is a constant M so that for all $x, y \in S$,

$$\varrho\big(f(x), f(y)\big) \leq M\varrho(x, y)^p. \tag{1}$$

We may say $f \in \operatorname{Lip}(p)$. This is also called a **Hölder condition** of order p.

Proposition 6.8.9. *Let $f \colon S \to T$, $f \in \operatorname{Lip}(p)$. Then* (a) $p \dim f[S] \leq \dim S$ *and* (b) $p \operatorname{Dim} f[S] \leq \operatorname{Dim} S$.

Proof. Let M satisfy (1). (a) If $A \subseteq S$, then $\operatorname{diam} f[A] \leq M (\operatorname{diam} A)^p$. Let $\delta > 0, s > 0$, and let $\{A_n\}$ be a cover of S by sets of diameter $\leq \delta$. Then $\{f[A_n]\}$ is a cover of $f[S]$ by sets of diameter $\leq M\delta^p$. So

$$\overline{\mathcal{H}}_{M\delta^p}^s\big(f[S]\big) \leq \sum_{n \in \mathbb{N}} \big(\operatorname{diam} f[A_n]\big)^s \leq M^s \sum_{n \in \mathbb{N}} \big(\operatorname{diam} A_n\big)^{ps}.$$

Taking the infimum over all δ-covers, we get

$$\overline{\mathcal{H}}_{M\delta^p}^s\big(f[S]\big) \leq M^s \overline{\mathcal{H}}_\delta^{ps}(S).$$

Taking the limit as $\delta \to 0$, we get

$$\overline{\mathcal{H}}^s\big(f[S]\big) \leq M^s \overline{\mathcal{H}}^{ps}(S).$$

If $s > (1/p)\dim S$, then $\overline{\mathcal{H}}^{ps}(S) = 0$ so $\overline{\mathcal{H}}^s(f[S]) = 0$, which means $s \geq \dim f[S]$. Therefore $\dim f[S] \leq (1/p)\dim S$ as required.

(b) Let $\delta > 0$. Define $\delta' = M\delta^p$. Then $\delta' \to 0$ as $\delta \to 0$. Also, $\log(2/\delta') = p\log(2/\delta) + C$ for a certain constant C. Now let $E \subseteq S$. If $u_1, \cdots, u_N \in f[E]$ have $\varrho(u_i, u_j) \geq \delta'$ for all $i \neq j$, then there exist $x_i \in E$ with $f(x_i) = u_i$ and $\varrho(x_i, x_j) \geq \delta$ for all $i \neq j$. Therefore

$$\dot{N}_{\delta/2}(E) \geq \dot{N}_{\delta'/2}(f[E]),$$

and thus $\overline{\dim}_B E \geq p\,\overline{\dim}_B f[E]$.

Let $S = \bigcup_i E_i$ be a countable cover. Then

$$\sup_i \overline{\dim}_B E_i \geq p\sup_i \overline{\dim}_B f[E_i] \geq p\,\overline{\dim}_{MB} f[S].$$

Take the infimum over all covers to get $\overline{\dim}_{MB} S \geq p\,\overline{\dim}_{MB} f[S]$. And by Prop. 6.8.8, $\overline{\dim}_{MB} = \mathrm{Dim}$. $\qquad\Box$

Theorem 6.8.10. *Let $0 < p \leq 1$. Suppose $f\colon [a, b] \to \mathbb{R}$ satisfies a Lipschitz condition of order p. Then the graph*

$$G = \big\{\, (x, f(x)) : x \in [a, b] \,\big\}$$

satisfies $\dim G \leq 2 - p$.

Proof. We may assume that $[a, b] = [0, 1]$. Divide $[0, 1]$ into n sub-intervals of length $1/n$. On each of these intervals f can vary by no more than $M(1/n)^p$. Thus, the part of the graph over one of the sub-intervals can be covered by no more than $Mn^{1-p} + 1$ squares of side $1/n$. Thus

$$\overline{\mathcal{H}}^s_{\sqrt{2}/n}(G) \leq n(Mn^{1-p} + 1)\left(\frac{\sqrt{2}}{n}\right)^s = M2^{s/2}n^{2-p-s} + 2^{s/2}n^{1-s}.$$

If $s = 2 - p$, then this shows $\mathcal{H}^s(G) \leq M2^{(2-p)/2} + 2^{(2-p)/2}$, so $\dim G \leq s = 2 - p$. $\qquad\Box$

Besicovitch and Ursell [5, part V] gave examples of functions satisfying a Lipschitz condition of order p (and no better) where $\dim G = 2 - p$ and other examples where $\dim G < 2 - p$.

6.9 *Remarks

Felix Hausdorff [32] formulated the concepts that we call today the Hausdorff measures and the Hausdorff dimension. Almost all of the early work on the subject was done by A. S. Besicovitch [5]. Mandelbrot therefore uses the term "Hausdorff–Besicovitch dimension".

In fact, Hausdorff proposed a much more general class of measures than the ones discussed here. For example, he proposed using functions of the diameter other than a power: for example, the function $h(x) = x^s \left(1/\log(1/x)\right)^t$ corresponds to the construction in 6.8.2 when $s = t = 1$. He also proposed using characteristics of the covering sets other than the diameter. The seminal paper [32] is required reading for the aspiring expert on fractals. Computation of the Hausdorff dimension using self-similarity appears even in Hausdorff's paper. It was carefully worked out by P. A. P. Moran [51] for subsets of $\mathbb{R}$, and by John Hutchinson [36] for subsets of $\mathbb{R}^d$. The open set condition is used in both of these papers.

The similarity dimension agrees with the Hausdorff and packing dimensions also for iterated function systems in (complete separable) metric spaces other than Euclidean space. But the open set condition must—in general [59]—be replaced by a **strong open set condition**. In addition to the properties listed on p. 149, the closure $\overline{U}$ of the open set U must intersect the attractor K.

The packing measure was introduced by Claude Tricot (but see [34, Exercise (10.51), p. 145]), and advocated by Taylor & Tricot [61], Saint Raymond & Tricot [60], and Taylor [62]. Fractal dimensions in addition to those defined here can be found in [44, p. 357*ff*] and [45].

The generalization of self-similarity that we have called "graph self-similarity" has a complicated history. The version that is used here is based on the work of R. Daniel Mauldin and S. C. Williams [48]. The "two-part dust" was invented explicitly to illustrate the computation of the Hausdorff dimension for graph self-similar sets.

The pentadendrite was shown to me by my colleague W. A. McWorter. The terdragon comes from Chandler Davis and Donald Knuth [13].

Topological vs. Hausdorff Dimension

In Theorem 6.3.11, $\operatorname{Cov} S \leq \dim S$ was proved only for some spaces S, such as compact spaces. In fact, it is true for any metric space S. A complete proof is in [18, Sect. 3.1]. As noted, that proof used Lebesgue integration. Recently, a proof without integration was published by M. G. Charalambous [9]. The covering dimension of a space S is a topological property of the space. That is, if S is homeomorphic to T, then $\operatorname{Cov} S = \operatorname{Cov} T$. The Hausdorff dimension is not a topological property. The spaces $(E^{(\omega)}, \varrho_r)$, where $E = \{0, 1\}$ is a two-letter alphabet, are all homeomorphic, but the Hausdorff dimension varies as r varies. We know that $\operatorname{Cov} S \leq \dim S$. In fact, $\operatorname{Cov} S$ is the largest topologically invariant lower bound for $\dim S$:

Theorem 6.9.1. *Let S be a separable metric space. Then*

$$\operatorname{Cov} S = \inf \left\{ \dim T : T \text{ is homeomorphic to } S \right\}.$$

I will prove here only the simplest case:

Proposition 6.9.2. *Let S be a separable zero-dimensional metric space. Then*

$$0 = \inf \{ \dim T : T \text{ is homeomorphic to } S \}.$$

Proof. First, S is homeomorphic to a subset T of the string space $\{0,1\}^{(\omega)}$, by Theorem 3.4.4. With metric ϱ_r, the space $\{0,1\}^{(\omega)}$ has Hausdorff dimension $\log 2/\log(1/r)$. But $\lim_{r\to 0} \log 2/\log(1/r) = 0$. □

The general case may be proved in a similar way [35, Theorem VII 5]. For example, the Menger sponge is a universal 1-dimensional space, so metric spaces homeomorphic to the Menger sponge, but with Hausdorff dimension close to 1 should be exhibited. The approximation shown in Fig. 6.9.3 suggests the idea. It is self-affine, rather than self-similar, so our methods of computation will not evaluate its Hausdorff dimension, however.

Two-Dimensional Lebesgue Measure Compared to Two-Dimensional Hausdorff Measure

According to Theorem 6.3.6 there is a positive constant c such that $\mathcal{H}^2(B) = c\mathcal{L}^2(B)$. We will show here that $c = 4/\pi$.

For the lower bound on $\mathcal{H}^2$, we need an interesting fact from two-dimensional geometry: Among all sets with a given diameter, the disk has the largest area. That is, if A is a set with diameter t, then $\overline{\mathcal{L}}^2(A) \le \pi t^2/4$.[*] The proof requires some knowledge concerning convexity in two dimensions. First, A has the same diameter as its convex hull, so we may assume that A is convex. Similarly we may assume that A is closed. Choose any boundary point of A; let it be the origin of coordinates. A has a support line there,

Fig. 6.9.3. Homeomorph of the Menger Sponge

---

[*] The corresponding fact for higher dimensions can be proved from Steiner's symmetrization construction. See, for example, [21, p. 107].

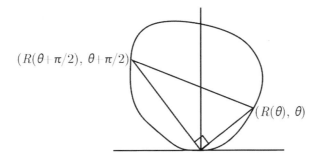

$(R(\theta + \pi/2), \theta + \pi/2)$

$(R(\theta), \theta)$

Fig. 6.9.4. Polar Coordinates

let it be the x-axis. Then the set A is given in polar coordinates (r, θ) by equations

$$0 \le r \le R(\theta), \quad 0 \le \theta \le \pi,$$

for some function R. Now A has diameter t. So the distance between the polar points $(R(\theta), \theta)$ and $(R(\theta + \pi/2), \theta + \pi/2)$ is at most t (Fig. 6.9.4). By the Pythagorean theorem, we may conclude

$$R(\theta)^2 + R(\theta + \pi/2)^2 \le t^2.$$

Then the area may be computed in polar coordinates:

$$\int_0^\pi \frac{R(\theta)^2}{2}\, d\theta = \int_0^{\pi/2} \frac{R(\theta)^2}{2}\, d\theta + \int_{\pi/2}^\pi \frac{R(\theta)^2}{2}\, d\theta$$
$$= \int_0^{\pi/2} \frac{R(\theta)^2 + R(\theta + \pi/2)^2}{2}\, d\theta$$
$$\le \int_0^{\pi/2} \frac{t^2}{2}\, d\theta = \frac{\pi t^2}{4}.$$

So: a set $A \subseteq \mathbb{R}^2$ with diameter t has area at most $\pi t^2/4$. Then the argument given in Theorem 6.3.6, with $a = 4/\pi$, will show that $(4/\pi)\mathcal{L}^2(B) \le \mathcal{H}^2(B)$ for any Borel set B.

For the upper bound, we use the Vitali covering theorem [23, Theorem 1.10] or [18, Theorem 1.3.7]. Let $b = \mathcal{H}^2(Q)$, where Q is the open unit square. Now a disk can be approximated inside and outside by little squares, so we have (by the argument in the proof of Theorem 6.3.6) $\mathcal{H}^2(B) = b\mathcal{L}^2(B)$ for all disks B. The collection of all closed disks with diameter $< \varepsilon$ and contained in the square Q satisfies the hypothesis of the Vitali theorem, so there is a countable disjoint set $\{B_i : i \in \mathbb{N}\}$ of them with $\mathcal{L}^2(Q \setminus \bigcup_{i \in \mathbb{N}} B_i) = 0$. But then, by the inequality $\mathcal{H}^2(B) \le b\mathcal{L}^2(B)$, we know that $\mathcal{H}^2(Q \setminus \bigcup_{i \in \mathbb{N}} B_i)$ is also 0, so $\mathcal{H}_\varepsilon^2(Q \setminus \bigcup_{i \in \mathbb{N}} B_i) = 0$. Now

$$\mathcal{H}_\varepsilon^2\left(\bigcup_{i\in\mathbb{N}} B_i\right) \le \sum_{i=1}^\infty (\operatorname{diam} B_i)^2$$

$$= \frac{4}{\pi} \sum_{i=1}^\infty \frac{\pi}{4} (\operatorname{diam} B_i)^2$$

$$= \frac{4}{\pi} \sum_{i=1}^\infty \mathcal{L}^2(B_i)$$

$$= \frac{4}{\pi} \mathcal{L}^2(Q) = \frac{4}{\pi}.$$

Therefore $b = \mathcal{H}^2(Q) = \mathcal{H}^2(\bigcup_{i\in\mathbb{N}} B_i) \le 4/\pi$.

So we have exactly $\mathcal{H}^2(B) = (4/\pi)\mathcal{L}^2(B)$ for all Borel sets B.

The same result is true in $\mathbb{R}^d$, namely $\mathcal{H}^d(B) = c_d \mathcal{L}^d(B)$, where c_d is the appropriate constant $1/\mathcal{L}^d(B_{1/2}(0))$.

The Sim-Value of a Mauldin–Williams Graph

The sim-value of a Mauldin–Williams graph exists and is unique. The proof of this fact will be given here. It requires some knowledge of linear algebra. In particular, it requires information from the Perron–Frobenius theorem (stated below).

Let A be a square matrix. The *spectral radius* of A is the maximum of the absolute values of all of the complex eigenvalues of A. We will write $\operatorname{sp rad} A$ for the spectral radius of A.

We will use some additional notation: $A \ge 0$ means all of the entries of A are nonnegative, and $A > 0$ means all of the entries of A are positive; $A \ge B$ means $A - B \ge 0$, and $A > B$ means $A - B > 0$. The matrix $A \ge 0$ is called *reducible* iff the rows and columns can be permuted (by the same permutation) so that A has the form

$$A = \begin{bmatrix} B & O \\ C & D \end{bmatrix},$$

where B and D are square matrices (with at least one row each), and O is a rectangular matrix of zeros. If A is not reducible, then it is *irreducible*. A column matrix with all entries 0 is $\mathbf{0}$, and a column matrix with all entries 1 is $\mathbf{1}$.

Here is (part of) the Perron-Frobenius theorem. See [27, Chap. XIII] for a proof.

Theorem 6.9.5. *Let $A \ge 0$ be an irreducible square matrix, and let $\lambda \in \mathbb{R}$. Then:*

(1) *If $\lambda = \operatorname{sp rad} A$, then there is a column matrix $\mathbf{x} > \mathbf{0}$ with $A\mathbf{x} = \lambda\mathbf{x}$.*
(2) *If there is a nonzero column matrix $\mathbf{x} \ge \mathbf{0}$ with $A\mathbf{x} = \lambda\mathbf{x}$, then $\lambda = \operatorname{sp rad} A$.*

(3) *If there is a nonzero column matrix* $\mathbf{x} \geq \mathbf{0}$ *with* $A\mathbf{x} < \lambda\mathbf{x}$, *then* $\lambda > \mathrm{sp\,rad}\,A$.

(4) *If there is a nonzero column matrix* $\mathbf{x} \geq \mathbf{0}$ *with* $A\mathbf{x} > \lambda\mathbf{x}$, *then* $\lambda < \mathrm{sp\,rad}\,A$.

Now we are in a position to prove that the dimension of a strongly connected, strictly contracting Mauldin-Williams graph exists and is unique.

Theorem 6.9.6. *Let* (V, E, i, t, r) *be a strongly connected, strictly contracting Mauldin–Williams graph. There is a unique number* $s \geq 0$ *such that positive numbers* q_v *exist satisfying*

$$q_u^s = \sum_{\substack{v \in V \\ e \in E_{uv}}} r(e)^s \, q_v^s$$

for all $u \in V$.

Proof. We will be using matrices with the rows (and columns) labeled by V. For each pair $u, v \in V$, and $s \geq 0$, let

$$A_{uv}(s) = \sum_{e \in E_{uv}} r(e)^s.$$

Let $A(s)$ be the matrix with entry $A_{uv}(s)$ in row u column v. Let $\Phi(s) = \mathrm{sp\,rad}\,A(s)$ be the spectral radius of the matrix $A(s)$. Now the matrix $A(s)$ has nonnegative entries. The entry $A_{uv}(s)$ is positive if and only if E_{uv} is not empty. Since the graph is strongly connected, the matrices $A(s)$ are irreducible. I will prove: (a) s-dimensional Perron numbers exist if and only if $\Phi(s) = 1$, and (b) the equation $\Phi(s) = 1$ has a unique solution in $[0, \infty)$.

First, suppose that s-dimensional Perron numbers exist, so that

$$q_u^s = \sum_{\substack{v \in V \\ e \in E_{uv}}} r(e)^s \, q_v^s$$

for all $u \in V$. Thus, if the column matrix $\mathbf{x}$ has entries q_v^s, then $\mathbf{x} > \mathbf{0}$ and $A(s)\mathbf{x} = \mathbf{x}$, so by the Perron-Frobenius theorem, 1 is the spectral radius of $A(s)$.

Conversely, suppose that $1 = \mathrm{sp\,rad}\,A(s)$. Then by the Perron-Frobenius theorem, there is a column matrix $\mathbf{x} > \mathbf{0}$ with $A(s)\mathbf{x} = \mathbf{x}$. If we write x_v for the entries of $\mathbf{x}$, then the numbers $q_v = x_v^{1/s}$ will be s-dimensional Perron numbers.

Next, I claim that the function Φ is continuous. Certainly the entries $A_{uv}(s)$ of the matrix are continuous functions of s. Fix a number s_0. Let $\mathbf{x} > \mathbf{0}$ be the Perron-Frobenius eigenvector: $A(s_0)\mathbf{x} = \Phi(s_0)\mathbf{x}$. Let the entries of $\mathbf{x}$ be x_v. Define positive numbers a, b by $a = \min x_v$, $b = \max x_v$. Suppose V has n elements, so the matrices are $n \times n$. Let $\varepsilon > 0$ be given. By the continuity of the entries A_{uv}, there exists $\delta > 0$ so that if $|s - s_0| < \delta$, then

$$|A_{uv}(s) - A_{uv}(s_0)| < \frac{a\varepsilon}{nb}$$

for all u, v. Now we have

$$\sum_v A_{uv}(s)x_v = \sum_v A_{uv}(s_0)x_v + \sum_v (A_{uv}(s) - A_{uv}(s_0))x_v$$

$$\leq \varPhi(s_0)x_u + n\frac{a\varepsilon}{nb}b \leq (\varPhi(s_0) + \varepsilon)x_u.$$

Therefore, by the Perron–Frobenius theorem, $\varPhi(s) = \text{sp rad } A(s) \leq \varPhi(s_0) + \varepsilon$. Similarly $\varPhi(s) \geq \varPhi(s_0) - \varepsilon$. This shows that $\varPhi$ is continuous.

Since the graph is strongly connected, each row has at least one nonzero entry. So for each u there is v with $A_{uv}(0) \geq 1$. Therefore $A(0)\mathbf{1} \geq \mathbf{1}$, so that $\varPhi(0) \geq 1$. The entries $A_{uv}(s) \to 0$ as $s \to \infty$, so for large enough s, we have $A_{uv}(s) \leq 1/(2n)$ for all u, v, so that $A(s)\mathbf{1} \leq (1/2)\mathbf{1}$, and thus $\varPhi(s) \leq 1/2$. Now by the intermediate value theorem, there is a solution s to the equation $\varPhi(s) = 1$.

Finally, to prove the uniqueness, we will show that $\varPhi$ is strictly decreasing. The derivative of A_{uv} is ≤ 0, and in fact < 0 unless A_{uv} is identically 0. Each row has at least one nonzero entry, so if $s > s_0$ and $\mathbf{x}$ is the Perron-Frobenius eigenvector for $A(s_0)$, we have $A(s)\mathbf{x} < A(s_0)\mathbf{x} = \varPhi(s_0)\mathbf{x}$. So $\varPhi(s) < \varPhi(s_0)$. Therefore the function $\varPhi$ is strictly decreasing. □

Exercise 6.9.7. Let (V, E, i, t, r) be a contracting, strongly connected Mauldin–Williams graph. Are the conclusions of Theorem 6.9.6 still correct?

To compute the dimension of a strictly contracting, strongly connected Mauldin–Williams graph, we would ordinarily find the numbers s such that 1 is an eigenvalue of the matrix $A(s)$. If that s is unique, it is the dimension. If not, then we find for each s the corresponding eigenvector for $A(s)$; only one of the values s will admit an eigenvector with all entries positive.

Remarks on the Exercises

Exercise 6.1.10: $f: S \to T$ is Lipschitz, and $A \subseteq S$. Say $\varrho(f(x), f(y)) \leq b\varrho(x, y)$. If $\mathcal{D}$ is a countable cover of A by sets with diameter at most ε, then $\mathcal{D}' = \{ f[D] : D \in \mathcal{D} \}$ is a countable cover of $f[A]$ by sets with diameter at most $b\varepsilon$. Now

$$\sum_{D \in \mathcal{D}} (\text{diam } f[D])^s \leq b^s \sum_{D \in \mathcal{D}} (\text{diam } D)^s,$$

so $\overline{\mathcal{H}}_{b\varepsilon}^s(f[A]) \leq b^s \overline{\mathcal{H}}_\varepsilon^s(A)$. Therefore $\dim f[A] \leq \dim A$. The case of inverse Lipschitz is similar.

Exercise 6.3.12: Suppose $\text{Cov } S \geq 1$. Then S does not have a base for the open sets consisting of clopen sets. So there is a point $a \in S$ and $\varepsilon > 0$ such that for $0 < r < \varepsilon$, the ball $B_r(a)$ is not clopen. The function $h: S \to \mathbb{R}$

Fig. 6.9.8. All edges have value $(3 - \sqrt{5})/2$

defined by $h(x) = \varrho(x, a)$ satisfies $|h(x) - h(y)| \leq \varrho(x, y)$. Its range includes the interval $(0, \varepsilon)$. So we have $\dim S \geq \dim h[S] \geq \dim(0, \varepsilon) = 1$.

For Exercise 6.5.11: The terdragon boundary is made up of two copies of the 120-degree dragon of Fig. 1.5.8. The open set condition (Plate 13) is satisfied by an open set the shape of the filled-in fudgeflake (Fig. 1.5.8); it may be thought of as the union of three terdragons. Exercise 6.7.11. Fig. 6.9.8.

Exercise 6.7.12. 1.22.

Exercise 6.7.5. [48].

Exercise 6.7.14. The graph of Fig. 4.3.13. This is a special case of the situation considered in [17].

The fractile lines of the sandstone.
—*Scribner's Magazine*, April, 1893
(quoted in the *Oxford English Dictionary*)

7

Additional Topics

This chapter includes additional examples of fractals, and hints at parts of the subject that we have not covered. In Sections 7.3 and 7.4, the computation of the fractal dimension requires more than just a simple application of the results of Chap. 6. So these examples show that there is more to the subject than we have seen in this book.

7.1 *Deconstruction

Let K be the attractor for an iterated function system (f_e) of similarities in $\mathbb{R}^d$, and let the ratio list have sim-value s. Then $\dim K \leq \operatorname{Dim} K \leq s$ (Theorem 6.4.10). If the parts $f_e[K]$ are disjoint, or have small overlap (specified by the open set condition), then $\dim K = \operatorname{Dim} K = s$ (Theorem 6.5.4). But if there   is too much overlap, then the dimension could be strictly smaller than s. Sometimes the fractal dimension can be still computed by "deconstructing" the attractor—interpreting it in a manner different from the one provided by the iterated function system.

Recall the Li's lace fractal on p. 84. It is made up of blocks of two kinds, P, Q, as described on p. 126. From the decompostions of the two sets, we get $P \supseteq Q$. Note that Fig. 4.3.3 not only describes the iterated function system, but also shows that the graph open set condition is satisfied. So we have $\dim P = \dim Q = \operatorname{Dim} P = \operatorname{Dim} Q = s$, where $s = \log(\sqrt{2}+2)/\log 2 \approx 1.7716$ is the sim-value for the M-W graph shown in Fig. 4.3.4(b).[†]

But that was not the way in which this fractal was originally defined. Example (c) in Jun Li's thesis [43] is described using an iterated function system in the plane consisting of four similarities:

[†] Answer for Exercise 4.3.11.

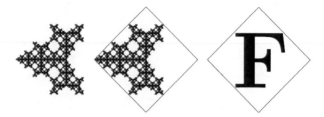

Fig. 7.1.1. A fractal and a surrounding square

Fig. 7.1.2. Four images

$$f_1 \begin{bmatrix} x \\ y \end{bmatrix} = \begin{bmatrix} 1/2 & 0 \\ 0 & 1/2 \end{bmatrix} \begin{bmatrix} x \\ y \end{bmatrix}, \qquad f_3 \begin{bmatrix} x \\ y \end{bmatrix} = \begin{bmatrix} 0 & 1/2 \\ -1/2 & 0 \end{bmatrix} \begin{bmatrix} x \\ y \end{bmatrix} + \begin{bmatrix} 1 \\ 0 \end{bmatrix},$$

$$f_2 \begin{bmatrix} x \\ y \end{bmatrix} = \begin{bmatrix} 1/2 & 0 \\ 0 & 1/2 \end{bmatrix} \begin{bmatrix} x \\ y \end{bmatrix} + \begin{bmatrix} 1/2 \\ 0 \end{bmatrix}, \qquad f_4 \begin{bmatrix} x \\ y \end{bmatrix} = \begin{bmatrix} 0 & -1/2 \\ 1/2 & 0 \end{bmatrix} \begin{bmatrix} x \\ y \end{bmatrix} + \begin{bmatrix} 1 \\ 0 \end{bmatrix}.$$

The attractor is shown in Fig. 7.1.1, and the iterated function system is illustrated in Fig. 7.1.2.

Deconstruction. We claim that if P, Q are the sets described on p. 126, and four of these are assembled into a square F as shown,

then F is the attractor of the iterated function system 7.1.2. But we know that the attractor is unique (Theorem 4.1.3), so we need only prove the self-referential equation $F = f_1[F] \cup f_2[F] \cup f_3[F] \cup f_4[F]$. We will prove this using pictures, which can sometimes be more convincing than words.

We should show that

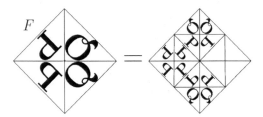

is the union of four sets:

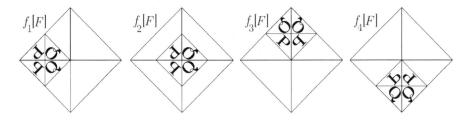

The outer triangles match as required. The triangles in $f_2[F]$ do not align with the others, so to see how the union behaves we require another level of subdivision. We need to prove that the inner square

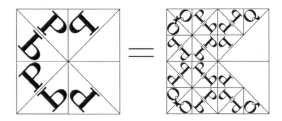

is the union of four sets:

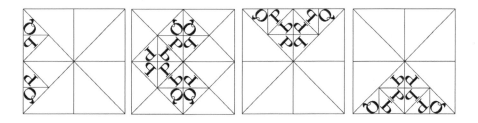

Everything now matches, taking into account $P \cup P = P$ and $P \cup Q = P$.

In this example, then, the iterated function system 7.1.1 has sim-value 2, but the attractor it defines has fractal dimension $\log(\sqrt{2} + 2)/\log 2 \approx 1.7716$.

More Self-Similar Sets with Overlap

Now we consider a "similarity dimension" example with overlap in $\mathbb{R}$, and leave the details to the reader. We will use the ratio list $(1/5, 1/5, 1/5)$. Let a, b, c be three real numbers. Consider the three dilations f_1, f_2, f_3 of $\mathbb{R}$ with fixed points a, b, c, respectively. For certain choices of the points a, b, c, this realization satisfies the open set condition, and the invariant fractal K has Hausdorff dimension equal to the sim-value of the ratio list. For certain other choices of a, b, c (such as two or three of them coincident) the Hausdorff dimension of K is not equal to the dimension of the ratio list. There is, nevertheless, always an inequality.

Let us normalize things by assuming $a = 0$, $c = 1$, $0 < b < 1$. (Any choice of three distinct points can be reduced to this case.) All three of the maps send $[0, 1]$ into itself, so the invariant set K is a subset of $[0, 1]$; in fact K may be constructed in the usual way by the contraction mapping theorem starting with $[0, 1]$.

Exercise 7.1.3. For what values of b are the three images of the open interval $(0, 1)$ disjoint?

Exercise 7.1.4. Compute the Hausdorff dimension when $b = 1/10$.

Exercise 7.1.5. Compute the Hausdorff dimension when $b = 1/5$.

There is a result of Falconer [24] that is relevant in situations like this. In this case it asserts that the Hausdorff dimension of the invariant set K is equal to the similarity dimension $\log 3/\log 5$ for *almost all* choices of $b \in [0, 1]$. That is, the set of all $b \in [0, 1]$ for which $\dim K = \log 3/\log 5$ fails is a set of Lebesgue measure 0.

Exercise 7.1.6. Give an example of an iterated function system of similarities in $\mathbb{R}^d$ where the Hausdorff dimension of the invariant set coincides with the similarity dimension, but Moran's open set condition fails.

Consider the Barnsley leaf fractal B defined on p. 26. It is the attractor of an iterated function system with three maps, and sim-value $2\log(1 + \sqrt{2})/\log 2$. This is > 2, so it is certainly not the fractal dimension of the fractal B itself. Plate 10 shows the set B. The three images are in three colors cyan, magenta, and yellow. Where cyan and magenta overlap, it is blue. Where all three overlap it is black.

Exercise 7.1.7. Deconstruct Barnsley's leaf, and determine its fractal dimension.

7.2 *Self-Affine Sets

The idea of an iterated function system makes good sense even when the maps are not similarities. One possibility that comes up often involves affine maps. The invariant set is then said to be **self-affine**. In the general self-affine case the evaluation of the Hausdorff dimension is not completely understood. It has even been argued [45] that the Hausdorff dimension is not the proper dimension to use at all. We will present a few examples in this section.

A Self-Affine Dust

As a reference, take the **unit square** in $\mathbb{R}^2$:

$$S = \{\,(x,y) :\ 0 \le x \le 1, 0 \le y \le 1\,\}.$$

The images of S under the two maps will be two rectangles:

$$R_1 = \{\,(x,y) :\ 0 \le x \le 1/2, 0 \le y \le 2/3\,\}$$
$$R_2 = \{\,(x,y) :\ 1/2 \le x \le 1, 1/3 \le y \le 1\,\}.$$

The function f_1 is an affine map of $\mathbb{R}^2$ onto itself, and sends the vertices of S to the corresponding vertices of R_1. The function f_2 is an affine map of $\mathbb{R}^2$ onto itself, and sends the vertices of S to the corresponding vertices of R_2.

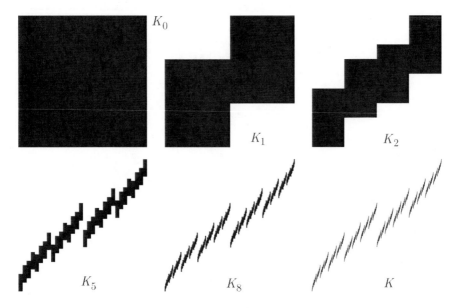

Fig. 7.2.1. Self-Affine Dust

---

* Optional section.

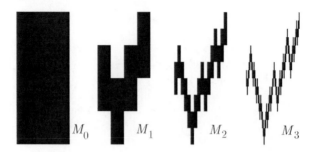

Fig. 7.2.4. Kiesswetter's curve

Exercise 7.2.2. There is a unique compact nonempty set $K \subseteq \mathbb{R}^2$ such that $K = f_1[K] \cup f_2[K]$.

Exercise 7.2.3. Compute the Hausdorff dimension of the set K.

Kiesswetter's Curve

This is illustrated in two different ways. The set can be decomposed into four subsets, which are affine images of the whole thing. The four affine maps may be written in matrix notation. A point (x, y) in the plane is identified with a 2×1 column matrix.

$$f_1 \begin{bmatrix} x \\ y \end{bmatrix} = \begin{bmatrix} 1/4 & 0 \\ 0 & -1/2 \end{bmatrix} \begin{bmatrix} x \\ y \end{bmatrix}$$

$$f_2 \begin{bmatrix} x \\ y \end{bmatrix} = \begin{bmatrix} 1/4 & 0 \\ 0 & 1/2 \end{bmatrix} \begin{bmatrix} x \\ y \end{bmatrix} + \begin{bmatrix} 1/4 \\ -1/2 \end{bmatrix}$$

$$f_3 \begin{bmatrix} x \\ y \end{bmatrix} = \begin{bmatrix} 1/4 & 0 \\ 0 & 1/2 \end{bmatrix} \begin{bmatrix} x \\ y \end{bmatrix} + \begin{bmatrix} 1/2 \\ 0 \end{bmatrix}$$

$$f_4 \begin{bmatrix} x \\ y \end{bmatrix} = \begin{bmatrix} 1/4 & 0 \\ 0 & 1/2 \end{bmatrix} \begin{bmatrix} x \\ y \end{bmatrix} + \begin{bmatrix} 3/4 \\ 1/2 \end{bmatrix} .$$

The first construction starts with the rectangle $M_0 = [0, 1] \times [-1, 1]$, and at each stage replaces the current set M_n with $M_{n+1} = f_1[M_n] \cup f_2[M_n] \cup f_3[M_n] \cup f_4[M_n]$. Because each of the maps f_j sends M_0 to a subset of M_0, this results in a decreasing sequence of compact sets. Kiesswetter's curve is the intersection $\bigcap_{n \in \mathbb{N}} M_n$.

The second construction starts with the line segment from $(0, 0)$ to $(1, 1)$, and makes the same transformation as before. Since $f_j\big((1, 1)\big) = f_{j+1}\big((0, 0)\big)$ for $j = 1, 2, 3$, these sets are all polygons. They are graphs of a sequence of continuous functions defined on $[0, 1]$; this sequence converges uniformly. The limit g is called **Kiesswetter's function**. Its graph $G = \{ (x, y) : y = g(x) \}$ is Kiesswetter's curve.

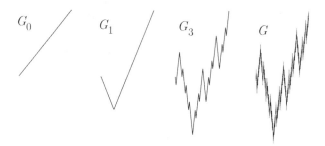

Fig. 7.2.5. Kiesswetter's curve

Exercise 7.2.6. Let g be Kiesswetter's function. Then for any integers $k \geq 0$ and $0 \leq j < 2^k$, prove

$$\left| g\left(\frac{j}{4^k}\right) - g\left(\frac{j+1}{4^k}\right) \right| = \frac{1}{2^k}.$$

Exercise 7.2.7. Kiesswetter's curve is the graph of a continuous but nowhere differentiable function $g \colon [0,1] \to \mathbb{R}$.

Exercise 7.2.8. Find the Hausdorff and packing dimensions of Kiesswetter's curve.

Besicovitch–Ursell Functions

Besicovitch and Ursell investigated the dimension of the graphs of non-differentiable functions. The most famous examples of these functions, dating back to Weierstrass, have a form

$$f(x) = \sum_{k=0}^{\infty} a_k \sin(b_k x),$$

for appropriate choices of a_k and b_k. A simpler variant was used by Besicovitch and Ursell, which will now be described.

Define a "sawtooth" function $g \colon \mathbb{R} \to \mathbb{R}$ by:

$$
\begin{aligned}
g(x) &= x && \text{for } -1/2 \leq x \leq 1/2 \\
g(x) &= 1 - x && \text{for } 1/2 \leq x \leq 3/2 \\
g(x+2) &= g(x) && \text{for all } x.
\end{aligned}
$$

If $0 < a < 1$, the **Besicovitch–Ursell** function with parameter a is:

$$f(x) = \sum_{k=0}^{\infty} a^k g(2^k x).$$

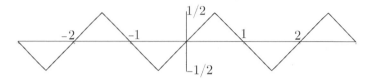

Fig. 7.2.9. Sawtooth function

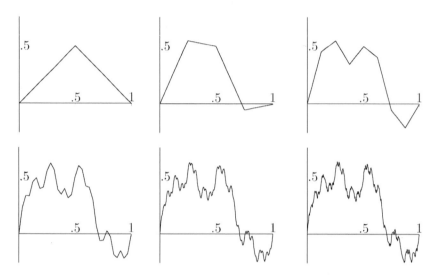

Fig. 7.2.10. Partial sums $\sum_{k=0}^{n} a^k g(2^k x)$ with $n = 0, 1, 2, 5, 6, 12$

Partial sums $\sum_{k=0}^{n} a^k g(2^k x)$ of the series are illustrated in Fig. 7.2.10, with $a = 0.6$. The pictures show only $0 \leq x \leq 1$, but the rest of the graph is simply related to this part.

Exercise 7.2.11. The function $f(x)$ exists and is continuous.

Exercise 7.2.12. Is the graph of f the invariant set for some iterated function system?

Pictures for various values of a are shown in Fig. 7.2.15.

Exercise 7.2.13. For what values of a is f a Lipschitz function?

Exercise 7.2.14. Compute the Hausdorff dimension of the graph of the Besicovitch-Ursell function with parameter $a = 3/5$.

Hironaka's Curve

Pictured (Fig. 7.2.16) are some approximations to **Hironaka's curve**. The first approximation consists of two vertical line segments, one unit long, one

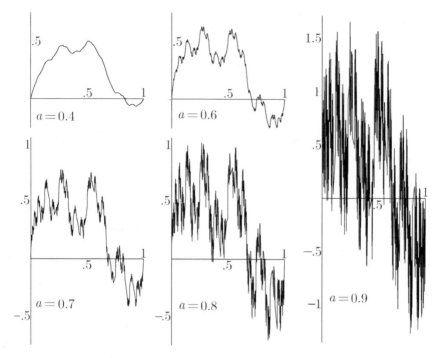

Fig. 7.2.15. Besicovitch-Ursell Functions

unit apart. For each subsequent approximation, additional line segments are
added. The length of the new line segments is decreased by a factor of 1/2 at
each stage. The distance between the line segments is decreased by a factor

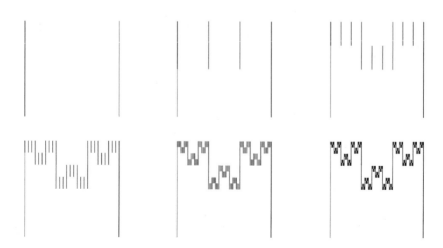

Fig. 7.2.16. Hironaka's curve

of $1/3$ at each stage. The position of the line segments is determined by the pattern illustrated. Hironaka's curve is the limit set.

Exercise 7.2.17. Find topological and Hausdorff dimensions for Hironaka's curve.

Number Systems

Here is a way to generalize the "number systems" of Sect. 1.6. Elements of $\mathbb{R}^d$ should be identified with $d \times 1$ column matrices. Let D be a finite set in $\mathbb{R}^d$, including 0, and let B be a $d \times d$ matrix. What conditions should B satisfy so that all of the following vectors exist?

$$\sum_{j=1}^{\infty} B^j a_j,$$

where the "digits" $a_j \in D$. The set F of all these vectors is the invariant set of an iterated function system of affine maps.

7.3 *Self-Conformal

An affine transformation that is not a similarity changes distances by different ratios in different directions. Here we will talk about non-affine transformations that change distances by the same ratio in all directions, but only in the limit near a point.

Let S be a metric space, let $f\colon S \to S$ be a transformation, let $r > 0$, and let $a \in S$. We say that f is **conformal** at a with ratio r if:

$$\lim_{\substack{x,y \to a \\ x \neq y}} \frac{\varrho(f(x), f(y))}{\varrho(x, y)} = r.$$

More technically stated: for every $\varepsilon > 0$ there is $\delta > 0$ such that for all $x, y \in S$, if $\varrho(x, a) < \delta$ and $\varrho(y, a) < \delta$, then

$$(1 - \varepsilon)r\varrho(x.y) \leq \varrho(f(x), f(y)) \leq (1 + \varepsilon)r\varrho(x, y).$$

We say that f is conformal on a set E if a is conformal at every point of E, but not necessarily with the same ratio.

Proposition 7.3.1. *Let $f\colon \mathbb{R} \to \mathbb{R}$ be continuously differentiable, let $a \in \mathbb{R}$, and assume $f'(a) \neq 0$. Then f is conformal at a with ratio $|f'(a)|$.*

---

* Optional section.

Proof. Let $\varepsilon > 0$ be given. Then since f' is continuous, there is $\delta > 0$ so that if $|x - a| < \delta$, then

$$(1 - \varepsilon)|f'(a)| < |f'(x)| < (1 + \varepsilon)|f'(a)|.$$

Now let x, y satisfy $|x - a| < \delta$, $|y - a| < \delta$, and $x \neq y$. Applying the Mean Value Theorem on the interval from x to y, we conclude there is z between x and y so that

$$\frac{f(x) - f(y)}{x - y} = f'(z).$$

Now $|z - a| < \delta$, so we have

$$\frac{|f(x) - f(y)|}{|x - y|} = |f'(z)| < (1 + \varepsilon)|f'(a)|,$$

so that

$$|f(x) - f(y)| < (1 + \varepsilon)|f'(a)| \, |x - y|.$$

Similarly,

$$|f(x) - f(y)| > (1 - \varepsilon)|f'(a)| \, |x - y|.$$

Thus f is conformal at a with ratio $|f'(a)|$. $\qquad\square$

Is continuity of the derivative required?

Exercise 7.3.2. Give an example where $f \colon \mathbb{R} \to \mathbb{R}$ is differentiable at a point a with $f'(a) \neq 0$, but f is not conformal at a.

If you have studied multi-dimensional calculus, you can attempt the next exercise.

Exercise 7.3.3. Let $f \colon \mathbb{R}^d \to \mathbb{R}^d$ be continuously differentiable, and let $a \in \mathbb{R}^d$. If the derivative $Df(a)$, interpreted as as $d \times d$ matrix, defines a similarity on $\mathbb{R}^d$, then f is conformal at a, and the ratio of f at a is the same as the ratio of the similarity $Df(a)$.

In Euclidean space $\mathbb{R}^d$, examples of conformal maps (where they are defined) are: translation, rotation, reflection in a hyperplane, reflection in a sphere. In particular, in $\mathbb{R}^2$, reflection in a circle (p. 28) is conformal. Of course the ratio is not the same everywhere.

In the mathematical subject of complex analysis, you can find this:

Proposition 7.3.4. *Let $f \colon \mathbb{C} \to \mathbb{C}$ be continuously differentiable in the complex sense, let $a \in \mathbb{C}$, and assume $f'(a) \neq 0$. Then f is conformal at a with ratio $|f'(a)|$.*

If a function f (defined for complex numbers z) has the form

$$f(z) = \frac{az + b}{cz + d},$$

where $a, b, c, d \in \mathbb{C}$, then f is called a **linear fractional transformation**. If $ad - bc = 0$, then f is constant, so we will assume $ad - bc \neq 0$. Then f is defined everywhere in $\mathbb{C}$ except $z = -b/a$, and

$$f'(z) = \frac{ad - bc}{(cz + d)^2}$$

is never zero. So f is conformal. An important property of a linear fractional transformation is that it maps circles to circles (provided a line is considered to be a circle).

The attractor of an iterated function system consisting of conformal maps is known as a **self-conformal** set. Of course self-similar sets are self-conformal. Pharaoh's breastplate (p. 30) is self-conformal but not self-similar.

Appolonian Gasket

Figure 7.3.5 shows a subset of the plane. The first approximation is obtained by taking three mutually tangent circles with radius 1. The set C_0 is the region enclosed by three arcs (including the arcs themselves). Each approximation will consist of some regions bounded by three mutually tangent circular arcs. To obtain C_{k+1}, remove from each region of C_k the circle in the region tangent to all three of the arcs. (The boundary of the circle remains.) The Appolonian gasket is the "limit" (intersection) of the sets C_k.

The gasket is is self-coinformal. It is not self-similar or self-affine.

Exercise 7.3.6. The Appolonian gasket is an invariant set for an iterated function system of linear fractional transformations.

Exercise 7.3.7. Discuss the topological dimension and fractal dimension of the Appolonian gasket.

Fig. 7.3.5. Appolonian Gasket

Julia Set

Recall the Julia set described on p. 28 for the function $\varphi(z) = z^2 + c$, where $c = -0.15 + 0.72i$. The two branches f_0, f_1 of the function $\sqrt{z - c}$ are not continuous on the set J that we construct, so it is not so simply interpreted as the attractor of an iterated function system.

The Julia set J is the union of two sets U and $L = -U$. See Fig. 7.3.8. There are choices of inverse maps $f_0(z), f_1(z) = -f_0(z)$ for φ that are continuous on U and choices of inverse maps $g_0(z), g_1(z) = -g_0(z)$ for φ that are continuous on L so that

$$U = f_0[U] \cup g_0[L], \qquad V = f_1[U] \cup g_1[L].$$

See Fig. 7.3.9. The point c does not belong to J, and $\sqrt{z - c}$ is conformal except at the point c. All four maps are conformal on their appropriate domains, since they are continuous branches of $\sqrt{z - c}$. Sets U and L are isometric; U is the attractor of the iterated function system consisting of conformal maps $f_0(z), g_0(-z)$. So U is **self-conformal**.

Fig. 7.3.8. J made up of U and L

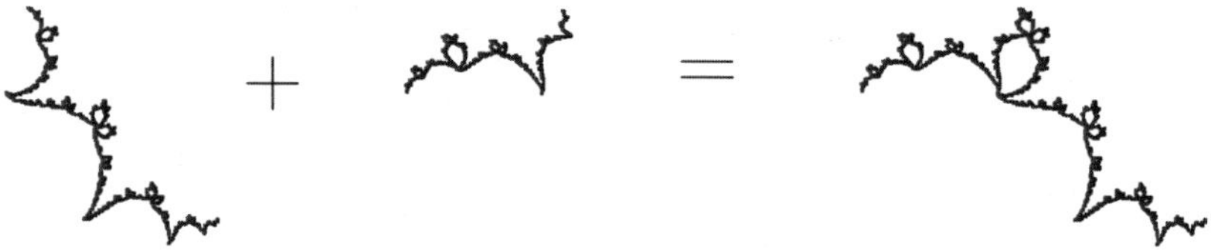

Fig. 7.3.9. $f_0[U]$ and $g_0[L]$ make up U

7.4 *A Multifractal

Most of the fractals that have been considered in this book are closed sets (or even compact sets). We will discuss now an example that is not closed. Mandelbrot calls it the **Besicovitch fractal**. It was studied by Besicovitch and Eggleston; more recently it occurs in the physics literature in connection with "multifractals" or "fractal measures". The proof will require some knowledge of probability theory, however.

Given $x \in [0, 1]$, consider its binary expansion, $x = \sum_{i=1}^{\infty} a_i 2^{-i}$, where each a_i is 0 or 1. We are interested in the frequency of the occurrence of the digit 0. More precisely, let $K_n^{(0)}(x)$ be the number of 0's and $K_n^{(1)}(x)$ the number of 1's occurring among the first n digits, $(a_1, a_2, \cdots, a_n)$. The **frequencies** in question are

$$F^{(0)}(x) = \lim_{n \to \infty} \frac{K_n^{(0)}(x)}{n},$$

$$F^{(1)}(x) = \lim_{n \to \infty} \frac{K_n^{(1)}(x)}{n}.$$

(Of course, the limits in question exist for only some $x \in [0, 1]$.)

Fix a number p, with $0 < p < 1$. We are interested in the set

$$S_p = \left\{ x \in [0, 1] : F^{(0)}(x) \text{ exists, and } F^{(0)}(x) = p \right\}.$$

We will compute the Hausdorff dimension of the set S_p. If we write $q = 1 - p$, so that $F^{(0)}(x) = p$ implies $F^{(1)}(x) = q$, then we will show that

$$\dim S_p = \frac{-p \log p - q \log q}{\log 2}.$$

The proof will use a string model, as usual. But it will also use the "strong law of large numbers", an important result from probability theory.

Before we turn to the proof, let us consider the set S_p more carefully. Note that $[0, 1]$ is not equal to $\bigcup_{0 \leq p \leq 1} S_p$, since the limit $F^{(0)}(x)$ does not exist for many x.

I will next prove that the set S_p is a Borel set. (This is the first example we have seen where measurability is not immediately obvious.) First, given a, b, n, the set

$$\left\{ x : a \leq K_n^{(0)}(x) \leq b \right\}$$

---

* Optional section.

is a Borel set, since it consists of a finite number of intervals of length 2^{-n}. Then

$$S_p = \left\{ x : \lim_{n \to \infty} \frac{K_n^{(0)}(x)}{n} = p \right\}$$

$$= \bigcap_{k \in \mathbb{N}} \bigcup_{N \in \mathbb{N}} \bigcap_{n \geq N} \left\{ x : p - \frac{1}{k} \leq \frac{K_n^{(0)}(x)}{n} \leq p + \frac{1}{k} \right\},$$

so S_p is a Borel set.

Next, note that if x and y agree except in the first k digits, then $F^{(0)}(x) = F^{(0)}(y)$. So any open interval in $[0,1]$ intersects S_p. That is, S_p is dense in $[0,1]$. Certainly $S_p \neq [0,1]$, so of course S_p is not closed.

If the digits of x are all shifted to the right, and a new digit is added on the left, then the frequencies are unchanged. So S_p exhibits a natural self-similarity: If $x \in [0,1]$, then $F^{(0)}(x) = F^{(0)}(x/2) = F^{(0)}(1/2 + x/2)$. Thus the two similarities

$$f_0(x) = \frac{x}{2},$$

$$f_1(x) = \frac{x+1}{2},$$

have the property

$$S_p = f_0[S_p] \cup f_1[S_p],$$

with no overlap. The similarity dimension of the iterated function system (f_0, f_1) is 1. The conclusion is: similarity dimension may be misleading for non-closed sets.

Theorem 7.4.1. *The Hausdorff dimension of the set S_p is*

$$s = \frac{-p \log p - q \log q}{\log 2}.$$

Proof. Let $E = \{0, 1\}$ be our two-letter alphabet, and recall the "base 2" model map $h \colon E^{(\omega)} \to [0, 1]$ defined on p. 14. Then $h[E^{(\omega)}] = [0, 1]$. Also, $\text{diam}[\alpha] = \text{diam}\, h[[\alpha]] = 2^{-n}$ if $\alpha \in E^{(n)}$. We define frequencies for strings in the same way as for numbers: For $\alpha \in E^{(*)}$, let $K^{(0)}(\alpha)$ be the number of 0's in α, let $K^{(1)}(\alpha)$ be the number of 1's in α. For $\sigma \in E^{(\omega)}$ let

$$F^{(0)}(\sigma) = \lim_{n \to \infty} \frac{K^{(0)}(\sigma \restriction n)}{n},$$

$$F^{(1)}(\sigma) = \lim_{n \to \infty} \frac{K^{(1)}(\sigma \restriction n)}{n}.$$

These limits are defined for some strings $\sigma \in E^{(\omega)}$, and not for others. Let

$$T_p = \left\{ \sigma \in E^{(\omega)} : F^{(0)}(\sigma) = p \right\}.$$

Then clearly $S_p = h[T_p]$.

Now consider a measure $\mathcal{M}_p$ defined on $E^{(\omega)}$ as follows. Let $\alpha \in E^{(n)}$. If $k = K^{(1)}(\alpha)$, $n - k = K^{(0)}(\alpha)$, let $w_\alpha = p^{n-k}q^k$. Then $w_\alpha = w_0 + w_1$, so these numbers define a metric measure $\mathcal{M}_p$ on $E^{(\omega)}$ with $\mathcal{M}_p([\alpha]) = w_\alpha$ for all $\alpha \in E^{(*)}$.

Now we require the result from probability theory. According to the measure $\mathcal{M}_p$ just defined, the "digits" of σ constitute independent Bernoulli trials, with probability p of outcome 0 and probability $q = 1 - p$ of outcome 1. So by the strong law of large numbers (for example, [7, Example 6.1]), we have

$$\mathcal{M}_p(T_p) = 1, \qquad \text{or, equivalently,} \qquad \mathcal{M}_p(E^{(\omega)} \setminus T_p) = 0.$$

We will take the case $p < 1/2$. The case $p > 1/2$ is similar, and the case $p = 1/2$ is the usual measure $\mathcal{M}_{1/2}$ and dimension 1 computed before (Proposition 6.3.1).

We begin with the upper bound, $\dim S_p \le s$. Let $\varepsilon > 0$ be given, and let $N \in \mathbb{N}$ satisfy $2^{-N} < \varepsilon$. Let $q' < q$. We will show that $\dim S_p \le s'$, where $s' = (-q' \log q - (1 - q') \log p) / \log 2$.

Consider the set $G \subseteq E^{(*)}$ defined as follows: if $\alpha \in E^{(n)}$, and $k = K^{(1)}(\alpha)$, then $\alpha \in G$ iff $k/n > q'$. For such α, we have $\operatorname{diam}[\alpha] = 2^{-n}$ and

$$\mathcal{M}_p([\alpha]) = p^{n-k}q^k = p^n \left(\frac{q}{p}\right)^k$$

$$> p^n \left(\frac{q}{p}\right)^{q'n} = p^{(1-q')n}q^{q'n}$$

$$= \left(2^{-n}\right)^{s'} = \left(\operatorname{diam}[\alpha]\right)^{s'}.$$

Let G' be the set of all $\alpha \in G$ with length $|\alpha| \ge N$ but $\alpha{\restriction}n \notin G$ for $N \le n < |\alpha|$. That is, α belongs to G, but no ancestors of α (except possibly ancestors before generation N) belong to G. If $\sigma \in T_p$, then $\lim_{n \to \infty} K^{(1)}(\sigma{\restriction}n)/n = q > q'$, so for some $n \ge N$ we have $\sigma{\restriction}n \in G$, and therefore for some $n \ge N$ we have $\sigma{\restriction}n \in G'$. So

$$\{ [\alpha] : \alpha \in G' \}$$

is a disjoint cover of T_p. But

$$\sum_{\alpha \in G'} \left(\operatorname{diam}[\alpha]\right)^{s'} < \sum_{\alpha \in G'} \mathcal{M}_p([\alpha])$$

$$= \mathcal{M}_p\left(\bigcup_{\alpha \in G'} [\alpha]\right) \le 1.$$

Therefore $\overline{\mathcal{H}}_\varepsilon^{s'}(T_p) \leq 1$. Let $\varepsilon \to 0$ to conclude $\mathcal{H}^{s'}(T_p) \leq 1$, and therefore $\dim T_p \leq s'$. Now when $q' \to q$ we have $s' \to s$, so $\dim T_p \leq s$.

Now the model map h has bounded decrease, so $\dim S_p \leq s$. (Or, cover S_p with the sets $h[[\alpha]], \alpha \in G'$.)

Next I must prove the lower bound, $\dim S_p \geq s$. Let $q' > q$, and define $s' = (-q' \log q - (1 - q') \log p)/ \log 2$. I will show $\dim S_p \geq s'$. Now

$$\mathcal{M}_p \left\{ \sigma \in E^{(\omega)} : \lim \frac{K^{(1)}(\sigma \restriction n)}{n} = q \right\} = \mathcal{M}_p(T_p) = 1,$$

and therefore

$$\mathcal{M}_p \left\{ \sigma : \text{there exists } N \in \mathbb{N} \text{ such that for all } n \geq N, \frac{K^{(1)}(\sigma \restriction n)}{n} < q' \right\}$$
$$= 1.$$

So by countable additivity,

$$\lim_{N \to \infty} \mathcal{M}_p \left\{ \sigma : \sup_{n \geq N} \frac{K^{(1)}(\sigma \restriction n)}{n} < q' \right\} = 1.$$

Choose N so that $\mathcal{M}_p(F) > 1/2$, where

$$F = \left\{ \sigma : \sup_{n \geq N} \frac{K^{(1)}(\sigma \restriction n)}{n} < q' \right\}.$$

Let $\varepsilon = 2^{-N}$.

Suppose $\mathcal{A}$ is a countable cover of S_p by sets A with $\operatorname{diam} A \leq \varepsilon$. First, we reduce to a cover by intervals of the form $h[[\alpha]]$. Each set $A \in \mathcal{A}$ is covered by (at most) three of the intervals $h[[\alpha]]$, where the length $|\alpha|$ is the integer n with $2^{-n} < \operatorname{diam} A \leq 2^{-n+1}$. Let $G \subseteq E^{(*)}$ be the set of all these α. (We may assume that the sets $[\alpha]$ are disjoint, since if two of them intersect, then one is a subset of the other, so we may delete the smaller one.) Thus

$$\sum_{\alpha \in G} (\operatorname{diam}[\alpha])^{s'} < 3 \sum_{A \in \mathcal{A}} (\operatorname{diam} A)^{s'}$$

and $T_p \subseteq \bigcup_{\alpha \in G}[\alpha]$, so $\mathcal{M}_p(\bigcup_{\alpha \in G}[\alpha]) = 1$.

For $\alpha \in G$ we have $|\alpha| \geq N$. If $[\alpha] \cap F \neq \emptyset$, $|\alpha| = n$ and $K^{(1)}(\alpha) = k$, then

$$\mathcal{M}_p([\alpha]) = p^{n-k} q^k = p^n \left(\frac{q}{p}\right)^k$$
$$< p^n \left(\frac{q}{p}\right)^{q'n} = p^{(1-q')n} q^{q'n}$$
$$= (2^{-n})^{s'} = (\operatorname{diam}[\alpha])^{s'}.$$

Now if $G' = \{ \alpha \in G : [\alpha] \cap F \neq \varnothing \}$, then

$$\frac{1}{2} < \mathcal{M}_p(F) \leq \mathcal{M}_p \left(\bigcup_{\alpha \in G'} [\alpha] \right)$$

$$= \sum_{\alpha \in G'} \mathcal{M}_p([\alpha]) < \sum_{\alpha \in G'} \left(\operatorname{diam}[\alpha] \right)^{s'}$$

$$\leq \sum_{\alpha \in G} \left(\operatorname{diam}[\alpha] \right)^{s'} < 3 \sum_{A \in \mathcal{A}} (\operatorname{diam} A)^{s'}.$$

Now $\mathcal{A}$ is any cover of S_p by sets of diameter $\leq \varepsilon$, so $\overline{\mathcal{H}}_\varepsilon^{s'}(S_p) > 1/6$. Therefore

$\mathcal{H}^{s'}(S_p) > 1/6$, so $\dim S_p \geq s'$. Now let $q' \to q$ to obtain $\dim S_p \geq s$.   □

Exercise 7.4.2. Compute the box dimension (the lower entropy index) of the set S_p.

Exercise 7.4.3. Compute the packing dimension $\operatorname{Dim} S_p$.

Exercise 7.4.4. Let E be a finite alphabet, let $\mathcal{M}$ be a metric measure on the space $E^{(\omega)}$ of infinite strings, and let ϱ be a metric on $E^{(\omega)}$. Suppose t is a positive real number, and let

$$S = \left\{ \sigma \in E^{(\omega)} : \lim_{n \to \infty} \frac{\log \mathcal{M}([\sigma \restriction n])}{\log \operatorname{diam}[\sigma \restriction n]} = t \right\}.$$

If $0 < \mathcal{M}(S) < \infty$, does it follow that $\dim S = t$?

7.5 *A Superfractal

Examples called "Kline curves" were included in the first edition of this book. It was included as extra material that could be assigned to students for independent investigation. The Kline curves provide examples of parametric curves in the plane where the Lipschitz classes of the two coordinate functions and the fractal dimension of the curve itself can be controlled independently. Kline's paper [40] was published in 1945.

Around 2005, motivated by their study of fractal methods for picture generation, Barnsley, Hutchinson, and Stenflo developed the **superfractal** formalism (see Barnsley's book [4]). Unexpectedly, the Kline curves give us an interesting example of a superfractal.

---

* Optional section.

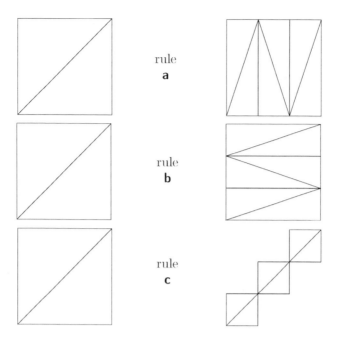

Fig. 7.5.1. Kline rules

Kline curves

The Kline curves are subsets of the plane $\mathbb{R}^2$. They are constructed using approximation by "Kline polygons". We begin with the line segment from the point $(0,0)$ to the point $(1,1)$; it is the diagonal of the rectangle (actually a square) $[0,1] \times [0,1]$. There are three rules used to build more complicated Kline polygons. Each of them replaces each of the line segments by three line segments. Rule **a** is implemented by subdividing the horizontal dimension of the containing rectangle in thirds, and replacing the diagonal by a three-part zig-zag, as illustrated. Rule **b** is implemented by subdividing the vertical dimension of the containing rectangle in thirds, and replacing the diagonal by a three-part zig-zag. Rule **c** is implemented by subdividing both the horizontal and vertical dimensions by three, and replacing the line segment by three parts of itself, inside the three diagonal subrectangles.

Each Kline polygon is obtained by applying these three rules in some order. Each finite string built from the alphabet $\{a, b, c\}$ may be considered a "program" for the construction of a polygon. Several examples are illustrated in Fig. 7.5.3. We will write Kline[α] for the Kline polygon corresponding to the string α.

Now let $\sigma \in \{a, b, c\}^{(\omega)}$ be an infinite string. The **_Kline curve_** Kline[σ] is the limit of the Kline polygons Kline[$\sigma{\restriction}k$] as k increases.

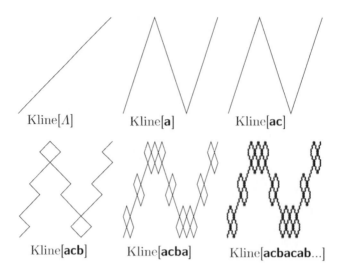

Fig. 7.5.3. Kline polygons, Kline curve

Exercise 7.5.2. If σ is an infinite string from the alphabet $\{a, b, c\}$, then Kline$[\sigma{\restriction}k]$ converges in the Hausdorff metric.

What is the justification of the use of the word "curve"? The "natural" parameterization of a polygon Kline$[\alpha]$ with 3^n segments of equal length ($n = |\alpha|$) is obtained by subdividing $[0, 1]$ into 3^n subintervals of equal length, and mapping each of the subintervals affinely onto the corresponding segment of the polygon.

Exercise 7.5.4. Determine necessary and sufficient conditions on the string σ for the natural parameterizations of the polygons Kline$[\sigma{\restriction}n]$ to converge uniformly to a parameterization of the Kline curve Kline$[\sigma]$ (which will again be called the natural parameterization).

There are two periodic strings σ that deserve special mention. They are cases where Kline$[\sigma]$ specializes to curves which we have seen before. For the constant string ccc$\cdots$, the Kline curve is a line segment. It has $\mathrm{Cov} = \dim = \mathrm{Dim} = 1$. For the period-two string abab$\cdots$, the Kline curve is the Peano space-filling curve (p. 70). It has $\mathrm{Cov} = \dim = \mathrm{Dim} = 2$. Clearly for a general string σ, the topological dimension $\mathrm{Cov}\,\mathrm{Kline}[\sigma]$ is either 1 or 2. And the fractal dimension satisfies $1 \le \dim \mathrm{Kline}[\sigma] \le 2$.

Exercise 7.5.5. Prove necessary and sufficient conditions on the string σ for $\mathrm{Cov}\,\mathrm{Kline}[\sigma] = 1$.

Exercise 7.5.6. Let $\sigma \in \{a, b, c\}^{(\omega)}$ be the program for a Kline curve. For $n \in \mathbb{N}$, let $a_n(\sigma)$ be the number of times the letter a occurs in the restriction

$\sigma \restriction n$. Similarly, let $b_n(\sigma)$ be the number of times the letter b occurs and $c_n(\sigma)$ the number of times the letter c occurs. Assume the limits

$$\alpha = \lim_{n\to\infty} \frac{a_n(\sigma)}{n}, \qquad \beta = \lim_{n\to\infty} \frac{b_n(\sigma)}{n}, \qquad \gamma = \lim_{n\to\infty} \frac{c_n(\sigma)}{n}$$

exist and $\alpha \geq \beta$. Show that the Hausdorff dimension of Kline$[\sigma]$ is

$$\frac{2\alpha + \gamma}{\alpha + \gamma} = 2 - \frac{\gamma}{\alpha + \gamma} = 2 - \frac{\gamma}{1 - \beta}.$$

Exercise 7.5.7. Discuss the packing dimension of a Kline curve.

Exercise 7.5.8. Let $\sigma \in \{a, b, c\}^{(\omega)}$, assume α, β, γ exist as in Exercise 7.5.6, and assume $1 > \alpha \geq \beta > 0$. Let $(\varphi(t), \psi(t))$, $t \in [0, 1]$, be the natural parameterization of Kline$[\sigma]$. Show that

$$\varphi \in \mathrm{Lip}\left(\frac{1}{1 - \alpha}\right), \qquad \psi \in \mathrm{Lip}\left(\frac{1}{1 - \beta}\right).$$

SuperIFS

Let us review the definition of a hyperbolic iterated function system and its attractor. We have a complete metric space S. We have a finite index set, an alphabet, E. For each letter $e \in E$ we have a contractive Lipschitz map $f_e \colon S \to S$. The data of the iterated function system let us define a map $F \colon \mathbb{H}(S) \to \mathbb{H}(S)$ by

$$F(A) = \bigcup_{e \in E} f_e[A]. \tag{1}$$

Sometimes we use the same letter (here F) to refer either to the iterated function system $(f_e)_{e \in E}$ itself or to the corresponding map (1) of the hyperspace. The attractor for the iterated function system F is the fixed point, the unique $K \in \mathbb{H}(S)$ satisfying the self-referential equation $F(K) = K$. The attractor K may be described using a string model. The addressing function $h \colon E^{(\omega)} \to S$ defined on the string space $E^{(\omega)}$ is defined by

$$h(\sigma) = \lim_n f_{\sigma \restriction n}(x),$$

where the limit is independent of the point $x \in S$. The range of h is K.

A superIFS is, roughly speaking, an IFS where the space is a hyperspace $\mathbb{H}(S)$ and the maps are themselves IFSs on S. The space S itself is where we are interested in describing sets, but there are two layers of data used to do it.

A more precise description: Let S be a complete metric space. Let E be a finite set, an alphabet. For each $e \in E$, let F_e be a hyperbolic IFS on the space S, so that the corresponding map $F_e \colon \mathbb{H}(S) \to \mathbb{H}(S)$ is a contractive Lipschitz map. (The IFSs F_e all act on the same space S, and each one has an alphabet and set of maps. Their alphabets may or may not be the same as each other, and probably are not the same as the master alphabet E.) An IFS $(F_e)_{e \in E}$ constructed in this way is known as a ***superIFS*** on S. So we define a map $\mathbf{F} \colon \mathbb{H}(\mathbb{H}(S)) \to \mathbb{H}(\mathbb{H}(S))$ by

$$\mathbf{F}(\mathcal{A}) = \bigcup_{e \in E} F_e[\mathcal{A}]. \tag{2}$$

Sometimes we use the same letter (here $\mathbf{F}$) to refer either to the SuperIFS $(F_e)_{e \in E}$ itself or to the corresponding map (2) of the hyperhyperspace. And there is an attractor: a unique $\mathcal{K} \in \mathbb{H}(\mathbb{H}(S))$ satisfying the self-referential equation $\mathbf{F}(\mathcal{K}) = \mathcal{K}$. This is called a ***superfractal***.[*] There is, as usual, a string model. For any string $\sigma \in E^{(\omega)}$ define

$$\mathbf{h}(\sigma) = \lim_n F_{\sigma \restriction n}(A).$$

This limit is taken according to the Hausdorff metric in $\mathbb{H}(S)$, and it does not depend on the starting set $A \in \mathbb{H}(S)$. So $\mathbf{h} \colon E^{(\omega)} \to S$ is continuous, and its range is the superfractal $\mathcal{K}$.

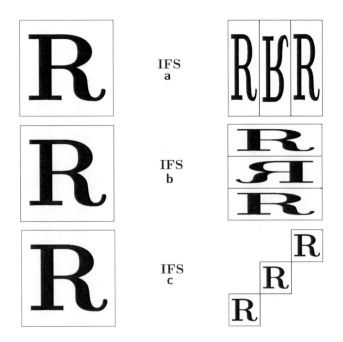

Fig. 7.5.9. Kline IFSs

---

[*] In [4] this is called a "1-variable superfractal", and that book also discusses V-variable superfractals.

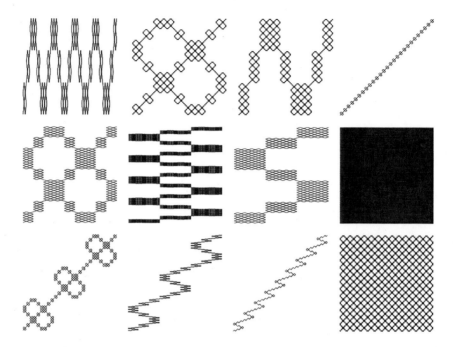

Fig. 7.5.10. Kline superfractal

The Kline curve construction described above is an example of a superfractal. Space S is the unit square. The master alphabet is $E = \{a, b, c\}$. The three IFSs are shown in Fig. 7.5.9. The superfractal $\mathcal{K}$ attractor is made up of all the Kline curves. The addressing function is $\sigma \mapsto \text{Kline}[\sigma]$. Some Kline curves are shown in Fig. 7.5.10. A family resemblance among the images reflects the fact that they all arise from a single superfractal $\mathcal{K}$.

See Barnsley [4, Chap. 5] and the references there for additional material on superfractals.

7.6 *Remarks

Robert Strichartz took the modern literary term "deconstruction" for use with iterated function systems. Deconstruction of a literary or philosophical text may mean finding meanings that were not intended by the original author. So deconstruction of the invariant set of an IFS means decomposing it in a way different from the one provided by the original iterated function system.

Karl Kiesswetter's curve is from [39]. It was proposed as a particularly elementary example of a continuous but nowhere differentiable function.

Theorem 7.4.1 is due to A. S. Besicovitch [5, Part II] and H. G. Eggleston [21]. Another proof is given by Patrick Billingsley [6, Section 14].

Comments on the Exercises

Exercise 7.1.4. The open set condition is satisfied, but not by the open set $(0, 1)$.

Exercise 7.1.5. The set K is contained in $[0, 1]$. Consider 4 parts of the set:

$$A = K \cap [4/5, 1]$$
$$B = K \cap [1/5, 4/5]$$
$$C = K \cap [4/25, 1/5]$$
$$D = K \cap [0, 4/25],$$

and observe that $(1/5)A \subseteq C$. Show that the graph similarity obeys Fig. 7.6.1, and the open set condition is satisfied for the corresponding realization. The dimension is $\log \left((\sqrt{5} + 3)/2 \right) / \log 5 \approx 0.59799$; compare it to the upper bound obtained from the the ratio list $(1/5, 1/5, 1/5)$, namely $\log 3 / \log 5 \approx 0.6826$.

Exercise 7.1.7. This is more like a class project than a homework assignment. Of course there are many possible answers. My deconstruction involves four isosceles right triangluar blocks A, D, H, K and their reflections A', D', H', K' with the graph iterated function system and open set condition shown in Fig. 7.6.2. The fractal dimensions are ≈ 1.92926.

Exercise 7.2.7. If g is differentiable at a point a, and $x_n \leq a \leq y_n$, $\lim x_n = a = \lim y_n$, $x_n < y_n$, then

$$\lim_{n \to \infty} \frac{g(y_n) - g(x_n)}{y_n - x_n} = g'(a).$$

This is false by Exercise 7.2.6.

Exercise 7.2.8: $3/2$.

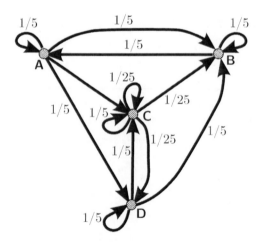

Fig. 7.6.1. A graph

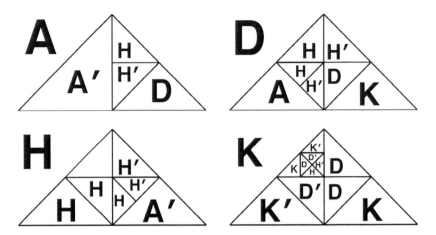

Fig. 7.6.2. Leaf deconstruction

Exercise 7.2.14: [5, Part V].
Exercise 7.2.17: [49].
Exercise 7.3.7: [23, Section 8.4].
Exercise 7.4.4: [6, Section 14].
Exercise 7.5.6: [40].

Life is a fractal in Hilbert space.
—Rudy Rucker, *Mind Tools*

I am a strange loop.
—Douglas Hofstadter

Appendix

Terms

Terms are listed here with the page number of an appropriate definition. In most cases, there is a reminder of the definition here, but for the complete definition see the page cited.

σ-**algebra of sets**, 147. Collection of subsets, contains the empty set and the whole space, closed under complements and countable unions.

accumulation point of a set, 46. Every ball centered at the point meets the set.

address, 125. Inverse image for the model map.

affine, 67. $f(tx + (1 - t)y) = tf(x) + (1 - t)f(y)$.

at most n-to-one, 111. The inverse image of any point consists of at most n points.

base for the open sets, 47.

bicompact, 61. Every family of closed sets with the finite intersection property has nonempty intersection.

Borel set, 147. Belongs to the σ-algebra generated by the open sets.

boundary point of a set, 54. Belongs to the closure of the set and to the closure of the complement.

boundary of a set, 54. The collection of all boundary points of the set.

Cauchy sequence, 52. $\varrho(x_n, x_m) \to 0$.

clopen, 86. Closed and open.

closed ball, 45. $\overline{B_r}(x) = \{\, y \in S : \varrho(y, x) \le r \,\}$.

closed set, 46. Contains all of its accumulation points.

closure of a set, 53. The set together with all of its accumulation points.

cluster point of a sequence, 51. Every ball centered at the point contains infinitely many terms of the sequence.

compact set, 61.

complete metric space, 52. Every Cauchy sequence converges.

concatenation of strings, 11. One string followed by the other.

topological property, 50. A property preserved by homeomorphism.
trema, 2. part removed in construction.
two-dimensional Lebesgue measure, 152.
two-dimensional Lebesgue outer measure, 152.
ultrametric space, 47. $\varrho(x, z) \leq \max\{\varrho(x, y), \varrho(y, z)\}$.
uniform convergence, 66.
uniformly continuous, 64.
zero-dimensional, 86. Every open cover has a clopen partition refinement

Notation

An index of the notations used in the book.

$\overline{\mathcal{L}}^2$, two-dimensional Lebesgue outer measure: 152.

$\mathcal{L}^2$, two-dimensional Lebesgue measure: 152.

$\mathcal{L}^d$, d-dimensional Lebesgue measure: 154.

$\overline{\mathcal{H}}^s$, Hausdorff outer measure: 167.

$\mathcal{H}^s$, Hausdorff measure: 167.

dim S, Hausdorff dimension: 168.

$\mathcal{P}^s$, packing measure: 173.

Dim S, packing dimension: 174.

sp rad A, spectral radius: 220.

Examples

Reading

The mathematical theory of fractal geometry, as I understand it, rests on three mathematical foundations: metric topology, measure theory, and probability. The metric topology in Chap. 2 is almost all that is generally needed for the study of fractal geometry. (Some additional "descriptive set theory" may be useful.) The measure theory of Chap. 5 is, however, only a part of what is needed for the deeper study of fractal geometry. Additional topics related to measure theory, such as integration, potential theory, and harmonic analysis, will be needed by the serious student of fractal geometry. (Some of the texts are [11], [34], [58].) Probability is an important branch of mathematics that is used often in fractal geometry. In order to reduce the background required here, I have almost entirely avoided using it explicitly. But further study of fractal geometry almost surely will require study of probability theory. And by "probability theory" I mean the modern subject that depends on measure theory, rather than the older version that gets by with calculus alone. (Some of the texts are [7], [10].)

Here are some suggestions for further reading on the topics considered in this book.

Benoit B. Mandelbrot, *The Fractal Geometry of Nature* [44]. This is the basic reference on fractals, together with a discussion of the applications of fractal sets in many branches of science. But it is not as mathematical as mathematicians would like, while it is too mathematical for many others. It contains many computer-generated pictures.

James Gleick, *Chaos: Making a New Science* [31]. This is a non-technical account by a New York *Times* reporter. It is concerned with the scientific phenomena governed by chaotic dynamical systems. "Chaos" and "fractals" are not the same field of study, but there are many relations between them.

Michael Barnsley, *Fractals Everywhere* [3]. This book is an introduction to fractal geometry from a somewhat different point of view than the one I have used here. A sequel, *Superfractals*, was published in 2006.

W. Hurewicz and H. Wallman, *Dimension Theory* [35]. This is a book on topological dimension. It is a bit out of date by now, but it has everything you need to know when you restrict attention to separable metric spaces. It requires background in metric topology, such as our Chap. 2. Other references for topological dimension are [22],[53],[54],[55].

K. J. Falconer, *Fractal Geometry* [26]. This is a modern text on the subject. It contains much more material than I have included here.

C. A. Rogers, *Hausdorff Measures* [57]. This book is concerned with the more technical aspects of the subject. It was written before Mandelbrot's sudden popularity, illustrating the fact that mathematicians had been doing this sort of thing ever since Hausdorff. But before Mandelbrot, scientists (including many mathematicians) considered it to be of only minor importance.

H.-O. Peitgen and P. H. Richter, *The Beauty of Fractals* [56]. This book is a discussion of Julia sets, the Mandelbrot set, and related topics, mostly without proofs. It contains many computer-generated color pictures; also an essay on whether they should be considered to be "art".

Chandler Davis and Donald E. Knuth, "Number representations and dragon curves" [13]. Two papers discuss the Heighway dragon (and some other dragons), and how they are related to representations of numbers in complex bases.

Michel Dekking, Michel Mendès France, and Alf van der Poorten, "Folds!" [14]. (The title was inspired by a movie current at the time.) Heighway's dragon can be generated by folding a strip of paper. A reader who can get past the "humor" will learn about this, and many other interesting topics.

William A. McWorter and Jane M. Tazelaar, "Creating Fractals" [50]. Instructions for drawing a wide variety of dragon curves, with code in BASIC. This is the source for McWorter's pentigree.

References

1. Abelson, H., diSessa, A.: Turtle Geometry: The Computer as a Medium for Exploring Mathematics. MIT Press, Cambridge (1981)
2. Barnsley, M. F., Demko, S: Iterated function systems and the global construction of fractals. Proc. R. Soc. London, A **399** (1985) 243–275
3. Barnsley, M. F.: Fractals Everywhere. Academic Press, Boston (1988); Second edition (1993)
4. Barnlsey, M. F.: Superfractals. Cambridge University Press, Cambridge (2006)
5. Besicovitch, A.S.: Sets of fractional dimensions. Part I: Math. Ann. **101** (1929) 161–193; Part II: Math. Ann. **110** (1934) 321–329; Part III: Math. Ann. **110** (1934) 331–335; Part IV: J. London Math. Soc. **9** (1934) 126–131; Part V: J. London Math. Soc. **12** (1934) 18–25
6. Billingsley, P: Ergodic Theory and Information. John Wiley & Sons, New York (1965)
7. Billingsley, P: Probability and Measure. John Wiley & Sons, New Tork (1979); Third edition (1995)
8. Blumenthal, L. M., Menger, K.: Studies in Geometry. Freeman, San Francisco (1970)
9. Charalambous, M. G.: A note on the relations between fractal and topological dimensions, Questions Answers Gen. Topology **17** (1999) 9–16; Correction: Questions Answers Gen. Topology **18** (2000) 121.
10. Chung, K. L.: A Course in Probability. Academic Press, New York (1974).
11. Cohn, D. L.: Measure Theory. Birkhäuser, Boston (1980).
12. Coxeter, H. S. M.: Introduction to Geometry. John Wiley & Sons, New York (1961); Second edition (1969)
13. Davis, C., Knuth, D. E.: Number representations and dragon curves, Part I: J. Recreational Math. **3** (1970) 66–81; Part II: J. Recreational Math. **3** (1970) 133–149
14. Dekking, M., Mendès France, M, van der Poorten, A.: Folds! Part I: Math. Intelligencer **4** (1982) 130–138; Part II: Math. Intelligencer **4** (1982) 173–181; Part III: Math. Intelligencer **4** (1982) 190–195
15. Dekking, M: Recurrent sets. Advances in Math. **44** (1982) 78–104
16. Devaney, R. L.: An Introduction to Chaotic Dynamical Systems. Benjamin/Cummings Publishing Company, Menlo Park (1986); Second edition, Addison–Wesley, Redwood City (1989)
17. Drobot, V., Turner, J: Hausdorff dimension and Perron-Frobenius Theory. Ill. J. Math. **33** (1989) 1–9
18. Edgar, G.: Integral, Probability, and Fractal Measures. Springer, New York (1998)
19. Edgar, G: Classics on Fractals. Westview Press, Boulder (1992)
20. Eggleston, H. G.: Convexity. Cambridge University Press, Cambridge (1958)
21. Eggleston, H. G.: The fractional dimension of a set defined by decimal properties. Quarterly J. Math. **20** (1949) 31–36
22. Engelking, R: Dimension Theory. North-Holland, Amsterdam (1978)
23. Falconer, K. J.: The Geometry of Fractal Sets. Cambridge University Press, Cambridge (1985)
24. Falconer, K. J.: The Hausdorff dimension of some fractals and attractors of overlapping construction. J. Statist. Phys. **47** (1987) 123–132

258   References

25. Falconer, K. J.: The Hausdorff dimension of self-affine fractals. Math. Proc. Camb. Phil. Soc. **103** (1988) 339–350

26. Falconer, K. J.: Fractal Geometry: Mathematical Foundations and Applications. Wiley, Chichester, 2003

27. Gantmacher, F. R.: The Theory of Matrices. Chelsey, New York (1959); Volume II, Chapter XIII: "Matrices with non-negative elements"

28. Gardner, M.: Mathematical Games. Scientific American (March, April, and July, 1967)

29. Gilbert, W. J.: Fractal geometry derived from complex bases. Math. Intelligencer **4** (1982) 78–86

30. Gilbert, W. J.: The fractal dimension of sets derived from complex bases. Canad. Math. Bull. **29** (1986) 495–500

31. Gleick, J.: Chaos: Making a New Science. Viking, New York (1987)

32. Hausdorff, F.: Dimension und äußeres Maß. Math. Ann. **79** (1918) 157–179. English translation in [19]

33. Hausdorff, F.: Mengenlehre, 3 Auflage. Dover Publications, New York (1944)

34. Hewitt, E., Stromberg, K.: Real and Abstract Analysis. Springer-Verlag, New York (1965)

35. Hurewicz, W., Wallman, H.: Dimension Theory. Princeton University Press, Princeton (1941)

36. Hutchinson, J. E.: Fractals and self similarity. Indiana Univ. Math. J. **30** (1981) 713–747

37. Jackson, A., Shapiro, D.: Arnold Ross (1906–2002). With contributions by: Prakash Bambah, Thomas Banchoff, Felix Browder, David Pollack, Peter Roquette, Karl Rubin, Paul J. Sally Jr., James Schultz, Alice Silverberg, Glenn Stevens, Bert K. Waits, Max Warshauer and Gloria Woods. Notices Amer. Math. Soc. **50** (2003) 660–665.

38. Kelley, J. L.: General Topology. Van Nostrand, Princeton (1955)

39. Kießwetter, K.: Ein einfaches Beispiel für eine Funktion, welche überall stetig und nicht differenzierbar ist. Math. Phys. Semesterber. **13** (1966) 216–221

40. Kline, S. A.: On curves of fractional dimensions. J. London Math. Soc. **20** (1945) 79–86

41. Kuratowski, K.: Topology, Volume I. Academic Press, New York (1966)

42. Lapidus, M. L., Frankenjuijsen, M. v.: Fractal Geometry, Complex Dimensions and Zeta Functions. Springer, New York (2006)

43. Li, J.: Dimensions d'objets géométriques et récurences vectorielles. Thése Ph.D., Université de Montréal (1999)

44. Mandelbrot, B. B.: The Fractal Geometry of Nature. W. H. Freeman and Company, New York (1982)

45. Mandelbrot, B. B.: Self-affine fractal sets. In: Pietronero, L., Tosatti, E.: Fractals in Physics. Elsevier Science Publishers, Amsterdam (1986)

46. Mandelbrot, B. B.: Multifractals and $1/f$ Noise. Springer-Verlag, New York (1999)

47. Marion, J.: Mesure de Hausdorff d'un fractal à similitude interne. Ann. Sc. Math. Québec **10** (1986) 51–84

48. Mauldin, R. D., Williams, S. C.: Hausdorff dimension in graph directed constructions. Trans. Amer. Math. Soc. **309** (1988) 811–829

49. McMullen, C.: The Hausdorff dimension of general Sierpiński carpets. Nagoya Math. J. **96** (1984) 1–9

50. McWorter, W. A., Tazelaar, J. M.: Creating Fractals. Byte (August 1987) 123–134

51. Moran, P. A. P.: Additive functions of intervals and Hausdorff measure. Proc. Cambr. Philos. Soc. **42** (1946) 15–23

52. Morgan, F.: Geometric Measure Theory. Academic Press, Boston (1988)

53. Nagami, K.: Dimension Theory. Academic Press, Boston (1970)

54. Nagata, J.: Modern Dimension Theory. Noordhoff, Groningen (1965)

55. Pears, A.: Dimension Theory of General Spaces. Cambridge University Press, Cambridge (1975)

56. Peitgen, H.-O., Richter, P. H.: The Beauty of Fractals. Springer-Verlag, Berlin (1986)

57. Rogers, C. A.: Hausdorff Measures. Cambridge University Press, Cambridge (1970)

58. Royden, R. L.: Real Analysis, Second edition. Macmillan, London (1968)

59. Schief, A.: Separation properties for self-similar sets. Proc. Amer. Math. Soc. **122** (1994) 111–115

60. Saint Raymond, X., Tricot, C.: Packing regularity of sets in n-space. Math.l Proc. Cambr. Philos. Soc. **103** (1988) 133–145

61. James Taylor, S. J., and Tricot, C.: Packing measure, and its evaluation for a Brownian path. Trans. Amer. Math. Soc. **288** (1985) 679–699

62. Taylor, S. J.: The measure theory of random fractals. Math. Proc. Camb. Phil. Soc. **100** (1986) 383–406.

Index

Undergraduate Texts in Mathematics

Irving: Integers, Polynomials, and Rings: A Course in Algebra.

Isaac: The Pleasures of Probability. *Readings in Mathematics.*

James: Topological and Uniform Spaces.

Jänich: Linear Algebra.

Jänich: Topology.

Jänich: Vector Analysis.

Kemeny/Snell: Finite Markov Chains.

Kinsey: Topology of Surfaces.

Klambauer: Aspects of Calculus.

Knoebel, Laubenbacher, Lodder, Pengelley: Mathematical Masterpieces: Further Chronicles by the Explorers.

Lang: A First Course in Calculus. Fifth edition.

Lang: Calculus of Several Variables. Third edition.

Lang: Introduction to Linear Algebra. Second edition.

Lang: Linear Algebra. Third edition.

Lang: Short Calculus: The Original Edition of "A First Course in Calculus."

Lang: Undergraduate Algebra. Third edition.

Lang: Undergraduate Analysis.

Laubenbacher/Pengelley: Mathematical Expeditions.

Lax/Burstein/Lax: Calculus with Applications and Computing. Volume 1.

LeCuyer: College Mathematics with APL.

Lidl/Pilz: Applied Abstract Algebra. Second edition.

Logan: Applied Partial Differential Equations, Second edition.

Logan: A First Course in Differential Equations.

Lovász/Pelikán/Vesztergombi: Discrete Mathematics.

Macki-Strauss: Introduction to Optimal Control Theory.

Malitz: Introduction to Mathematical Logic.

Marsden/Weinstein: Calculus I, II, III. Second edition.

Martin: Counting: The Art of Enumerative Combinatorics.

Martin: The Foundations of Geometry and the Non-Euclidean Plane.

Martin: Geometric Constructions.

Martin: Transformation Geometry: An Introduction to Symmetry.

Millman/Parker: Geometry: A Metric Approach with Models. Second edition.

Moschovakis: Notes on Set Theory. Second edition.

Owen: A First Course in the Mathematical Foundations of Thermodynamics.

Palka: An Introduction to Complex Function Theory.

Pedrick: A First Course in Analysis.

Peressini/Sullivan/Uhl: The Mathematics of Nonlinear Programming.

Prenowitz/Jantosciak: Join Geometries.

Priestley: Calculus: A Liberal Art. Second edition.

Protter/Morrey: A First Course in Real Analysis. Second edition.

Protter/Morrey: Intermediate Calculus. Second edition.

Pugh: Real Mathematical Analysis.

Roman: An Introduction to Coding and Information Theory.

Roman: Introduction to the Mathematics of Finance: From Risk management to options Pricing.

Ross: Differential Equations: An Introduction with Mathematica®. Second Edition.

Ross: Elementary Analysis: The Theory of Calculus.

Samuel: Projective Geometry. *Readings in Mathematics.*

Saxe: Beginning Functional Analysis

Scharlau/Opolka: From Fermat to Minkowski.

Schiff: The Laplace Transform: Theory and Applications.

Sethuraman: Rings, Fields, and Vector Spaces: An Approach to Geometric Constructability.

Shores: Applied Linear Algebra and Matrix Analysis.

Sigler: Algebra.

Silverman/Tate: Rational Points on Elliptic Curves.

Simmonds: A Brief on Tensor Analysis. Second edition.

Singer: Geometry: Plane and Fancy.

Singer: Linearity, Symmetry, and Prediction in the Hydrogen Atom.

Singer/Thorpe: Lecture Notes on Elementary Topology and Geometry.

Smith: Linear Algebra. Third edition.

Smith: Primer of Modern Analysis. Second edition.

Stanton/White: Constructive Combinatorics.

Stillwell: Elements of Algebra: Geometry, Numbers, Equations.

Stillwell: Elements of Number Theory.

Stillwell: The Four Pillars of Geometry.

Stillwell: Mathematics and Its History. Second edition.

Stillwell: Numbers and Geometry. *Readings in Mathematics.*

Strayer: Linear Programming and Its Applications.

Toth: Glimpses of Algebra and Geometry. Second Edition. *Readings in Mathematics.*

Troutman: Variational Calculus and Optimal Control. Second edition.

Valenza: Linear Algebra: An Introduction to Abstract Mathematics.

Whyburn/Duda: Dynamic Topology.

Wilson: Much Ado About Calculus.

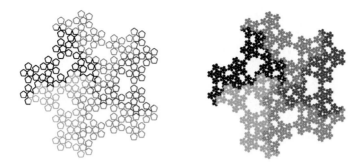

Plate 1. McWorter pentigree

Plate 2. Pentigree self-similar

Plate 3. Pentadendrite

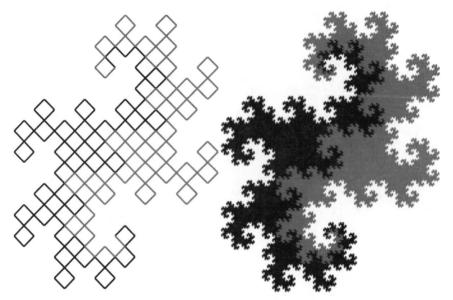

Plate 4. Two dragons = twindragon

Plate 5. Twindragon self-similar

Plate 6. Eisenstein fractions

Plate 7. Bagula double V

Plate 8. Koch curve

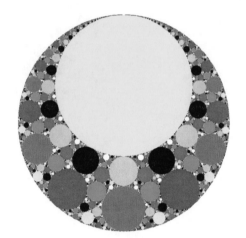

Plate 9. Pharaoh

Plate 10. Leaf overlap

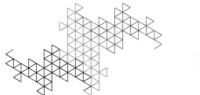

Plate 11. Terdragon

Plate 12. Six terdragons

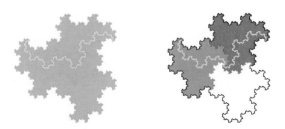

Plate 13. Open set condition

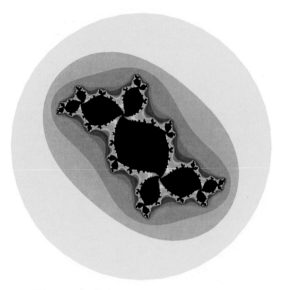

Plate 14. Julia set, $c = -0.15 + 0.72i$

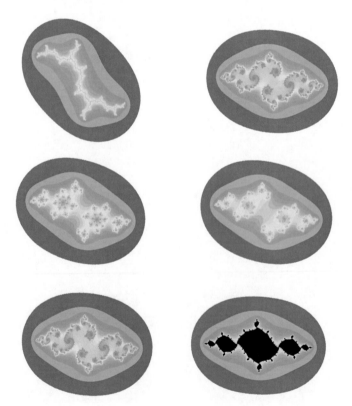

Plate 15. i, $-0.8 + 0.2i$, $-0.7 + 0.5i$, $-0.8 + 0.4i$, $-0.8 + 0.2i$, $-1.0 + 0.1i$

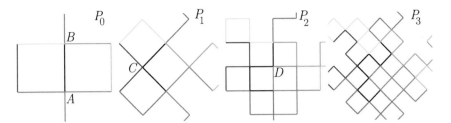

Plate 16. Heighway tiling

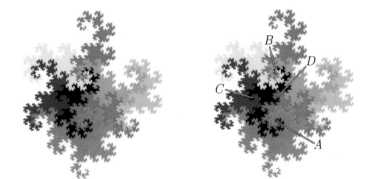

Plate 17. Heighway tiling

Plate 18. Heighway OSC

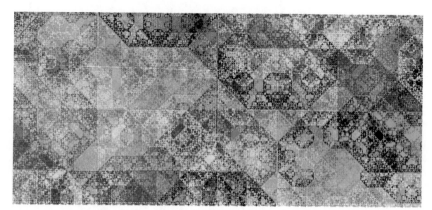

Plate 19. Levy tiling

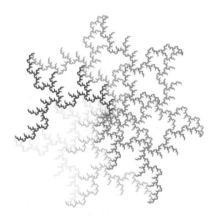

Plate 20. McWorter Lucky Seven

Printed by Books on Demand, Germany